OSMOTICALLY DRIVEN MEMBRANE PROCESSES - APPROACH, DEVELOPMENT AND CURRENT STATUS

Edited by **Hongbo Du, Audie Thompson**
and **Xinying Wang**

Osmotically Driven Membrane Processes - Approach, Development and Current Status
http://dx.doi.org/10.5772/intechopen.68607
Edited by Hongbo Du, Audie Thompson and Xinying Wang

Contributors
Murat Eyvaz, Serkan Arslan, Ebubekir Yüksel, İsmail Koyuncu, Derya İmer, Jimmy Roach, Mandy Bondaruk, Zain Ali Burney, Amira Abdelrasoul, Huu Doan, Ali Lohi, Albert S. Kim, Sangwoo Shin, Martha Chollom, Sudesh Rathilal, Achisa Mecha, Machawe Motsa, Bhekie Mamba, Julio Romero, Humberto Estay, Mohtada Sadrzadeh, Amrit Bhinder, Simin Shabani, Torleif Holt, Edvard Sivertsen, Willy R. Thelin, Geir Magnar Brekke, Audie Key Thompson, Felecia Nave, Raghava Kommalapati

Published by InTech
Janeza Trdine 9, 51000 Rijeka, Croatia

Publishing Process Manager Martina Usljebrka
Technical Editor SPi Global
Cover InTech Design team

First published March, 2018
Printed in Croatia
Legal deposit, Croatia: National and University Library in Zagreb

Additional hard copies can be obtained from orders@intechopen.com

Osmotically Driven Membrane Processes - Approach, Development and Current Status
Edited by Hongbo Du, Audie Thompson and Xinying Wang
p. cm.
Print ISBN 978-953-51-3921-8
Online ISBN 978-953-51-3922-5

For EU product safety concerns:
IN TECH d.o.o., Prolaz Marije Krucifikse Kozulić 3, 51000 Rijeka, Croatia,
info@intechopen.com or visit our website at intechopen.com.

Meet the editors

Dr. Hongbo Du received his BSc and MSc degrees in Mechanical Engineering from Beijing Forestry University, China, and his PhD degree in Mechanical Engineering from Colorado State University, USA. He is currently a researcher in the NSF CREST Center for Energy and Environmental Sustainability, Prairie View A&M University, a member of the Texas A&M University System. He worked as a postdoctoral researcher in the Department of Chemical Engineering and the Department of Biomedical Engineering, University of Arkansas, for 3 years. Dr. Du has published 18 journal papers and 2 book chapters and made 49 presentations at regional, national, and international conferences. His research interests are shale gas and oil-produced water treatment with photocatalysis and forward osmosis membranes, membrane surface modifications, thermo-responsive membranes, carbon capture with polyamine-functionalized nanomaterials, and environmental sustainability.

Dr. Audie Key Thompson is a research assistant professor in the Department of Chemical Engineering, Roy G. Perry College of Engineering, Prairie View A&M University (PVAMU), a member of the Texas A&M University System. Prior to joining PVAMU in 2011, Thompson served as an assistant professor in the Department of Chemistry and Physics at Alcorn State University. She is a recipient of several awards and honors. Dr. Thompson was recognized by the Thurgood Marshall College Fund on behalf of the Department of Defense for her work as a coprincipal investigator on the project, PVAMU REU in Membrane Separation, and recognized as a Southern Regional Education Board scholar from 2004 to 2010. She has also served as a reviewer for the National Science Foundation Graduate Research Fellowship and USDA.

Dr. Xinying Wang is currently working on his PhD degree studies in Electrical and Computer Engineering at the University of Illinois Urbana-Champaign since 2014. He is currently conducting multidisciplinary research, which is on applying Internet of Thing (IoT)/wireless sensor network to process monitoring in reverse osmosis desalination plant and other industrial environments. From 2010 to 2014, he had been working as a chemical engineer at the Illinois Sustainable Technology Center on forward osmosis including high-performance forward osmosis membrane development and novel draw solution recycling process development. In 2010, he was awarded his PhD degree in Chemical Engineering from Colorado State University. He has coauthored 18 publications including journal papers, conference proceedings, patents, and book chapters. He is a recipient of multiple outstanding research and teaching awards.

Contents

Preface

Osmotically driven membrane processes (ODMPs) including forward osmosis (FO) and pressure-retarded osmosis (PRO) have attracted increasing attention in fields such as water treatment, desalination, power generation, and life science. In contrast to pressure-driven membrane processes, e.g., reverse osmosis, which typically employs applied high pressure as driving force, ODMPs take advantages of naturally generated osmotic pressure as the sole source of driving force. In light of this, ODMPs possess many advantages over pressure-driven membrane processes. The advantages include low energy consumption, ease of equipment maintenance, low capital investment, high salt rejection, and high water flux.

A decade ago, there was an excellent review paper on FO written by Cath et al. published in *Journal of Membrane Science*. Since then, numerous studies on ODMPs have been emerging and evolving. In the past decade, over 300 academic papers on ODMPs have been published in a variety of fields. The number of such publication is still rapidly growing. This book intends to provide readers a comprehensive overview about ODMPs, from fundamentals, history, and challenges to current state-of-the-art applications, technology commercialization status, and so on. Chapter 1 overviews ODMPs as a whole. In this chapter, various aspects of ODMP technologies are briefly discussed. Chapters 2 and 3 explore manufacturing methods and application fields of FO membranes. Chapter 4 discusses the effects of internal and external concentration polarizations on the performance of FO. Chapter 5 focuses on the temperature effects on the performance of FO. Chapter 6 introduces pressure dependency of the membrane structure parameter and implications in PRO. Chapter 7 discusses nonideal solution behavior in FO. Chapters 8 and 9 elucidate fouling mechanisms in ODMPs and introduce different membrane cleaning technologies. Chapters 10 and 11 present applications of FO in wastewater treatment and seawater desalination. Chapter 12 summarizes membrane gas absorption theory and applications.

ODMPs' approach, fabrications, recent development and applications in wastewater treatment, power generation, seawater desalination, and gas absorption are presented in this book. We can clearly predict the rapid development of ODMPs with high performance in the next few years due to the enormous demands of complex wastewater treatment such as shale gas and oil-produced water and high saline water. Dr. Du would like to acknowledge the support from the USA, for his research through the NSF CREST Center for Energy and Environmental Sustainability (Award No. #1036593).

Dr. Hongbo Du
Prairie View A&M University, USA

Dr. Audie Thompson
Prairie View A&M University, USA

Dr. Xinying Wang
University of Illinois, USA

Chapter 1

Introductory Chapter: Osmotically Driven Membrane Processes

Felecia Nave, Raghava Kommalapati and
Audie Thompson

Additional information is available at the end of the chapter

http://dx.doi.org/10.5772/intechopen.72569

1. Forward osmosis fundamentals

Global climate patterns and urban growth are two of the many factors that have affected the world's water resources. During the twentieth century, the population of the world tripled, and it is predicted to increase by another 15–20% in the next 50 years [1, 2]. The demand for fresh potable water correlates with the increase in the world's population, thus access to safe and sufficient drinking water is now an international aim. Sadly, over 1 billion people across the world currently have limited to no access to drinking water [3]. In particular, the demand for water drastically outweighs the availability of water in some Middle Eastern countries and even within the United States, in states such as California that has recently experienced droughts [4]. Further, urbanization throughout the world has also impacted groundwater resources [5], and this controversy has led to surging interest in the efficiency and practicality of ocean water desalination [6].

Desalination is the process of obtaining drinking water by removing salt ions, minerals, and other undesired contaminants from seawater [7], and currently, there is an increasing interest in using FO in desalination. In arid regions of the world, such as the Mediterranean and the Middle East, desalination research has made great strides over the past 30 years [8]. In fact, there are approximately 14,000 desalination plants in 150 countries with a production of millions of gallons per day [8]. In countries, such as Saudi Arabia and the United Arab Emirates, 70% of water supplies are dependent on desalination. Hence, energy production is concurrently linked to the production of freshwater, as desalination of seawater requires more energy than transportation of water from a lake or river [9]. It is also important to note that nuclear plants and other energy sources (coal or oil) require 20–50 K gallons of water per megawatt-hour of electricity produced [10]. Furthermore, gasoline vehicles, plug-in vehicles, ethanol-running vehicles and hydrogen-fuel cell vehicles all consume gallons of water to operate. Thus, the demand for water is intrinsically tied to energy and sustainable practices and processes must be used. Discovering energetically efficient methods to produce and reuse water is pertinent in providing strategies to combat the energy consumption demands. Additionally, industrial plants consume a drastic amount of water for their industrial processes, and 70% of fresh water is utilized in agricultural processes [11]. Therefore, water shortages will hinder many areas of human daily activity and existence.

Most water-related technologies are based on advanced materials, advanced manufacturing technologies, biotechnology, and integrated filtration systems. Therefore, research and development of new materials with tailored properties and nanomaterials are necessary to meet the water demands and provide connections between eco-efficiency, performance, processing, recyclability, costs, and water reuse. Although the development of membrane technology for producing clean water in wastewater treatment and desalination is vital, there are challenges that must be further addressed in all water filtration processes [12, 13]. Water-selective membranes have gained vast interest for their advantages like high energy efficiency, reasonable cost, and environmental sustainability. The ideal water-selective membranes are fabricated to have high water permeability, selectivity, as well as stability [14]. However, major constraints include operational fouling, waste residue disposal, cost, and acceptance by utility organizations and the public.

The current and most widely used water purification is reverse osmosis (RO)—a membrane-based separation process that removes salts, microbial constituents, both organic and inorganic compounds from water and has been used extensively in a variety of fields including desalination of seawater, ultrapure water production, and wastewater treatment [15, 16]. RO goes against the laws of nature and uses pressure to force a solvent through the membrane, which retains the solute on one side and allows the pure solvent to pass to the other side. Since its discovery, RO has become a very useful process when it comes to removing salt ions from a solution.

There has been an increased focus on membrane technology research because of the high efficiency and low-cost solutions for water purification. Currently, forward osmosis (FO) systems are seen as favorable alternatives to RO systems, as they have been also utilized in electricity generation, food processing [11], industrial wastewater, and add produced water treatment [17–19]. In nature, when two solutions are separated by a semipermeable membrane, the solvent molecules will tend to move through the membrane into the region of higher solute concentration until equilibrium is reached. FO separates two solutions with different concentrations using the natural osmotic pressure difference. The osmotic gradient is the driving force instead of externally applied pressure.

Even though RO systems have dominated the water purification arena for decades, FO systems offer an advantage of rejecting a wide range of contaminants. FO systems experience less fouling than RO systems; therefore, a membrane with anti-fouling properties could be efficient and beneficial. Within the RO process, the saline water, which has a high salt concentration, is forced through a membrane to a region of low solute concentrate by applying pressure in excess of osmotic pressure [20, 21], where the osmotic pressure is the minimum pressure needed to prevent the water molecules from moving back to the feed side from the permeate side. This occurs when the hydrostatic pressure differential resulting from the concentration changes on both sides of the semipermeable membrane is equal to the osmotic pressure of the solute [21]. The semipermeable membrane allows the passage of water but not salt ions. The feed water must pass through a very narrow passage as a result of the way the membrane is packaged. This causes for an initial treatment phase, where fine particulates or suspended solids must be removed to prevent fouling. In contrast, the FO system will have higher productivity and be considered an energy saving device since no external pressure is required. However, a major and unresolved challenge in FO remains an efficient draw solution that could result in high flux and reconstituted using a low-energy separation process which will be discussed later.

Two key factors in FO utilization are selecting the membrane and appropriate draw solute (DS). The DS should be non-toxic, generate high osmotic pressure, and be easily regenerated [22]. Continuous reconcentration is required to sustain the FO driving force to purify water. NaCl, $MgCl_2$, $CaCl_2$, and $MgSO_4$ are commonly used DSs; however, they are energy intensive and consequently costly [22, 23]. Alternatively, the DS can be treated wastewater effluent brine or seawater; the diluted DS will lower the energy demand [22]. Other limitations are the diffusion of the DS into the feed solution, low water flux compared to RO, membrane fouling, and concentration polarization. Therefore, many researchers are investigating alternative DSs.

1.1. Wastewater and water recycling

Wastewater sources include municipal and industrial plants and consume a drastic amount of water for their industrial processes. Some plants also produce oily wastewater end products. The industries that account for oil in water emulsions are petroleum, pharmaceutical, polymer, leather, polish, cosmetic, food, polymer, textile, agriculture, prints, and paper [24]. Helen Wake reports that oil refineries in European and Middle Eastern countries alone produce over 2 billion tons of wastewater [25]. This strikes as a major ecological problem, due to the discharge of oily wastewater into the ecosystem [25]. Furthermore, a principal fraction of oil/water emulsions' treatment technologies is often ineffective and expensive [24].

Produced water (PW) is generated during oil and gas production and is the biggest waste stream in the energy industries [26, 27]. Therefore, PW is contaminated with oils and salts of organic and inorganic compounds [27]. Releasing PW onto nature has an environmental impact and is a noteworthy issue of ecological concern. Ordinarily, PW is treated through various physical, chemical, and biological strategies. In offshore stages, as a result of space imperatives, minimal physical and substance frameworks are utilized. Unfortunately, current advances cannot dislodge these minute suspended oil particles. In addition, natural pretreatment of wastewater can be financially expensive. As high salt fixation and varieties of influent qualities have an impact on PW, it is suitable to fuse a physical treatment (e.g., film) to refine the material. Hence, future research endeavors are concentrating on the streamlining of flow innovations, utilization of consolidated methodology, organic treatment of delivered water, and review of reuse and release limits.

Agricultural wastewater, which comes from all animal farms and food processing, requires unique treatment before disposal or reuse [28]. Untreated agricultural wastewater results in pollution of groundwater, rivers, and lakes, thereby disrupting ecosystems and resulting in a chain of negative effects. However, with proper treatment and filtration, this wastewater can become a valuable resource. Primary treatment involves separating solids from the liquids and producing "sludge." The secondary treatment removes contaminants and dissolved solids from the effluent. Ultraviolet light, specialized enzymes, and microbes are often used for further treatment [29, 30]. After which, the "safe" water is returned to a waterway (ocean or river) or reused in agriculture [31]. Thus, treated wastewater can be reused in a sustainable fashion.

Where efficient irrigation methods and collection of run-off are in place, there is little wastewater [tailwater] to be treated for reuse. However, when bountiful tailwater is available, it often contains large amounts of salt and nutrients which makes it non-permissible for irrigation [31]. Innovative effluent treatment permits water reuse for irrigation and animal needs, making the "sludge" and subsequent effluent suddenly valuable. Additionally, collecting and reusing tailwater can benefit a farm through fertilization, and it can protect the environment by avoiding salt and nutrient discharge. Thus, utilizing tailwater and food processing wastewater could be profitable for farmers and positive for our environment.

1.2. Membrane fouling

Most membrane technologies experience reduction in performance as a result of various types of fouling. Therefore, designing and investigating membranes to combat fouling is imperative in creating proficient systems. Membrane fouling is the accumulation of unwanted matter such as colloids, salts, and microorganisms during the water purification process. Foulants accumulating on the surface reduces the water flow either temporarily or possibly permanently. Unfortunately, this is a common problem, and these foulants deteriorate and increase the ineffectiveness of the system.

During mass transport, various aspects lead to adsorption of particles within and onto the membrane surface, causing membrane fouling [22]. Contaminated feed water results in compounds and unwanted material adhering to the membrane, resulting in fouling, which is a major problem for most membrane-based systems and often results in a decline in flux [23]. Therefore, minimizing fouling is the key to optimal membrane operation and keeping costs down. Depending upon the polymer utilized for membrane fabrication, additional characteristics can be optimized to prevent fouling. Regardless of the membrane system, biofouling is a long-term problem [32]. All types of fouling (biofouling, organic, colloidal, and scaling) can be damaging [32]. It has been noted that FO is less likely to foul and less complicated than pressure-driven membrane processes like RO [23, 32]. This is because applied hydraulic pressure causes compact foulant layers, which diminish the effectiveness of cleaning the membranes.

Biofouling is considered to be the most difficult and detrimental to water filtration processes and decreases the durability of membranes. Therefore, membranes that are resistant to the accumulation of microorganisms are a necessity for water purification. Ultimately, biofouling causes higher than necessary energy consumption, deterioration of system performance, and water production. Due to the aforementioned issues, it is technologically essential to find efficient methods to minimize membrane biofouling. Studies have shown that FO membranes are more effective in preventing foulant permeation into the draw solute and reducing fouling in the downstream RO membrane [23].

Organic foulants are dominant and precursors to biofouling when using membrane bioreactor (MBR) for wastewater treatment [22, 33]. Therefore, biofouling can be prevented by controlling the organic matter. Hydrophobic and hydrophilic polysaccharides and transphilic organic macromolecules are all found in the feed water and may lead to organic fouling. Of these examples, polysaccharides are three times more likely than other humic acid contaminants to cause fouling [33].

1.3. Membrane selection

Material selection for membrane fabrication is significant in developing a system with optimal flux, as flux decline is directly connected to membrane fouling. Regardless of the polymeric material, asymmetric membranes are preferred during liquid separation due to their thin top layer on top of a porous support layer. FO asymmetric membranes consist of a dense active layer and a loosely bound support layer. The dense top layer is selective and the large pores in the support layer reduce hydraulic resistance [34]. Thin-film composite (TFC) and polysulfone are currently the most widely used materials for membrane fabrication due to their stability and high-pressure tolerance. However, Poly [vinyl alcohol] (PVA) hydrogels have been shown to be a suitable membrane used for water treatment, and PVA is an excellent surface modifier. Their hydrophilicity, water permeability, and anti-fouling potential make them ideal candidates in the further development of composite membranes [35, 36]. Research continues to investigate ways to optimize PVA hydrogel membranes based on their degree of polymerization and incorporation of nanoparticles [37]. Furthermore, studies have proven that ideal membranes should have high water permeability, selectivity, and stability [14].

1.4. Concentration polarization

As many are investigating FO for wastewater treatment and desalination, one of the major weaknesses of FO is internal concentration polarization (ICP). The configuration of the membrane contributes to the aforementioned fouling possibility and other complications such as ICP which minimized flux efficiency [33]. Traditionally, the support layer faces the feed in normal mode and faces the active layer in the reverse mode. The inability of the salt to pass easily through the active layer results in a concentration increase within the support layer. Amid the process, fouling such as scaling contributes to concentrative ICP [22, 33]. In the normal mode, the support layer diminishes water transport hydraulic resistance, and the solute freely enters, leading to minimum ICP [38]. Just as fouling leads to lower water flux, ICP within asymmetric thin-film composite (TFC) FO membranes does the same. Contrarily, in reverse mode, the active layer faces the feed solution contributing to ICP. The concentration is increased in the support as the active layer prevents the passage of salt. Thus, ICP greatly reduces the driving force for transport. However, a thin low porosity support minimizes ICP [33] and surface modifications, such as coating with another polymer, has been one of the most effective methods [21]. Studies have been conducted to improve membrane design for new-generation FO membranes and mitigate the ICP effect. Researchers have explored membrane structures to prevent salt leakage and minimize ICP in FO [39]. Altering phase inversion fabrication protocol by examining different casting substrate, consequently, results in an open structure with increased porosity in the middle support layer. During desalination, the FO system showed decreased salt leakage with mitigated ICP [21]. The ICP and ECP (external concentration polarization) structural value of the double dense-layer membrane is much smaller than those reported in the literature [21]. Moreover, lower CP values were seen after an intermediate solvent/water immersion was performed before complete immersion in water [39]. Additionally, Tang et al. [33] investigated ICP and fouling during humic acid filtration. They reported that despite initial ICP, the active facing orientation resulted in stable flux in contrast to flux diminution when facing foulant humic acid feed water.

2. Pressure retarded osmosis

Most water purification processes are known to consume energy. However, using the salinity differences between two bodies of water, pressure retarded osmosis (PRO) generates power. PRO is based on membrane technology similar to FO but results in sustainable osmotic power energy. During PRO, additional back pressure is applied to the draw solute, creating chemical potential between seawater and fresh water. As a result, electricity is produced from the conversion of flux into mechanical energy [22], and the net flux is similar to FO in the direction of the DS [40]. Unfortunately, membrane fouling consequently reduces the permeate flux and osmotic power generation, thus increasing overall cost similar to other membrane technologies. Research has been conducted on different quality feed waters to identify the main foulants on the surface in the PRO processes, and silica has been shown to cause severe scaling [41]. Again, structural parameters, material choice, pH of FS and/or DS played a critical role in mitigating IC of silica scaling [41]. Furthermore, organic and inorganic salt water was used to investigate cleaning methods to resolve fouling issues [32]. Using salt water as the DS, iron, aluminum, calcium, sodium, and silica were the inorganic foulants discovered [32]. Also, humic substances, polysaccharides, and proteins were the organic foulants identified [32]. Sequential acidic and basic cleaners were proven to be successful with a flux recovery above 95% [32]. PRO processes and consequently osmotic power generation can be enhanced by decreasing membrane fouling via chemical cleaning [32].

3. Summary

In summary, many researchers have compared FO, PRO, and RO as shown in **Figure 1** [22]. The most noted comparisons are the necessary pressure difference, fouling tendencies, and application. All three systems have advantages but require necessary improvements for expansion of utilization in various applications. Although fouling is a challenge for membrane technologies, research has demonstrated various ways to diminish its effects on flux [22, 32, 41]. With the increasing water demands, FO is certainly a viable option to meet the water and energy challenges of a growing global population as PRO has the potential to be widely used for sustainable energy. With polymer chemistry and membrane innovations, FO will advance for continuous use in producing safe water for irrigation, pharmaceuticals, and human

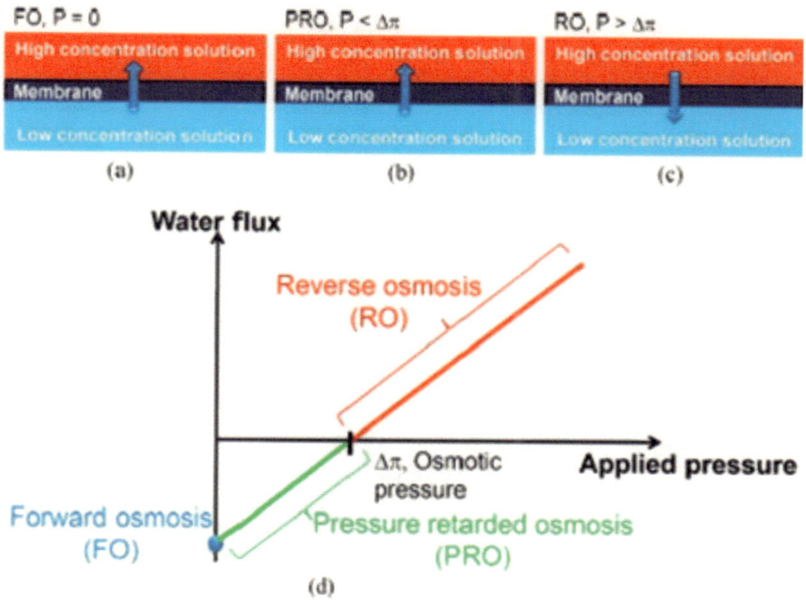

Figure 1. Illustration of FO, PRO, and RO processes [22].

consumption. This book will further discuss the headway in osmotically driven membrane processes (ODMP) research, findings, and contributions to membrane processes.

Author details

Felecia Nave, Raghava Kommalapati and Audie Thompson*

*Address all correspondence to: akthompson@pvamu.edu

Prairie View A&M University, Prairie View, TX, USA

References

[1] Kang G-D, Cao Y-M. Development of antifouling reverse osmosis membranes for water treatment: A review. Water Research. 2011;**46**(3):584-600

[2] Roser M, Ortiz-Ospina E. 'World Population Growth'. 2017. Published online at OurWorldInData.org

[3] Arnal JM, Garcia-Fayos B, Sancho M, Verdu G, Lora J. Design and installation of a decentralized drinking water system based on ultrafiltration in Mozambique. Desalination. 2010;**250**(2):613-617

[4] Wuertz G. Drought requires joint effort, not finger pointing. Rural Cooperatives. 2 May/June, 2015;**82**(4):40. USDA Rural Developments. https://www.rd.usda.gov/files/publications/RuralCoopMayJune2015.pdf

[5] McDonald RI, Weber K, Padowski J, Flörke M, Schneider C, Green PA, et al. Water on an urban planet: Urbanization and the reach of urban water infrastructure. Global Environmental Change. 2014;**27**:96-105

[6] Voutchkov N. Desalination Engineering Planning and Design. New York: McGraw Hill; 2013

[7] Bodzek M, Konieczny K, Kwiecinska A. Application of membrane processes in drinking water treatment-state of art. Desalination and Water Treatment. 2011;**35**(1–3):164-184

[8] Shatat M, Worall M, Riffat S. Opportunities for solar water desalination worldwide: Review. Sustainable Cities and Society. 2013;**9**:67-80

[9] Holland RA, Scott KA, Florke M, Brown G, Ewers RM, Farmer E, et al. Global impacts of energy demand on the freshwater resources of nations. Proceedings of the National Academy of Sciences of the United States of America. 2015;**112**(48):E6707-E6716

[10] Macknick J, Newmark R, Heath G, Hallett KC. Operational water consumption and withdrawal factors for electricity generating technologies: A review of existing literature. Environmental Research Letters. 2012;**7**(4):1-10

[11] Ronald P. Plant genetics, sustainable agriculture and global food security. Genetics. 2011;**188**(1):11-20

[12] Ho J, Low J, Sim L, Webster R, Rice S, Fane A, et al. In-situ monitoring of biofouling on reverse osmosis membranes; detection and mechanistic study using electrical impedance spectroscopy. Journal of Membrane Science. 2016;**518**:229-242

[13] Bereschenko LA, GHJ H, Nederlof MM, van Loosdrecht MCM, Stams AJ, GJW E. Molecular characterization of the bacterial communities in the different compartments of a full-scale reverse-osmosis water purification plant. Applied Environmental Microbiology. 2008;**26**:5297-5304

[14] Yang H, Wu H, Fusheng P, Li Z, Ding H, Guanhua L, et al. Highly water-permeable and stable hybrid membrane with asymmetric covaelent organic framework distribution. Journal of Membrane Science. 2016;**520**:583-595

[15] Zhao Y, Zhang Z, Dai L, Mao H, Zhang S. Enhanced both water flux and salt rejection of reverse osmosis membrane through combining isophphaloyl dichloride with binphenyl tetraacyl chlorise as organic phase monomer for seawater desalination. Journal of Membrane Science. 2017;**522**:175-182

[16] Sala-Comorera L, Blanch A, Vilaro C, Galofre B, Garcia-Aljaro C. Pseudomonas-related populations associated with revers osmosis in drinking water treatment. Journal of Environmental Management. 2016;**182**:335-341

[17] Coday BD, Cath TY. Forward osmosis: Novel desalination of produced water and fracturing flowback. Journal – American Water Works Association. 2014;**106**:E55-E66

[18] Lutchmiah K, Verliefde AR, Roest K, Rietveld LC, Cornelissen ER. Forward osmosis for application in wastewater treatment: A review. Water Research. 2014;**58**:179-197

[19] Alturki AA, Tadkaew N, McDonald JA, Khan SJ, Price WE, Nghiem LD, Combining MBR. NF/RO membrane filtration for the removal of trace organics in indirect potable water reuse applications. Journal of Membrane Science. 2010;**365**(1–2):206-215

[20] Geise GM, Park HB, Sagle AC, Freeman BD, McGrath JE. Water permeability and water/salt selectivity tradeoff in polymers for desalination. Journal of Membrane Science. 2011;**369**(1–2):130-138

[21] Low ZX, Liu Q, Shamsaei E, Zhang X, Wang H. Preparation and characterization of thin-film composite membrane with nanowire-modified support for forward osmosis process. Membranes (Basel). 2015;**5**(1):136-149

[22] Chun Y, Mulcahy D, Zou L, Kim ISA. Short review of membrane fouling in forward osmosis processes. Membranes (Basel). 2017;**7**(2):30

[23] Boo C, Elimelech M, Hong S. Fouling control in a forward osmosis process integrating seawater desalination and wastewater reclamation. Journal of Membrane Science. 2013; **444**:148-156

[24] Zolfaghari R, Fakhru'l-Razi A, Abdullah LC, SSEH E, Pendashteh A. Demulsification techniques of water-in-oil and oil-in-water emulsions in petroleum industry. Separation and Purification Technology. 2016;**170**:377-407

[25] Wake H. Oil refineries: A review of their ecological impacts on the aquatic environment. Estaurine Coastal and Shelf Science. 2004;**62**(1–2):131-140

[26] Kumar S, Guria C, Mandal A. Synthesis, characterization and performance studies of polysulfone/bentonite nanoparticles mixed-matrix ultra-filtration membranes using oil field produced water. Separation and Purification Technology. 2015;**150**:145-158

[27] Ebrahimi M, Ashaghi KS, Engel L, Willershausen D, Mund P, Bolduan P, et al. Characterization and application of different ceramic membranes for the oil-field produced water treatment. Desalination. 2009;**245**(1–3):533-540

[28] Coskun T, Debik E, Kabuk HA, Demir NM, Basturk I, Yildirim B, Temizel D, Kucuk S. Treatment of poultry slaughterhouse wastewater using a membrane process, water reuse, and economic analysis. Desalination and Water Treatment. 2015;**57**:4944-4951

[29] Bustillo-Lecompte CF, Mehrvar M. Slaughterhouse wastewater characteristics, treatment, and management in the meat processing industry: A review on trends and advances. Journal of Environmental Management. 2015;**161**:287-302

[30] Yordanov D. Preliminary study of the efficiency of ultrafiltration treatment of poultry slaughterhouse wastewater. Bulgarian Journal of Agricultural Science. 2010;**16**(6):700-704

[31] Kovacs K, Xu Y, West G, Popp M. The tradeoffs between market returns from agricultural crops and non-market ecosystem service benefits on an irrigated agricultural landscape in the presence of groundwater overdraft. Water. 2016;**8**(11):501

[32] Abbasi-Garravand E, Mulligan CN, Laflamme CB, Clairet G. Identification of the type of foulants and investigation of the membrane cleaning methods for PRO processes in osmotic power application. Desalination. 2017;**421**:135-148

[33] Tang CY, She Q, Lay WCL, Wang R, Fane AG. Coupled effects of internal concentration polarization and fouling on flux behavior of forward osmosis membranes during humic acid filtration. Journal of Membrane Science. 2010;**354**(1–2):123-133

[34] McCabe W, Smith J, Harriott P. Unit Operations of Chemical Engineering. New York: McGraw-Hill; 2005

[35] Arena JT, McCloskey B, Freeman BD, McCutcheon JR. Surface modification of thin film composite membrane support layers with polydopamine: Enabling use of reverse osmosis membranes in pressure retarded osmosis. Journal of Membrane Science. 2011;**375**(1–2):55-62

[36] Yin J, Fan H, Zhou J. Cellulose acetate/poly(vinyl alcohol) and cellulose acetate/crosslinked poly(vinyl alcohol) blend membranes: Preparation, characterization, and antifouling properties. Desalination and Water Treatment. 2015;**57**(23):10572-10584

[37] Zimpel A, Preiß T, Röder R, Engelke H, Ingrisch M, Peller M, et al. Imparting functionality to MOF nanoparticles by external surface selective covalent attachment of polymers. Chemistry of Materials. 2016;**28**(10):3318-3326

[38] Cath T, Childress A, Elimelech M. Forward osmosis: Principles, applications, and recent developments. Journal of Membrane Science. 2006;**281**(1–2):70-87

[39] Zhang S, Wang KY, Chung T-S, Chen H, Jean YC, Amy G. Well-constructed cellulose acetate membranes for forward osmosis: Minimized internal concentration polarization with an ultra-thin selective layer. Journal of Membrane Science. 2010;**360**(1–2):522-535

[40] Rastogi NK. Opportunities and challenges in application of forward osmosis in food processing. Critical Reviews in Food Science and Nutrition. 2016;**56**(2):266-291

[41] Wang Y-N, Li X, Wang R. Silica scaling and scaling control in pressure retarded osmosis processes. Journal of Membrane Science. 2017;**541**:73-84

Chapter 2

Forward Osmosis Membranes – A Review: Part I

Murat Eyvaz, Serkan Arslan, Derya İmer,
Ebubekir Yüksel and İsmail Koyuncu

Additional information is available at the end of the chapter

http://dx.doi.org/10.5772/intechopen.72287

Abstract

Forward osmosis (FO) is a technical term describing the natural phenomenon of osmosis: the transport of water molecules across a semi-permeable membrane. The osmotic pressure difference is the driving force of water transport, as opposed to pressure-driven membrane processes. A concentrated draw solution (DS) with osmotic pressure draws water molecules from the feed solution (FS) through a semi-permeable membrane to the DS. The diluted DS is then reconcentrated to recycle the draw solutes as well as to produce purified water. As a major disadvantage, nature of FO membranes (asymmetrical structure) causes international concentration polarization (ICP) which promotes the decrease in water flux. Therefore, the number of studies related to improving both active and support layers of FO membranes is increasing in the applications. The purpose of the chapter is to bring an overview on the FO membrane manufacturing, characterizing and application area at laboratory or full scales. This chapter is published in two parts. In the first part, which appears here, the overview of membrane technologies and the definition of forward osmosis process are stated. The manufacturing methods of support and active layers forming FO membranes are described with common and/or new modification procedures.

Keywords: forward osmosis, water flux, reverse salt flux, active layer, support layer, thin film composite, structural parameter, porosity, internal concentration polarization

1. Introduction

Membrane separation processes have been widely applied for many years in environmental, industrial applications, and domestic use such as water/wastewater treatment, desalination, specific industrial purposes and energy recovery. Among the concentration-driven processes, FO has recently attained many attractions due to its advantages such as less energy requirement, lower fouling tendency or easier fouling removal and higher water recovery. However, all drawbacks of FO process such as; (i) membrane fouling originated from ICP, (ii) lower flux, and (iii) reverse salt diffusion (RSD) limit the performance of the FO applications in environmental studies. Moreover, "necessity of concentrate management" and "meeting of discharge standards with high amounts of product water" oblige the developing new processes, membrane materials or modifications and finding new DS. In recent studies, developing new FO membranes by optimization of thickness, porosity, tortuosity of active/support layer of FO membrane to increase water flux and decrease ICP are mainly focused.

FO is a technical term describing the natural phenomenon of osmosis: the transport of water molecules across a semi-permeable membrane. The osmotic pressure difference is the driving force of water transport, as opposed to pressure-driven membrane processes A concentrated DS with osmotic pressure draws water molecules from the FS through a semi-permeable membrane to the DS. The diluted DS is then reconcentrated to recycle the draw solutes as well as to produce purified water. As the driving force is only the osmotic pressure difference between two solutions which means that there is no need to apply an external energy, this results in low fouling propensity of membrane and minimization of irreversible cake forming which are the main problems controverted by membrane applications -especially- in biological treatment systems (e.g. FO-MBR). However, nature of FO membranes (asymmetrical structure) causes ICP which promotes the decrease in water flux. Therefore, the number of studies related to improving both active and support layers of FO membranes is increasing in the applications.

In this first part of chapter, advantages of FO over conventional membrane processes and main drawbacks originated from the nature of FO membranes are thoroughly stated by considering both review and research articles in the available literature. The book chapter consists of three main titles firstly including introduction section, the second of which states a literature survey on early definitions of diaphragm, membrane and selectivity phenomenon by considering about last two centuries. Basic principles of FO phenomenon is also expressed in this part. However, the special aspects of FO process are discussed in the third part in view of FO membrane properties. The water flow is mainly determined by the support layer, while the selectivity is by the active layer of FO membrane. Therefore, both support and active layer designing are overemphasized by addressing new materials, manufacturing methods and modification steps to overcome the main challenges of FO processes such as low water flux and concentration polarization phenomena causing the membrane fouling.

2. General aspects of membrane processes

2.1. Membrane technology

Systematic studies of the membrane phenomenon can be traced to the eighteenth century philosopher scientists. For example, Abb'e Nolet prepared an 'osmosis' word in 1748 to describe water permeability through a diaphragm. Through the nineteenth century and beginning of the twentieth century, membranes were not used for industrial or commercial purposes, but they were used as laboratory tools to develop physical/chemical theories. For example, the measurement of solution osmotic pressure by membranes by Traube and Pfeffer was used by van't Hoff in 1887 to improve the limit law, which describes the behavior of ideal diluted solutions; this work led directly to the van't Hoff equation. At the same time, Maxwell and others used a perfectly selective semipermeable membrane concept in the development of the kinetic theory of gases [1].

Early membrane researchers have experimented with all sorts of diaphragms available for themselves, such as pigs, cattle or fish and sausage covers made of animal guts. Later, nitrocellulose membranes were preferred because they could be reproducibly produced. In 1907, Bechhold developed a technique for preparing nitrocellulose membranes of a graded pore size determined by the bubble test [2]. Other early workers, the technique of Bechhold and were introduced into the market of microporous nitrocellulose membranes at the beginning of the 1930s [3–5]. In the following 20 years, this early microfiltration (MF) membrane technology has expanded to other polymers, especially cellulose acetate. Membranes found their first important practice in drinking water testing at the end of the Second World War. Drinking water sources serving large communities in Germany and elsewhere in Europe were destroyed and filters were urgently needed to test water safety. The research effort to develop these filters, backed by the US Army, was later exploited by Millipore Corporation, the first and largest producer of US MF membranes. By 1960, elements of modern membrane science were developed, but membranes were used only in a few laboratories and small, specialized industrial applications. An effective membrane industry was not available and the total annual sales of the membranes for all industrial applications probably did not exceed US $ 20 million in 2003. There were four problems that prevented membranes from being widely used as a separation process: they were too unreliable, too slow, too unselective and too expensive. Solutions for each of these problems have been developed over the past 30 years and membrane separation systems have become more common [1].

The first discovery of the conversion of membrane separation into an industrial process from a laboratory appeared with defect-free, high-flux anisotropic reverse osmosis (RO) membranes produced by the Loeb-Sourirajan process at the beginning of the 1960s [6]. These membranes consisted of an ultra-thin, selective surface on a much thicker but more permeable microporous support providing mechanical strength. The flux of the first Loeb-Sourirajan RO membrane was 10 times higher than that of any available membrane, and this performance made the RO potentially a practical method for desalinating water. Loeb and Sourirajan's work and large-scale research and development by the US Department of Interior Office of Saline Water (OSW) have been a pioneer in the commercialization of RO and this has been a major factor in the development of MF and ultrafiltration (UF) membranes. In addition, the development of electrodialysis was supported by OSW funding. With the development of these industrial applications of membranes, the development of

-especially- artificial kidneys, has been provided for medical separation procedures. Kolff et al. [7] demonstrated the first successful artificial kidney in Holland in 1945. It took about 20 years for technology to be applied to large-scale works, but these developments were completed in the early 1960s. Since then, the use of membranes in artificial organs has become an important life-saving procedure [1].

Currently, more than 800,000 people are protected with artificial kidneys and about 1 million of people who have undergone open heart surgery every year through a possible procedure by developing a membrane blood oxygenator. The sales of these devices easily exceed the total industrial membrane separation market. Another important medical application of membranes is for controlled drug delivery systems. An important figure in this area was Alex Zaffaroni, who founded Alza, a company dedicated to developing these products in 1966. Membrane techniques developed by Alza and his competitors are widely used in the pharmaceutical industry to improve efficacy and safety of drug delivery. Significant stages were recorded in the membrane technology of 1960–1980 period. Using the original Loeb-Sourirajan technique, other membrane production methods including interface polymerization and multilayer composite casting and coating were developed to produce high performance membranes. Membranes produced using these methods and containing thin layers of 0.1 μm or less are now produced by many companies. Along with the membrane type, membrane modules have been developed by working on the packing volume and the number of studies for increasing the membrane stability has increased. In the 1980s, large-scale installations involving MF, UF, RO and electrodialysis began to become widespread all over the world [1].

2.2. Conventional membrane processes

RO is primarily used to remove salts from brackish water or seawater while it can reject synthetic organic compounds. One of the latest developed membrane process, nanofiltration (NF), is used to soften fresh water and clear disinfection by-products (DBP) precursors. Electrodialysis is used to demineralize brackish and sea water and to soften fresh water. UF and MF are used to remove turbidity, pathogens and particles in fresh water. In the broadest sense, a membrane, a common element of all of these processes, can be defined as any barrier to the flow of suspended, colloidal or dissolved species in any solvent. Applicable size ranges for membrane processes are shown in **Figure 1**. Generally, the cost of membrane processing increases when the size of the solute is reduced. The ionic range in **Figure 1** includes potable water solubles such as sodium, chloride, total hardness, maximum dissolved solids, and smaller DBP precursors. Macromolecular range includes large and small colloids, bacteria, viruses, and colors. The fine particle range includes particles that produce larger turbidity, most suspended solids, cysts, and larger bacteria. Membrane processes normally used in the ionic range can remove macromolecules and fine particles, but are not as cost effective as larger pore membranes due to some operational problems [9]. The comparison of the membrane properties with each other is given in **Table 1**.

RO is the tightest membrane process in liquid/liquid separation. In principle, water is the only substance passing through the membrane; essentially all dissolved and suspended material is rejected. The RO membranes with much larger pores are sometimes confused with

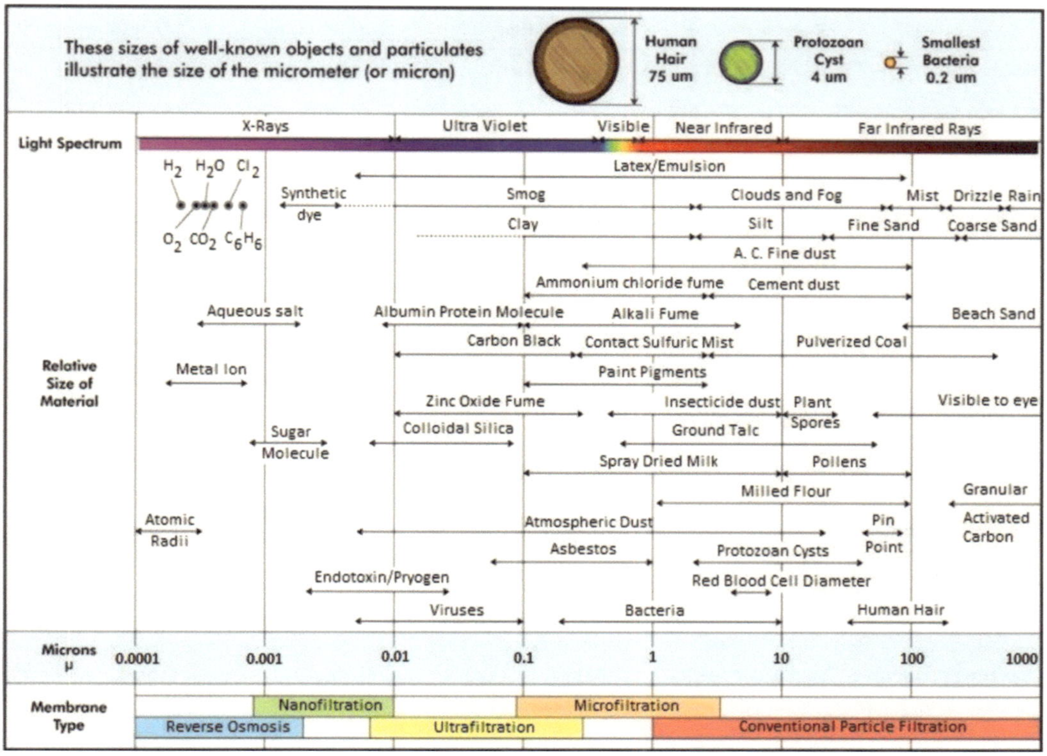

Figure 1. Membrane pore size compared with molecules, bacteria and virus [8].

NF membranes. True NF rejects ions with more than one negative charge, such as only sulfate or phosphate, while passing single charged ions. NF also rejects uncharged, dissolved materials and positively charged ions according to the molecular size and shape. Finally, the sodium chloride (NaCl) rejection varies from 0 to 50%, depending on NF and the rejected feed concentration. In contrast, "loose RO" is an RO membrane with reduced salt rejection. Such membranes are highly effective for a range of applications where moderate desalination is acceptable and, therefore, the operating pressure and power consumption are significantly reduced. Therefore, the costs are reduced in cases where complete desalination is not required.

UF is a process in which all low molecular weights compounds (LMWC) are freely permeable, while the high molecular weights compounds (HMWC), such as proteins, and suspended solids are rejected. Therefore, none of the mono- and di-saccharides, salts, amino acids, organics, inorganic acids or sodium hydroxide are rejected. Microfiltration (MF) is ideally a process where only suspended solids are rejected, and even proteins pass free through the membrane.

A wide range of products using membranes, but water desalination uses more than 80% of all membranes having ever been sold. The remaining 20% is used for -mostly- milk processing, while the rest is sold for use with many different liquids. Some liquids are waste products and some of which are very expensive pharmaceutical products. **Table 2** lists some typical applications and as seen in the table the permeate as well as the concentrate can be the desired product [10].

	Reverse osmosis	Nanofiltration	Ultrafiltration	Microfiltration
Membrane	Asymmetrical	Asymmetrical	Asymmetrical	Symmetrical Asymmetrical
Thickness	150 μm	150 μm	150–250 μm	10–150 μm
Thin film	1 μm	1 μm	1 μm	
Pore size	<0.002 μm	<0.002 μm	0.2–0.02 μm	4–0.02 μm
Rejection of	HMWC, LMWC Sodium chloride Glucose Amino acids	HMWC Mono-, di- and oligosaccharides Polyvalent neg. ions	Macromolecules, Proteins, Polysaccharides Vira	Particles Clay Bacteria
Membrane material(s)	CA Thin film	CA Thin film	Ceramic PSO, PVDF, CA Thin film	Ceramic PP, PSO, PVDF
Membrane module	Tubular, Spiral wound, Plate-and-frame	Tubular, Spiral wound, Plate-and-frame	Tubular, Hollow fiber Spiral wound, Plate-and-frame	Tubular, Hollow fiber
Operating pressure	15–150 bar	5–35 bar	1–10 bar	<2 bar

Table 1. Comparing membrane processes [10].

2.3. Forward osmosis process

van't Hoff's semipermeable membrane, which he assumed to promote the dilution of aqueous solution theory, is a permeable barrier to water (solvent), which is completely impermeable to dissolved solutes. For this reason, removal of the solvents results in a model barrier for all membrane filtration processes where the solutions are retained (concentrated). Like all joining properties, osmotic effects are limited to liquid solutions. Since we know nature is a watery system, the following solvent is water. When pure water and a random aqueous solution come in contact with the environment through a semi-permeable membrane, pure water is "drawn" into the solution, as if to dilute it: Osmosis. As is well known, osmosis is extremely important for the functioning of life when understood as a transport phenomenon at the molecular level. Live cell walls are osmotic barriers with improved selectivity towards inorganic and organic solutes ("biological membranes"). The direction of osmotic water transport, irrespective of the nature of the solution, indicates that the solution has a lower free energy (potential) than pure water. Specifically, the effectiveness of the solvent must be reduced by the effect of the solute, since the model barrier is assumed to communicate only through solvent [11].

Forward osmosis, an evolving separation/desalination process, has received increased interest in both academic research and industrial development in the past decade [12]. In FO, a semi-permeable membrane is placed between two solutions of different concentrations: a concentrated DS and a more dilute FS. Using the osmotic pressure differential to provide water permeation through the membrane, FO may respond to some of the deficiencies of hydraulic pressure driven membrane processes such as RO [13].

		Permeate	Concentrate
RO	Dyeing effluent	Clean water	BOD, salt, chemicals, waste products
	Water	Low salinity water	Salty water
	Whey	Low BOD permeate	Whey concentrate
	Antibiotics	Salty waste product	Desalted, concentrated antibiotics
NF	Dyeing effluent	Clean,salty water	BOD/COD, color
	water	softened water	waste product
	Whey	Salty waste water	Desalted whey concentrate
	Antibiotics	Clarified fermentation broth	Waste product
	Bio-gas waste	Clarified liquid for discharge	Microbes to be recycled
	Carrageenan	Waste product	Concentrated carrageenan
	Enzymes	Waste product	High value product
	Milk	Lactose solution	Protein concentrate for cheese production
UF	Oil emulsion	Oil free water (<10 ppm)	Highly concentrated oil emulsion
	Washing effluent	Clarified water	Dirty water (waste product)
	Water	Clarified water	Waste product
	Whey	Lactose solution	Whey protein concentrate
	Xantan	Waste product	Concentrated xantan

Table 2. Type of membrane process for several products (the shaded area representing the main product) [10].

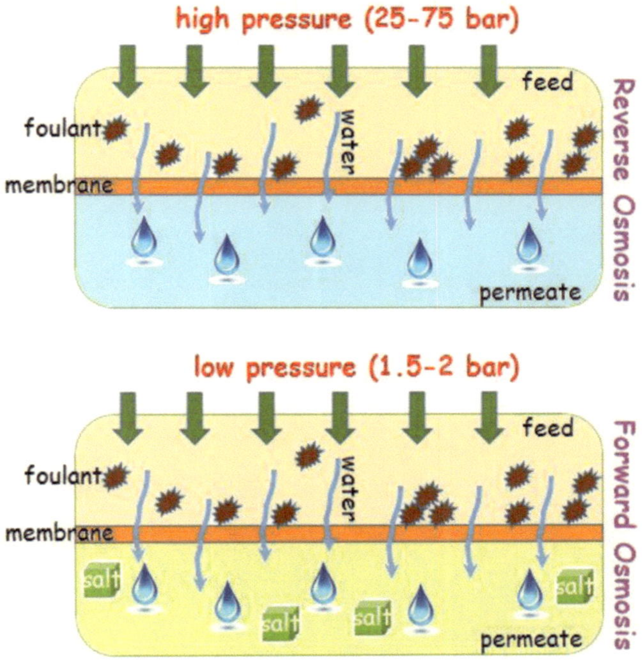

Figure 2. Illustration of comparison between FO and RO processes [16].

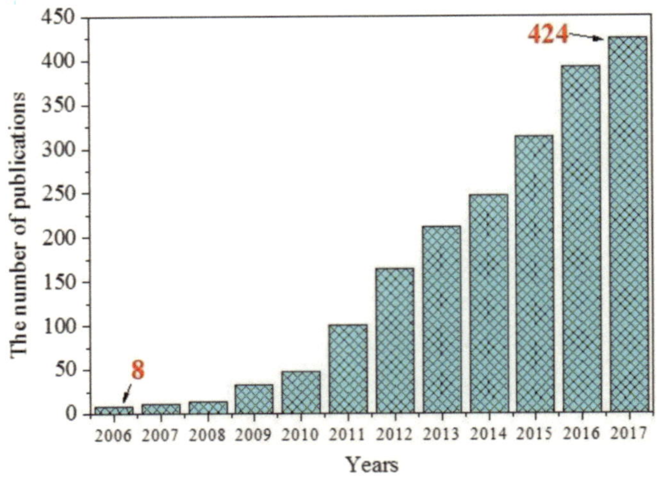

Figure 3. Annual number of publications on FO since 2006 until fourth quarter of 2017 (retrieved from science direct database search) (updated and adapted from Eyvaz et al. [16]).

The transport of water molecules from a semipermeable membrane to the concentrate/saline solution on the other side of the membrane is referred to a technical term as forward osmosis. Contrary to conventional pressure-driven membrane processes, no pressure is applied to the side of water or concentrated solution. The difference in osmotic pressure between the aqueous medium on both sides of the membrane serves as the driving force to transport the water. [14]. Concentrated solution (DS), which pulls water molecules, is diluted during the process. The diluted DS is then re-concentrated in order to separate the water from the DS with a suitable further process. Where appropriate, the re-concentrated solution can be used again as DS. [15]. The FO process is shown in **Figure 2**. The main advantages of FO are it is not operated

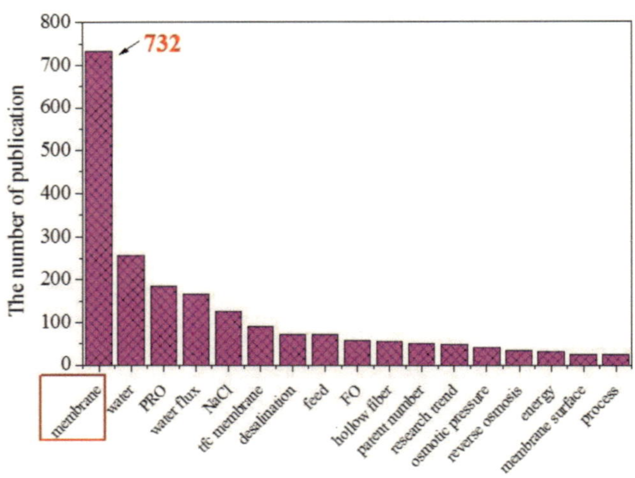

Figure 4. The number of publication about FO studies different research topics since 2006 until fourth quarter of 2017 (retrieved from science direct database search) (updated and adapted from Eyvaz et al. [16]).

under any hydraulic pressure, that a wide range of pollutants can be rejected at a high level and have lower irreversible pollution than pressure-based membrane processes [17].

As a method for water desalination, FO has been investigated for about four decades [18] and many researchers have found that (i) the selection or development of (new) membrane materials [19, 20], (ii) determining the suitable DS [21] The understanding of the mechanism of pollution [22], (iv) the characterization of concentration polarization (CP) [23]. In these pure academic publications, the FO survey and the increasing tendency of various special topics have been shown in **Figures 3** and **4** for the last 10 years. As seen in **Figure 3**, the number of researchers has been steadily increasing, and recent research has focused heavily on membrane properties and development [16].

3. Manufacturing of forward osmosis membranes

Currently used membranes are mostly asymmetric porous membranes [24, 25]. In asymmetric porous membranes, the structure and transport properties change across the membrane thickness. An asymmetric membrane normally consists of a dense layer of 0.1–1 μm thick and supported by a highly porous, 100–200 μm thick support layer [24]. The dense layer provides most of the selectivity for the membrane. The separation properties are determined by the chemical structure, the size of the pores (0.4–1 nm) and the thickness of the skin layer. It is believed that the porous substrate provides mechanical support for the thin and fragile selective layer and has little effect on the separation performance of the membrane. However, recently the effects of the chemical properties of the support layer (eg, hydrophilicity/hydrophobicity) and pore structure (e.g., pore size and porosity) on composite membrane transport have been reevaluated [26–28].

In thin film composite (TFC) membranes, the porous support layer is generally an integrally skinned membrane formed by a non-solvent induced phase separation (NIPS) process. The skin layer is typically formed by either interfacial polymerization (IP) or dip coating followed by crosslinking [24]. The most common thin film chemistry for RO membranes is based on a completely aromatic polyamide (PA) formed by the IP of meta-phenylenediamine (MDP) and trimesoyl chloride (TMC). In contrast, the popular PA NF membranes are formed by IP of piperazine and TMC [29]. It is assumed that the dense selective layer formed by the IP is heterogeneous (20–200 nm) throughout the thickness and is highly cross-linked. The surface properties of a PA film are different from those of the PA dense layer because the polymer density is not uniformly distributed [30]. The PA dense layer is extremely negatively charged because acyl chloride groups are not completely converted to amide during the formation process; however, recent direct titration experiments have demonstrated the presence of both positive and negative fixed charges in the dense layer of composite PA NF membranes [31]. According to [28], Freger and Srebnik suggested that the fixed charge is not uniform and that the film is actually a "sandwich" with two oppositely charged layers [32].

The dense coating layer has been treated as a non-porous film in the past. More advanced identification techniques such as atomic force microscopy, scanning electron microscopy, transmission electron microscopy (TEM), X-ray photoelectron spectroscopy, electron spin resonance, nuclear magnetic resonance (NMR), small angle X-ray scattering, and molecular dynamics simulations have been used to state the structure of the dense layer. In the literature [28], a highly cross-linked PA skin layer structure with sub-nanoscale pores (0.2–1 nm) and low porosity has been reported [33, 34].

Wang et al. [28] stated that mixed matrix membranes contain both organic and inorganic phases. The first mixed matrix membranes were produced to enhance the performance of the gas separation membranes by providing interconnected flow paths of materials with a high diffusion rate [35]. In aqueous separations, mixed matrix membranes are typically formed of a polymer matrix in which inorganic particles are dispersed. Classically, micron-scale inorganic fillers (eg, zeolites and silicalite) have been added to polymer membranes to create preferential flow paths for rapid transport of certain molecules [36]. When nanomaterials (eg, metal and zeolites nanoparticles) are used as the inorganic filler, these membranes are called nanocomposite membranes [37].

Inorganic particles may be present throughout the thickness of a symmetric or integrally-skinned membrane or exclusively in the coating film of a composite membrane. Theoretically, mixed matrices add an additional degree of freedom to membrane production because the advantages of a particular filler material can be imbued into a bulk membrane material [35]. Mixed matrices have been used to enhance the mechanical and chemical stability of organic membranes and to add specific functionality to the interface of polymer membranes, such as desired degradation, reduced fouling or increased selectivity [38, 39]. Organic and inorganic hybrid membranes are very interested in using it as a new generation of membrane materials for water treatment. According to Wang et al. [28], scientists have begun to use nanoparticles TiO_2 [40], carbon nanotubes [41], zeolites [42], clay [43], nonporous amorphous silica [44] and such as, to increase the water flux.

In another study [45] a new nanocomposite FO membrane is designed to perform oil/water separation and desalination at the same time. This nanocomposite FO membrane consists of an oil-repelling and salt-rejecting hydrogel selective layer on the top surface of graphene oxide (GO) nanosheets grafted into a polymeric support layer (**Figure 5**). This selective layer demonstrates strong underwater oleophobicity, which leads to superior anti-fouling properties under various oil/water emulsions and ICP can be decreased by and can be significantly reduced by GO in view of membrane structural parameter (S) decrease by about 20%. Compared to the commercial FO membrane, this new FO membrane has a markedly low membrane fouling tendency, having higher removal rates for oils and salts (>99.9% in oil and >99.7% in multivalent ions) for treatment of simulated shale gas wastewater (**Figure 6**). These combined advantages will endorse this new FO membrane in the treatment of highly saline and oily wastewater.

Xu et al. [46] reported that the availability of suitable FO membranes is crucial for the development of FO technology. Problems such as high RSD, high concentration polarization (CP) and poor mechanical strength are frequently encountered in FO processes. Meanwhile, although FO tends to exhibit a lower membrane fouling than pressure driven membrane processes, fouling is still the most serious problem that adversely affects FO performance. To overcome these problems, many new FO membranes have been fabricated or ready-made membranes have been modified by means of surface chemistry in recent years [47].

Chung et al. [48] stated that a few comprehensive reviews on FO membrane development are available in the literature [12, 18, 49]. Basically, most FO membranes are fabricated with conventional phase inversion [49] and TFC by IP processes [50]. Each layer of FO membranes have

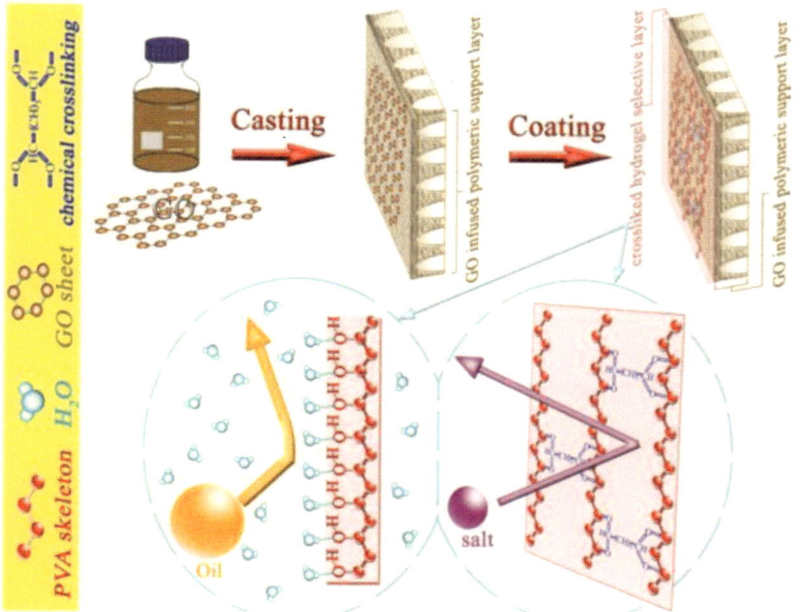

Figure 5. Illustration of the synthetic process and work mechanisms of hydrogel/GO FO membrane [45].

been investigated, but reverse solute diffusions (RSD) tend to be high [51]. Employing hydrophilic materials as substrates in FO membranes is crucial to increase water flux [52]. Recently, TFC-FO-membranes synthesized on nanofiber [53, 54] and multi-bore [20] surfaces with good mechanical properties have also been reported. Future R & D should focus on innovative membranes with low fouling and ICP. Until recently, double-skinned FO membranes with dense RO skin and a loose RO skin, have been promised reduced membrane fouling and ICP [55].

Xu et al. [46] declared that the FO membrane serves as a selective barrier to control the water transport and solute retention to maintain the separation efficiency. The initial attempt to use the RO membrane in the FO process faced with some operational limitations; such as low flux, due to the thick sponge-like substrate and compact support of the RO membrane hindering mass transfer and causing severe ICP within the support layer [28–30]. Hydration Technologies, Inc. (HTI) developed the first commercial FO membranes [56], one of which has a characteristic structure embedding a thin polyester mesh support in cellulose triacetate (CTA) (**Figure 7**). These membranes provide significantly better separation performance than commercially available RO membranes. In addition to the commercial CTA FO membranes, HTI has then introduced the TFC FO membrane to the market; The flux of the spiral element was twice that of existing CTA membranes. This is thought to be a new criterion in future studies on FO membranes [58]. However, FO membranes with superior water permeability and salt rejection are still subjects to be developed for commercialization of FO technology. SEM images of the other some commercial FO membranes are presented in **Figure 8**.

According to the research in literature, it is shown that the adjustment of the sub-layers is of great potential in tailoring PA-TFC membranes. In addition, the flexibility of PA-TFC membrane

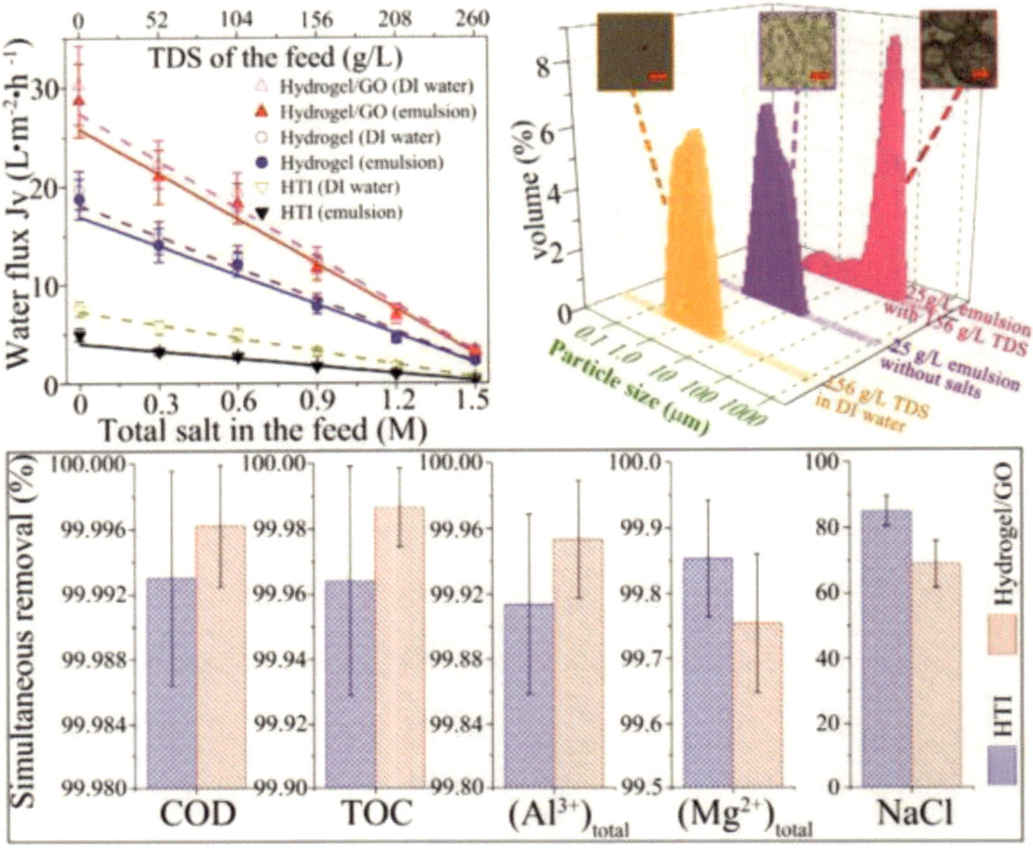

Figure 6. The study of simultaneously deoiling and desalting shale gas wastewaters [45].

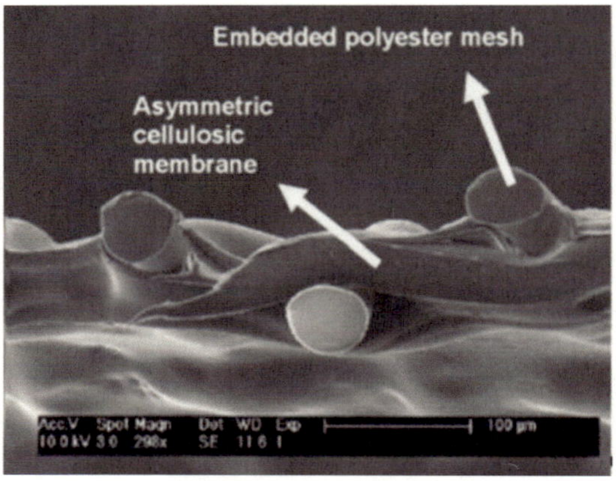

Figure 7. Cross-sectional SEM images of hydration technologies CA (the bar in each SEM image is 100 μm) (adapted from McCutcheon and Elimelech [57]).

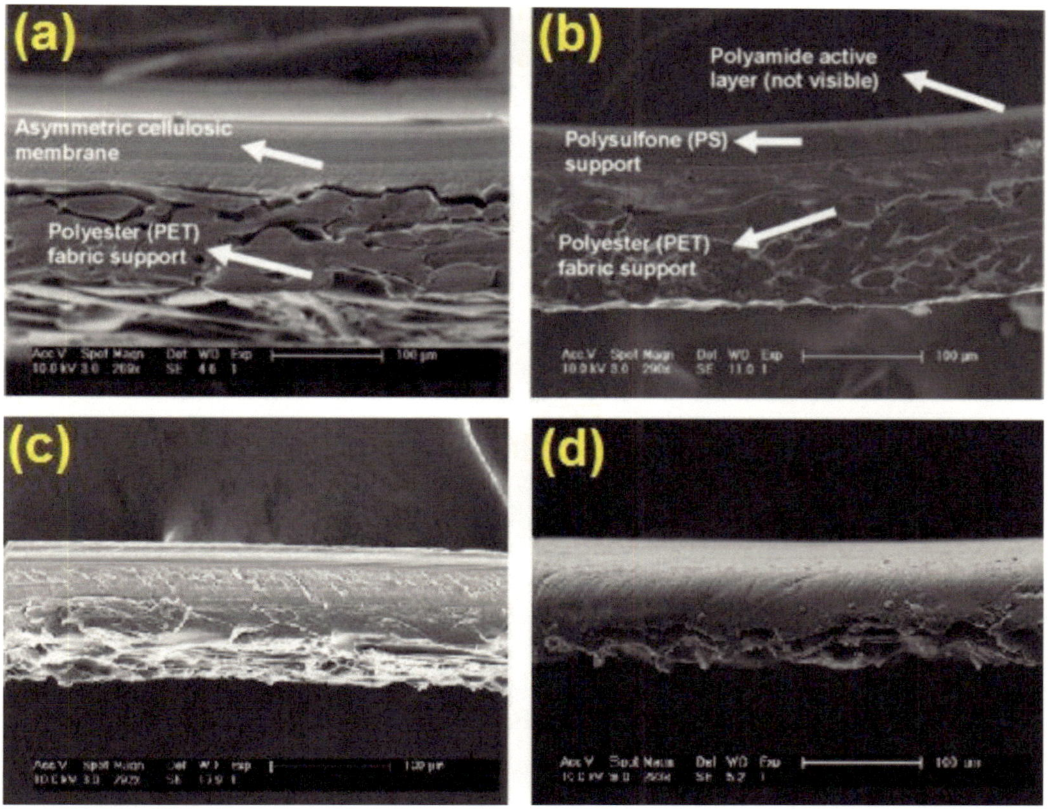

Figure 8. Cross-sectional SEM images for the (a) Osmonics CE, (b) Filmtec SW30 XLE and (c) Filmtec SW30 XLE, (d) Osmonics CE membrane with the fabric layer removed (the bar in each SEM image is 100 μm) (adapted from McCutcheon and Elimelech [58]).

structures has a positive effect on the improvement of the sub-layers, since each of the surface active and support layers can be individually constituted for a particular purpose (**Figure 9**) [59].

3.1. Support layer properties and manufacturing techniques

Li et al. [60] have recently reviewed recent researches on polymer and polymer composite membranes for RO and FO processes comprehensively. In one of these studies, a TFC FO flat membrane has a thin selective layer on top of a flat porous polymeric support that is produced by phase inversion with/without a thin nonwoven layer [50]. More recently, nanofiber mats with high porosity have been proposed as a support layer to reduce the ICP to a minimum [53, 54, 61]. Bucky-papers made from CNTs are also being tested as support layer candidates due to their flexibility, strength and high porosity, it is also recommended to investigate other low cost and high porosity materials such as metal oxide nanotubes [62]. Parallel to the studies with TFC flat plate modules, the number of research related to hollow fiber configuration is also increasing due to its advantages such as high packing density and enhanced flow

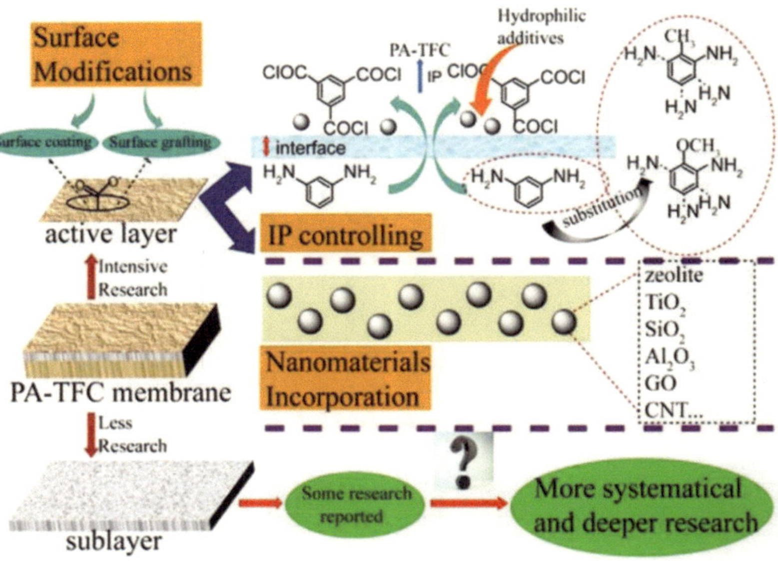

Figure 9. Flexibility in tailoring PA-TFC membranes by adjusting the surface active and support layers individually [59].

pattern and self-supporting structure [20, 63]. Under the same drive force effect, higher fluxes were obtained with hollow fiber membranes than with flat sheet membranes [64].

Similar to the characteristics of the support layers required to produce high performance PA TFC RO membranes, it is desirable that the support layers of the TFC FO membranes have high hydrophilicity, stability and mechanical strength, [65]. In addition, resistance to chemicals, temperature and oxidation, as well as low fouling tendencies, increase the potential use of FO membranes in harsh industrial environmental conditions. [66]. However, the thin support layer, high porosity, and low tortuosity will help reduce ICP [67]. For this reason, some of the research related to FO membranes have focused on support layer fabrication and modification. In these studies, fabrication parameters such as optimization of polymer concentration, solvent composition and functional additives have been considered in the synthesis of the support layer [68]. For example, a simultaneous casting of two polymer solutions with a co-casting technique, a synthesized support layer has played an important role in reducing ICP, improving water flux and reducing reverse salt flux [69]. Furthermore, even the selection of non-woven fabrics underneath the polymeric support layer significantly affects the adhesion of the support layer polymer to this non-woven sheet. For example, selecting a fabric with high tortuosity, large thickness, and low porosity leads to an undesirable decrease in the water flux of the FO membrane. Moreover, the addition of foreign components to the substrate casting solution can help improve the substrate properties. For example, when lignin content is incorporated into the polysulfone (PSf) substrate, the bulk porosity enhanced, shorter diffusion pathway is provided and TFC membrane performance is improved [70]. By using diethylene glycol as a pore forming agent in the PSf/sulfonated poly (ether ketone)/N-methylpyrrolidone (NMP) casting solution, greater porosity and wider pore size distribution were obtained which reduced the resistance of the support layer to the solution and the reverse salt flux was also relatively controlled [65]. Addition of PEG to the preparation of the CAP substrate increased the connectivity of the pores and prevented macrovoids, as well as positively interacted with the cast glass blade. The resulting support is suitable for high performance TFC FO membrane fabrication since it has a high porosity bottom and a tight top surface [52]. The result is that the TFC membrane with macrovoids (or finger-like structure) support facilitates mass transport and reduces ICP in FO [67]. However, these porous structures may become mechanically weak points in the membrane structure and may, in practice, worsen membrane separation under continuous water flow or backwash conditions [71]. The highly porous support can also increase the difficulty in forming an excellent active selective layer with the necessary mass transport properties [67]. However, the sponge-like structure of small pores surrounded by dense walls may be convenient to form an integral thin active layer and may exhibit better mechanical stability on the finger-like property; however, it increases resistance to mass transfer [64]. Experimental studies suggest that the ideal support with a sponge-like film on a finger-like substrate is very important in the fabrication of high-performance TFC FO membranes [50].

Recently, nanofiber support layers with scaffold-like and interconnected porous structures have been seen as promising alternatives to overcoming the disadvantages of sponge-like structures. The nanofiber supported PA TFC membranes exhibited much lower S (~80 μm) than a commercial HTI FO membrane (S = 620 μm) and thus with a low molarity (0.5 M NaCl) DS and a DI water as the FS, it has been observed that the water flux has increased by five times [72].

An FO membrane with tubular nanofiber support was manufactured for the first time in the study of Arslan et al. [53]. In the first stage of the manufacturing, the support layer (polyacrylonitrile (PAN) nanofiber) was coated on the hollow braided rope (backing layer) by electrospinning method. In the second step, the active layer called the TFC layer was coated on the formed nanofiber by the IP process. Schematic illustration of FO membrane manufacturing is shown in **Figure 10** and SEM images are presented in **Figure 11**. The TFC layer is the main selective barrier that prevents the transfer of the salt to the diluted side and allows the water molecules to diffusion into the DS side.

According to aforementioned review [60], Han et al. [65] pointed out that the hydrophilicity and support layer thickness are critical parameters in controlling of water transport. It is reported that the TFC membrane with support layer which is completely sponge-like and has a hydrophilic upper surface, provides a higher water flux than a TFC membrane with support layer which is completely hydrophobic and has finger-like structure. In order to prepare or modify the support layer, hydrophilic materials such as sulfonated polysulfone (sPSf), sulfonated copolymer made of polyethersulfone (PES) and polyphenylsulfone (PESU-co-sPPSU), sulfonated poly(ether ketone), poly- dopamine (PDA) or poly(vinyl alcohol) (PVA) have been explored [73–75]. Emadzadeh et al. improved the mass transfer and reduced ICP by applying TFC on the PSF support layer containing TiO_2, thus increasing the water flow of the FO membrane [76].

Liang et al. [77] for the first time in the production of TFC-FO membranes, it has been proposed to use vertical porous substrates as a support layer. The addition of acetone in aqueous phase promotes IP on vertical porous substrates. Positron annihilation lifetime analyses indicated that new FO membranes in the study have thicker and dense selective layers than conventional FO membranes with asymmetric substrates. These new FO membranes have a low structural parameter, indicating a greatly reduced ICP effect.

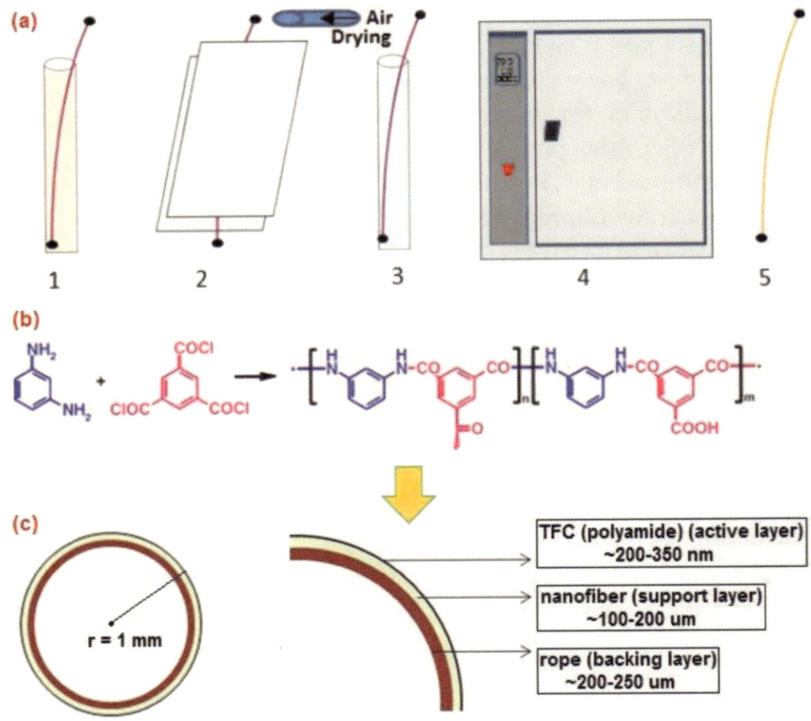

Figure 10. Demonstration of tubular nanofiber FO membrane manufacturing steps: (a) Steps of TFC process, 1— immersion of the tubular nanofiber membrane into the MPD solution, 2—air drying, 3—immersion of the tubular nanofiber membrane into the TMC solution, 4—heat treatment, 5— tubular nanofiber FO membrane. (b) Reaction mechanisms of the polyamide formation from MPD and TMC. (c) Schematic representation of cross-section of tubular nanofiber FO membrane [54].

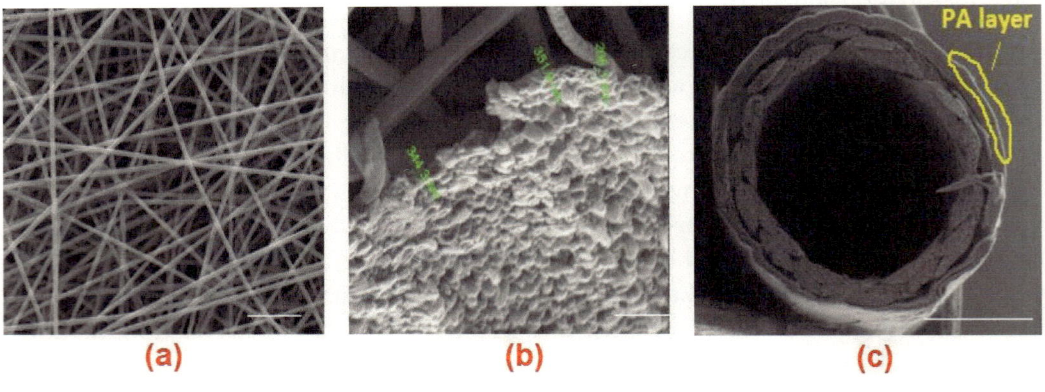

Figure 11. SEM images of (a) nanofibers in support layer at magnification of 10,000×, (b) gel-like formation of the polyamide layer on nanofibers at magnification of 50,000×, (c) cross-section of tubular nanofiber supported FO membrane at magnification of 10,000× (adapted from Arslan et al. [53]).

The selective layer exhibits an unprecedented water flux up to 93.6 L/m^2.h (LMH) (**Figure 12**) when driven by a 2 M NaCl as DS. This performance is evidenced by the FO membranes reported in the literature and commercially available. The authors suggested according to the results that substrates with vertically oriented porous structure are ideal supports for

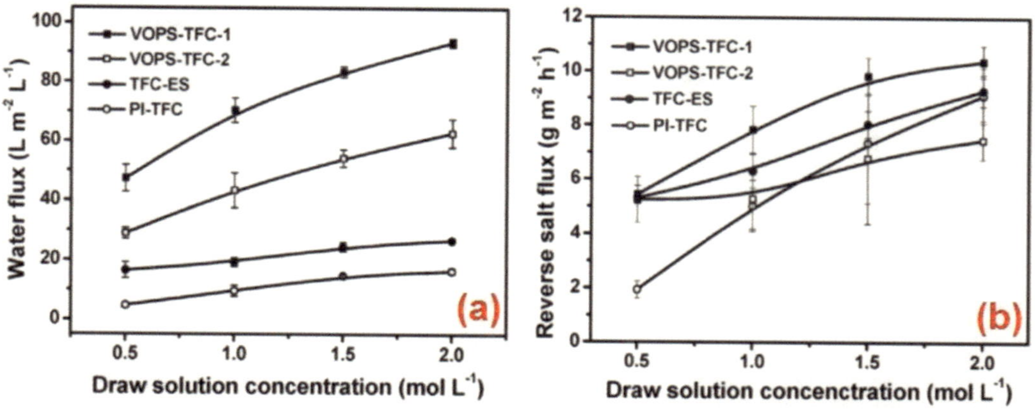

Figure 12. (a) Water flux and (b) reverse salt flux of the FO membranes in a process mode of the selective layer facing the FS (VOPS: vertically oriented porous substrates) [77].

developing FO membranes with lower ICP and ultra-high water flux. Proposed lower ICP in vertically oriented porous layer compared with a tortuous sponge-like structure in FO membranes by means of salt concentration profile are illustrated in **Figure 13** [77].

In a more recent study, Kwon et al. [78] produced a highly permeable and mechanically resistant TFC-FO membrane with a new support layer which has been already commercialized porous polyethylene (PE) membrane as the lithium ion battery separator. The very open and interconnected pore structure of the PE support, when combined with the thickness (~8 μm), is useful for alleviating the ICP, thus increasing the FO water flux. The use of a suitable plasma treatment and a surfactant in the PE support resulted in a stable formation of a PA permselective layer on the support by IP process. The prepared PE supported TFC (PE-TFC) membrane exhibited high water flux and low reverse salt flux performance due to its significantly low structural parameter. The performance values obtained in this study are also compared with other flux values in the literature (**Table 3**). The PE-TFC membrane has superior mechanical properties compared

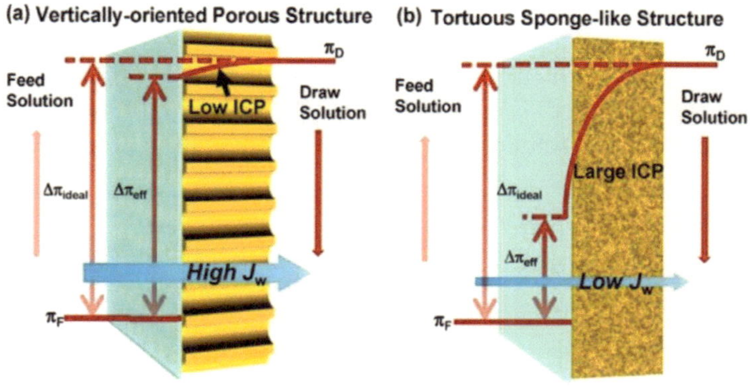

Figure 13. The salt concentration profiles of (a) VOPS-TFC and (b) PI-TFC FO membranes. πD and πF denote the osmotic pressure of draw solution and FS, respectively. $\Delta\pi_{ideal}$ indicates the osmotic pressure difference between the bulk feed and the bulk draw solution. $\Delta\pi_{eff}$ means the effective osmotic pressure driving force due to the presence of ICP effect [77].

Membranes	S (μm)	J_w (LMH)	J_s/J_w (g/L) (FO/PRO)	Refs.
PE-TFC	161	41.9/45.1	0.50/0.49	[78]
PES hollow fiber-TFC	219	26.5/37.6	0.17/0.14	[64]
PAN flat sheet-TFC	350	28.8/36.3	0.10/0.13	[79]
Cellulose ester flat sheet-TFC	32	56.9/89.5	0.14/0.12	[52]
PESU-co-sPPSU flat sheet-TFC	324	20.0/25.0	0.12/0.13	[80]
PTA-co-POD flat sheet-TFC	236	37.5/78.4	0.15/0.15	[81]
PSf/SPPO (50 wt% SPEK) flat sheet-TFC	381	16.0/32.0	0.28/0.19	[82]
PSU/SPEK (50 wt% SPEK) flat sheet-TFC	107	23.0/34.0	0.18/0.20	[65]
PES/SPES (50 wt% SPES) flat sheet-TFC	245	25.2/33.7	0.28/0.26	[83]
Polydopamine-coated PAI flat sheet-TFC	456	14.0/48.0	0.44/0.17	[84]
PSf-silica NPs flat sheet-TFC	216	31.0/60.5	0.24/0.26	[85]
PSf-zeolite flat sheet-TFC	340	33.0/65	0.55/0.47	[86]
PSf-LDH flat sheet-TFC	148	18.1/34.6	0.45/0.36	[87]
PES nanofiber-TFC	106	46.0/50.0	—	[72]
Nylon 6,6 nanofiber-TFC	—	21.0/27.0	0.24/0.44	[88]
PVDF nanofiber-TFC	193	22.0/31.0	0.17/0.43	[89]

Table 3. Comparison of the FO performance of RO-grade membranes (FS: DI water, DS: 1.0 M NaCl) [78].

to the much thicker commercial FO membrane due to the exceptionally high mechanical integrity of the PE support. The proposed strategy offers a new material platform for FO membranes with strong commercial potential and excellent performance and durability.

In another recent study on the support layer, Zhang et al. [90] manufactured hollow fiber FO membranes with improved thermal stability using IP process on the lumen side of the co-poly (phthalazinone biphenyl ether sulfone) (PPBES) substrate. The increase in water flow in the PPBES substrate also increased the flow in the FO membrane. IP preparation parameters such as solvent, monomer concentrations, reaction time and curing conditions have been shown to seriously affect the development of composite FO membrane properties. The water flux of composite FO membranes increased from 24.0 to 66.5 LMH without a significant change in salt flux/water flux (J_s/J_w) ratio when the draw solution temperature was raised from 23 to 85°C (**Figure 14**).

3.2. Active layer properties and manufacturing techniques

Li et al. [60] in their comprehensive review stated some recent applications on manufacturing or modifying active layer of FO membranes. Accordingly, the preparation of PA TF FO membranes is also similar to the preparation of TFC RO membranes. It is necessary to optimize the parameters such as the reaction time and the air drying duration and compositions of the monomers. Klaysom et al. [91] noted that both the surfactant additive and the drying of excess amine solution prior to the reaction are two critical parameters in PAN support preparation to control membrane properties. The addition of SDS increases polymerization and helps to form

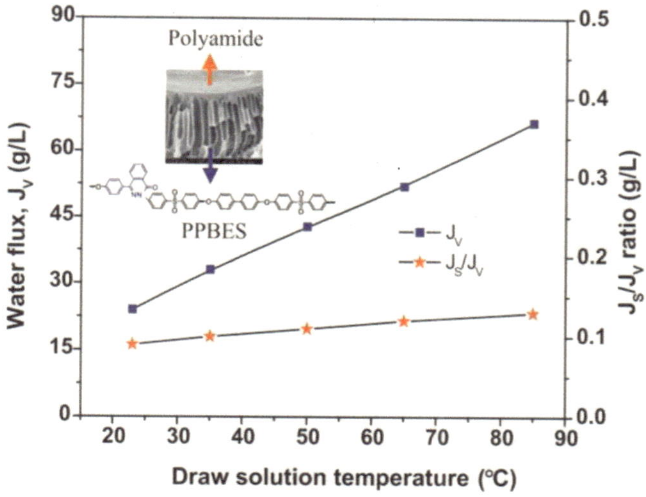

Figure 14. Water flux changing of thermally-stabile FO membrane with draw solution temperature [90].

a uniform and highly cross-linked PA film. Thus, the rate of salt uptake in this study increased from 57% to over 95%, and the decrease in permeability did not occur. On the other hand, the removal of the excess amine solution before interacting with TMC resulted in the formation of a less rough membrane with improved salt rejection [91]. Due to the ionic interaction between cetyltrimethylammonium chloride (CTAC) and m-phenylenediamine (MPD) in the aqueous solution, CTAC may alter the reaction of the monomers of the presence and polymer molecular aggregation. Increasing the CTAC content improves the formation of the linear PA structure and microcrystalline structure of the active layer, but consequently the water flux of the PA TFC hollow fiber membrane with PES support layer is reduced, despite the high reverse salt selectivity [92]. Thermal annealing after SDS/glycerol treatment on TFC FO membranes facilitates the removal of residual unreacted monomers from the surface of the active layer, increasing the free volume size/fractional free volume ratio and reducing the total membrane thickness; so that the water flux can be improved without losing the rejection performance of the membrane [52]. Another major problem encountered in FO processes is membrane fouling, although it is less severe and reversible compared to RO processes. The structures of the support layers also significantly affect the active layer properties and hence the fouling characteristics of the TFC FO membranes. Surfaces with high roughness and large leaf-like structures are more prone to foulant accumulation and exhibit a dramatic decline in flux through these membranes, making it more difficult to improve the flux by physical cleaning of the membrane [93]. When the TFC FO membrane surface is modified, for example by covalent attachment of PEG, the tendency of surface contamination is significantly reduced due to surface barriers that adsorb pollutants [74]. On the other hand, attachment of the functionalized silica on TFC membrane via covalent amide bonds between amine groups of functionalized nanoparticles and the carboxyl groups of the TFC surface improve the fouling resistance and reduce the BSA or alginate adhesion. This is explained by the presence of the tightly bound hydration layer and the reduction of the charged carboxyl groups on the TFC membrane surface [94].

A more recent research in the literature has produced nanoporous thin-film inorganic (TFI) FO membranes with a tetraethylorthosilicate-driven sol-gel process (**Figure 15(a)**). The produced

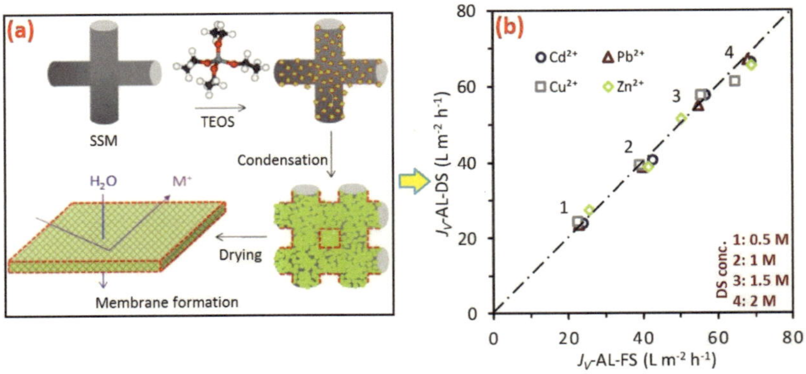

Figure 15. (a) Schematic diagram of TFI membrane formation (SSM: Stainless steel mesh, TEOS: Tetraethylorthosilicate), (b) FO water flux obtained under AL-FS and AL-DS mode at initial heavy metal concentration of 200 mg/L and pH 4.5 ± 0.5 for TFI membrane (adapted from You et al. [95]).

membrane was used for the removal of four typical ionic divalent heavy metals. In laboratory scale FO process, 69 LMH flux was obtained by using four heavy metal containing FS at pH 4.5 and 2 M NaCl as DS (**Figure 15(b)**). An average of 94% metal removal from the 200 mg/L FS solution was obtained. Since the hydrated ion diameters of the metals are smaller than the membrane pore size, the charge-interaction should be responsible for heavy metal rejection. Based on the classical Debye-Hückel theory and the Gouy-Chapman model, You et al. [95] have shown the importance of double-layer overlap in the membrane pore induced by electrostatic interaction between heavy metal ions and silica-made pore walls. Thus, the selectivity of the TFI membrane depends primarily on the function of the membrane pore size, the surface potential of the membrane pore wall, and Debye length (**Figure 16**). This study not only confirms the feasibility of the TFI membrane in the treatment of acidic heavy metal wastewater without pH adjustment, but it also suggests a simple theoretical scheme for better understanding and design of the charged membrane for FO applications.

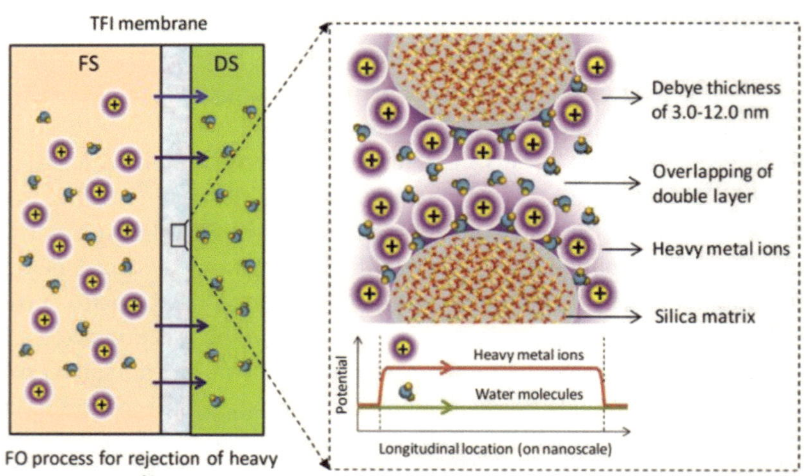

Figure 16. Schematic illustration of proposed mechanisms for rejection of heavy metal ions in FO process based on charge exclusion effect within the pores of TFI membrane [95].

Salehi et al. [96] in their work, fabricated a new and highly efficient FO membrane by using electrostatic interaction on a porous support layer employing layer-by-layer (LbL) assembly of positive chitosan (CS) and negative GO nanosheets. The support layer was prepared by mixing the hydrophilic sulfonated polyethersulfone (sPES) and PES using the wet phase inversion process (**Figure 17**).

Various characterization techniques have been used to confirm that the LbL membrane has been successfully fabricated. The number of layers formed in the SPES-PES support layer was easily adjusted by repeating the CS and GO deposition cycles. A TFC membrane with the same SPES-PES support layer and PA active layer was also prepared to compare membrane performances. Water permeability and salt rejection of the fabricated membranes were obtained with two types of DS (including Na_2SO_4 and sucrose) for two different membrane orientations. The results showed that the membrane coated by a CS/GO double layer had a flow rate of 2–4 orders of magnitude as much as the TFC. By increasing the number of CS/GO double layers, the selectivity of the LbL membrane was improved. The newly fabricated LbL membrane showed better fouling resistance than the TFC in the FS containing 200 ppm sodium alginate as the foulant model (**Figure 18**).

Xu et al. [60] reported that generally, the flux obtained in the active layer facing draw solution (AL-DS) (PRO) configuration is higher than in the active layer facing feed solution (AL-FS) (FO) mode, but more fouling may occur in the PRO mode if the FS containing scalants/foulants is easily transported to the porous support layer. Two active layered hollow fiber membranes, one at the top of the high porosity support layer and one at the bottom, have been proposed by Wang's group [97] so that scaling or fouling can be controlled without reducing water flux in the AL-DS mode. The hollow fiber membranes with RO and NF-like scales fabricated on a PAI support were subjected to high water flux and reverse salt flux values (41.3 and 5.2 LMH) using 2 M NaCl DS and DI as FS in AL-DS mode after IP reaction and polyethyleneimine (PEI) modification. In addition, the presence of the NF-like layer on the support layer can greatly increase the resistance to scaling in the AL-DS mode. A double-skinned hollow fiber membrane with $CaHPO_4$ scaling with a 2-hour backwash recovers 96% of the water flux while a hollow fiber membrane with a single RO selective layer has a recovery of 78% [97]. Recent studies have used polyelectrolyte LbL to form an NF-like skin in the support layer studies conducted without chemical modification, while PA-RO-like layers have also been formed. Since the resulting NF-like skin does not directly contact

Figure 17. Schematic illustration of CS/GO LbL assembly procedure [96].

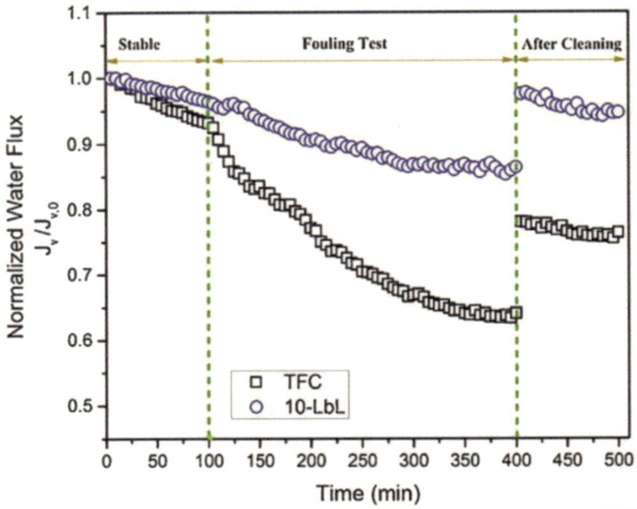

Figure 18. Fouling behavior and flux recovery of the TFC and 10-LbL membranes [96].

the FS and support layer, it prevents the transport of pollutants such as humic acid, dextran and lysozyme and thus the pore clogging. As a result, for a double-skinned hollow fiber membrane, the decrease in water flux was less than 30% during 4 hours operation, whereas for RO layer hollow fiber membranes, this reduction was 30–40% when 200 ppm foulant was used [98].

4. Conclusions

The active layer of an ideal FO membrane must be very thin and dense to achieve high salt retention. In order for the membrane to be able to be operated for a long time and the internal concentration polarization to be low, the support layer should be thin, hydrophilic, porous and exhibit mechanical strength as possible. The hydrophilicity must be high so that high flux and low fouling can be achieved. According to the current studies, utilizing novel nanomaterials, substrates and layer-by-layer assumptions in manufacturing of FO membrane undoubtedly enhance the water flux, rejection of the pollutants and minimize the membrane fouling but using synthetic wastewater -generally- containing one model foulant or DI water as feed solution makes it difficult to predict how FO membranes will act in real wastewaters or harsh environmental conditions. Therefore working with complex foulants and real wastewaters to better understanding of membrane behaviors, using modeling tools for fouling prediction and new cleaning strategies are essential to mitigate intrinsic challenges of the FO membranes.

In on-going researches, the developed new support layers appears continue to increase water flux slightly, however, lower water flux remains as a main challenge of the process when compared the conventional membrane systems. It is also a fact that the diffusion provided by draw solution in the process is not effective alone to increase product water volume, therefore, some promotive factors such as rehabilitated hydrodynamic behaviors or simultaneous filtration could be provided together with diffusion phenomena in further researches.

Acknowledgements

This work was supported by the Scientific and Technological Research Council of Turkey (TUBITAK), grant number: CAYDAG-113Y340.

Author details

Murat Eyvaz[1]*, Serkan Arslan[1], Derya İmer[2], Ebubekir Yüksel[1] and İsmail Koyuncu[2]

*Address all correspondence to: meyvaz@gtu.edu.tr

1 Environmental Engineering Department, Gebze Technical University, Gebze-Kocaeli, Turkey

2 Environmental Engineering Department, İstanbul Technical University, Maslak-İstanbul, Turkey

References

[1] Baker RW. Membrane Technology and Applications. 2nd ed. UK: John Wiley & Sons, Ltd.; 2004. 538 p. DOI: 10.1002/0470020393

[2] Bechhold H. Kolloidstudien mit der Filtrationsmethode. Zeitschrift für Physikalische Chemie. 1907;**60**:527-533. DOI: 10.1002/bbpc.19070133207

[3] Elford WJ. Principles governing the preparation of membranes having graded porosities. The properties of 'Gradocol' membranes as Ultrafilters. Transactions of the Faraday Society. 1937;**33**:1094-1104. DOI: 10.1039/TF9373301094

[4] Zsigmondy R, Bachmann W. Uber Neue Filter. Zeitschrift für Anorganische und Allgemeine Chemie. 1918;**103**:119-128. DOI: 10.1002/zaac.19181030107

[5] Ferry JD. Ultrafilter membranes and ultrafiltration. Chemical Reviews. 1936;**18**:373-455. DOI: 10.1021/cr60061a001

[6] Loeb S, Sourirajan S. Sea water demineralization by means of an osmotic membrane. In: Gould RF, editor. Saline Water Conversion–II, Advances in Chemistry Series Number 28. Washington, DC: American Chemical Society; 1963. pp. 117-132. DOI: 10.1021/ba-1963-0038.ch009

[7] Kolff WJ, Berk HTHJ, Welle NM, van der Ley AJW, van Dijk EC, van Noordwijk J. The artificial kidney: A dialyser with great area. Acta Medica Scandinavica. 1944;**117**:121-134. DOI: 10.1111/j.0954-6820.1944.tb03951.x

[8] McGowan W. Water Processing: Residential, Commercial, Light Industrial. 3rd ed. Water Quality Association: Illinois; 2001

[9] Taylor JS, Wiesner M. Membranes. In: Letterman RD, editor. Water Quality and Treatment: A Handbook of Community Water Supplies. 5th ed. New York: AWWA-McGraw-Hill; 1999. pp. 11.1-11.71

[10] Wagner J. Membrane Filtration Handbook Practical Tips and Hints. 2nd ed. Minnetonka, MN: Osmonics Inc.; 2001. p. 127

[11] Böddeker KW. Liquid Separations with Membranes: An Introduction to Barrier Interference. Berlin Heidelberg: Springer; 2007. p. 146

[12] Zhao S, Zou L, Tang CY, Mulcahy D. Recent developments in forward osmosis: Opportunities and challenges. Journal of Membrane Science. 2012;**396**:1-21. DOI: 10.1016/j.memsci.2011.12.023

[13] Shaffer DL, Yip NY, Gilron J. Elimelech M. Seawater desalination for agriculture by integrated forward and reverse osmosis: improved product water quality for potentially less energy. Journal of Membrane Science. 2012;**415-416**:1-8. DOI: doi.org/10.1016/j.memsci.2012.05.016

[14] Lutchmiah K, Lauber L, Roest K, Harmsen DJH, Post JW, Rietveld LC, van Lier JB, Cornelissen ER. Zwitter ions as alternative draw solutions in forward osmosis for application in wastewater reclamation. Journal of Membrane Science. 2014;**460**:82-90. DOI: 10.1016/j.memsci.2014.02.032

[15] Liu Y, Mi B. Combined fouling of forward osmosis membranes: Synergistic foulant interaction and direct observation of fouling layer formation. Journal of Membrane Science. 2012;**407-408**:136-144. DOI: 10.1016/j.memsci.2012.03.028

[16] Eyvaz M, Aslan T, Arslan S, Yuksel E, Koyuncu İ. Recent developments in forward osmosis membrane bioreactors: A comprehensive review. Desalination and Water Treatment. 2016;**57**(59):28610-28645. DOI: 10.1080/19443994.2016.1193448

[17] Holloway RW, Childress AE, Dennett KE Cath TY. Forward osmosis for concentration of anaerobic digester centrate. Water Research. 2007;**41**:4005-4014. DOI: 10.1016/j.watres.2007.05.054

[18] Alsvik IL, Hägg MB. Pressure retarded osmosis and forward osmosis membranes: Materials and methods. Polymer. 2013;**5**:303-327. DOI: 10.3390/polym5010303

[19] Flanagan MF, Escobar IC. Novel charged and hydrophilized polybenzimidazole (PBI) membranes for forward osmosis. Journal of Membrane Science. 2013;**434**:85-92. DOI: 10.1016/j.memsci.2013.01.039

[20] Luo L, Wang P, Zhang S, Han G, Chung TS. Novel thin-film composite tri-bore hollow fiber membrane fabrication for forward osmosis. Journal of Membrane Science. 2014;**461**:28-38. DOI: 10.1016/j.memsci.2014.03.007

[21] Ou R, Wang Y, Wang H, Xu T. Thermo-sensitive polyelectrolytes as draw solutions in forward osmosis process. Desalination. 2013;**318**:48-55. DOI: 10.1016/j.desal.2013.03.022

[22] Linares RV, Yangali-Quintanilla V, Li Z, Amy G. NOM and TEP fouling of a forward osmosis (FO) membrane: Foulant identification and cleaning. Journal of Membrane Science. 2012;**421-422**:217-224. DOI: 10.1016/j.memsci.2012.07.019

[23] Tan CH, Ng HY. Revised external and internal concentration polarization models to improve flux prediction in forward osmosis process. Desalination. 2013;**309**:125-140. DOI: 10.1016/j.desal.2012.09.022

[24] Mulder M. Basic Principles of Membrane Technology. London: Kluwer Academic Publishers; 1996. p. 564

[25] Loeb S, Sourirajan S. High Flow Semipermeable Membrane for Separation of Water from Saline Solutions, US Patent. 3,133,132. May 12, 1964

[26] Ghosh AK, Hoek EMV. Impacts of support membrane structure and chemistry on polyamide-polysulfone interfacial composite membranes. Journal of Membrane Science. 2009;**336**:140-148. DOI: 10.1016/j.memsci.2009.03.024

[27] Pendergast MM, Hoek EMV. A review of water treatment membrane nanotechnologies. Energy & Environmental Science. 2011;**4**:1946-1971. DOI: 10.1039/C0EE00541J

[28] Wang J, Dlamini DS, Mishra AK, Pendergast MTM, Wong MCY, Mamba BB, Freger V, Verliefde ARD, Hoek EMV. A critical review of transport through osmotic membranes. Journal of Membrane Science. 2014;**454**:516-537. DOI: 10.1016/j.memsci.2013.12.034

[29] Cadotte JE, Petersen RJ, Larson RE, Erickson EE. New thin-film composite seawater reverse-osmosis membrane. Desalination. 1980;**32**:25-31. DOI: 10.1016/S0011-9164(00)86003-8

[30] Freger V. Nanoscale heterogeneity of polyamide membranes formed by interfacial polymerization. Langmuir. 2003;**19**:4791-4797. DOI: 10.1021/la020920q

[31] Coronell O, Marinas BJ, Cahill DG. Depth heterogeneity of fully aromatic polyamide active layers in reverse osmosis and nanofiltration membranes. Environmental Science and Technology. 2011;**45**:4513-4520. DOI: 10.1021/es200007h

[32] Freger V, Srebnik S. Mathematical model of charge and density distributions in interfacial polymerization of thin films. Journal of Applied Polymer Science. 2003;**88**:1162-1169. DOI: 10.1002/app.11716

[33] Hilal N, Al-Zoubi H, Darwish NA, Mohammad AW. Characterisation of nanofiltration membranes using atomic force microscopy. Desalination. 2005;**177**:187-199. DOI: 10.1016/j.desal.2004.12.008

[34] Boussu K, De Baerdemaeker J, Dauwe C, Weber M, Lynn KG, Depla D, Aldea S, Vankelecom IFJ, Vandecasteele C, Van der Bruggen B. Physico-chemical characterization of nanofiltration membranes. Chemphyschem. 2007;**8**:370-379. DOI: 10.1002/cphc.200600512

[35] Zimmerman C, Singh A, Koros W. Tailoring mixed matrix composite membranes for gas separations. Journal of Membrane Science. 1997;**137**:145-154. DOI: 10.1016/S0376-7388(97)00194-4

[36] Jia MD, Peinemann KV, Behling RD. Molecular-sieving effect of the zeolite-filled silicone-rubber membranes in gas permeation. Journal of Membrane Science. 1991;**57**:289-296. DOI: 10.1016/S0376-7388(00)80684-5

[37] Brunet L, Lyon DL, Zodrow K, Rouch JC, Caussat B, Serp P, Remigy JC, Wiesner MR, Alvarez PJJ. Properties of membranes containing semi-dispersed carbon nanotubes. Environmental Engineering Science. 2008;**25**:565-575. DOI: 10.1089/ees.2007.0076

[38] Aerts P, Greenberg AR, Leysen R, Krantz WB, Reinsch VE, Jacobs PA. The influence of filler concentration on the compaction and filtration properties of Zirfon (R)-composite ultrafiltration membranes. Separation and Purification Technology. 2001;**22-23**(1-3):663-669. DOI: 10.1016/S1383-5866(00)00165-9

[39] Wu L, Shamsuzzoha M, Ritchie SMC. Preparation of cellulose acetate supported zero-valent iron nanoparticles for the dechlorination of trichloroethylene in water. Journal of Nanoparticle Research. 2005;**7**:469-476. DOI: 10.1007/s11051-005-4271-5

[40] Luo M, Wen Q, Liu J, Liu H, Jia Z. Fabrication ofSPES/nano-TiO_2 composite ultrafiltration membrane and its anti-fouling mechanism. Chinese Journal of Chemical Engineering. 2011;**19**:45-51. DOI: 10.1016/S1004-9541(09)60175-0

[41] Celik E, Park H, Choi H, Choi H. Carbon nanotube blended polyethersulfone membranes for fouling control in water treatment. Water Research. 2011;**45**:274-282. DOI: 10.1016/j.watres.2010.07.060

[42] Jeong BH, Hoek EMV, Yan YS, Subramani A, Huang XF, Hurwitz G, Ghosh AK, Jawor A. Interfacial polymerization of thin film nanocomposites: A new concept for reverse osmosis membranes. Journal of Membrane Science. 2007;**294**:1-7. DOI: 10.1016/j.memsci.2007.02.025

[43] Nandi BK, Uppaluri R, Purkait MK. Effects of dip coating parameters on the morphology and transport properties of cellulose acetate-ceramic composite membranes. Journal of Membrane Science. 2009;**330**:246-258. DOI: 10.1016/j.memsci.2008.12.071

[44] Pendergast MTM, Nygaard JM, Ghosh AK, Hoek EMV. Using nanocomposite materials technology to understand and control reverse osmosis membrane compaction. Desalination. 2010;**261**:255-263. DOI: 10.1016/j.desal.2010.06.008

[45] Qin D, Liu Z, Sun DD, Song X, Bai H. A new nanocomposite forward osmosis membrane custom-designed for treating shale gas wastewater. Scientific Reports. 2015;**5**(14530):1-14. DOI: 10.1038/srep14530

[46] Xu W, Chen Q, Ge Q. Recent advances in forward osmosis (FO) membrane: Chemical modifications on membranes for FO processes. Desalination. 2017;**419**:101-116. DOI: 10.1016/j.desal.2017.06.007

[47] Ulbricht M. Advanced functional polymer membranes. Polymer. 2006;**47**:2217-2262. DOI: 10.1016/j.polymer.2006.01.084

[48] Chung T, Luo L, Feng C, Cui Y, Amy G. What is next for forward osmosis (FO) and pressure retarded osmosis (PRO). Separation and Purification Technology. 2015;**156**:856-860. DOI: 10.1016/j.seppur.2015.10.063

[49] Chung TS, Li X, Ong RC, Ge QC, Wang HL, Han G. Emerging forward osmosis (FO) technologies and challenges ahead for clean water and clean energy applications. Current Opinion in Chemical Engineering. 2012;**1**:246-257. DOI: 10.1016/j.coche.2012.07.004

[50] Yip NY, Tiraferri A, Phillip WA, Schiffman JD, Elimelech M. High performance thin-film composite forward osmosis membrane. Environmental Science and Technology. 2010;**44**:3812-3818. DOI: 10.1021/es1002555

[51] Qiu CQ, Qi SR, Tang CY. Synthesis of high flux forward osmosis membranes by chemically crosslinked layer-by-layer polyelectrolytes. Journal of Membrane Science. 2011;**381**:74-80. DOI: 10.1016/j.memsci.2011.07.013

[52] Ong RC, Chung TS, de Wit JS, Helmer BJ. Novel cellulose ester substrates for high performance flat-sheet thin-film composite (TFC) forward osmosis (FO) membranes. Journal of Membrane Science. 2015;**473**:63-71. DOI: 10.1016/j.memsci.2014.08.046

[53] Arslan S, Aslan T, Eyvaz M, Güçlü S, Yüksekdağ A, Yüksel E, Koyuncu İ. Water and reverse salt flux performances of a novel nanofiber forward osmosis membrane. In: 5th IWA Regional Conference on Membrane Technology (IWA-RMTC2016); Kunming City, Yunnan Province China: IWA Publishing; Aug 22-24, 2016

[54] Arslan S, Eyvaz M, Aslan T, Koyuncu İ, Yüksel E. Manufacturing of high performance forward osmosis by innovative manufacturing methods and their applications in submerged membrane bioreactors (Poster Presentation in Turkish). In: Institute of Science and Technology, Graduate Researches-Promotion Days. Gebze Technical University, Department of Architecture, Kocaeli/Turkey. Abstracts Book. GTU Publishing; May 17-18, 2016. p. 46

[55] Wang KY, Ong RC, Chung TS. Double-skinned forward osmosis membranes for reducing internal concentration polarization within the porous sublayer. Industrial & Engineering Chemistry Research. 2010;**49**:4824-4831. DOI: 10.1021/ie901592d

[56] Chung TS, Zhang S, Wang KY, Su J, Ling MM. Forward osmosis processes: Yesterday, today and tomorrow. Desalination. 2012;**287**:78-81. DOI: 10.1016/j.desal.2010.12.019

[57] McCutcheon JR, Elimelech M. Influence of membrane support layer hydrophobicity on water flux in osmotically driven membrane processes. Journal of Membrane Science. 2008;**318**:458-466. DOI: 10.1016/j.memsci.2008.03.021

[58] Yuan H, Abu-Reesh IM, He Z. Enhancing desalination and wastewater treatment by coupling microbial desalination cells with forward osmosis. Chemical Engineering Journal. 2015;**270**:437-443. DOI: 10.1016/j.cej.2015.02.059

[59] Xu GR, Xu JM, Feng HJ, Zhao HL, Wu SB. Tailoring structures and performance of polyamide thin film composite (PA-TFC) desalination membranes via sublayers adjustment-a review. Desalination. 2017;**417**:19-35. DOI: 10.1016/j.desal.2017.05.011

[60] Li D, Yan Y, Wang H. Recent advances in polymer and polymer composite membranes for reverse and forward osmosis processes. Progress in Polymer Science. 2016;**61**:104-155. DOI: 10.1016/j.progpolymsci.2016.03.003

[61] Tian EL, Zhou H, Ren YW, Za M, Wang XZ, Xiong SW. Novel design of hydrophobic/hydrophilic interpenetrating network composite nanofibers for the support layer of forward osmosis membrane. Desalination. 2014;**347**:207-214. DOI: 10.1016/j.desal.2014.05.043

[62] Dumée L, Lee J, Sears K, Tardy B, Duke M, Gray S. Fabrication of thin film composite poly(amide)-carbon-nanotube supported membranes for enhanced performance in osmotically driven desalination systems. Journal of Membrane Science. 2013;**427**:422-430. DOI: 10.1016/j.memsci.2012.09.026

[63] Wang R, Shi L, Tang CY, Chou S, Qiu C, Fane AG. Characterization of novel forward osmosis hollow fiber membranes. Journal of Membrane Science. 2010;**355**:158-167. DOI: 10.1016/j.memsci.2010.03.017

[64] Sukitpaneenit P, Chung T-S. High performance thin-film composite forward osmosis hollow fiber membranes with macrovoid-free and highly porous structure for sustainable water production. Environmental Science and Technology. 2012;**46**:7358-7365. DOI: 10.1021/es301559z

[65] Han G, Chung T-S, Toriida M, Tamai S. Thin-film composite forward osmosis membranes with novel hydrophilic supports for desalination. Journal of Membrane Science. 2012;**423-424**:543-555. DOI: 10.1016/j.memsci.2012.09.005

[66] Yasukawa M, Mishima S, Shibuya M, Saeki D, Takahashi T, Miyoshi T, Matsuyama H. Preparation of a forward osmosis membrane using a highly porous polyketone microfiltration membrane as a novel support. Journal of Membrane Science. 2015;**487**:51-59. DOI: 10.1016/j.memsci.2015.03.043

[67] Tiraferri A, Yip NY, Phillip WA, Schiffman JD, Elimelech M. Relating performance of thin-film composite forward osmosis membranes to support layer formation and structure. Journal of Membrane Science. 2011;**367**:340-352. DOI: 10.1016/j.memsci.2010.11.014

[68] Huang L, McCutcheon JR. Impact of support layer pore size on performance of thin film composite membranes for forward osmosis. Journal of Membrane Science. 2015;**483**:25-33. DOI: 10.1016/j.memsci.2015.01.025

[69] Xiao P, Nghiem LD, Yin Y, Li X-M, Zhang M, Chen G, Song J, He T. A sacrificial-layer approach to fabricate polysulfone support for forward osmosis thin-film composite membranes with reduced internal concentration polarisation. Journal of Membrane Science. 2015;**481**:106-114. DOI: 10.1016/j.memsci.2015.01.036

[70] Vilakati GD, Wong MCY, Hoek EMV, Mamba BB. Relating thin film composite membrane performance to support membrane morphology fabricated using lignin additive. Journal of Membrane Science. 2014;**469**:216-224. DOI: 10.1016/j.memsci.2014.06.018

[71] Peng N, Widjojo N, Sukitpaneenit P, Teoh MM, Lipscomb GG, Chung T-S, Lai J-Y. Evolution of polymeric hollow fibers as sustainable technologies: Past, present, and future. Progress in Polymer Science. 2012;**37**:1401-1424. DOI: 10.1016/j.progpolymsci.2012.01.001

[72] Song X, Liu Z, Sun DD. Nano gives the answer: Breaking the bottleneck of internal concentration polarization with a nanofiber composite forward osmosis membrane for a high water production rate. Advanced Materials. 2011;**23**:3256-3260. DOI: 10.1002/adma.201100510

[73] Wang KY, Chung T-S, Amy G. Developing thin-film-composite forward osmosis membranes on the PES/SPSf substrate through interfacial polymerization. AICHE Journal. 2012;**58**:770-781. DOI: 10.1002/aic.12635

[74] Arena JT, McCloskey B, Freeman BD, McCutcheon JR. Surface modification of thin film composite membrane support layers with polydopamine: Enabling use of reverse osmosis membranes in pressure retarded osmosis. Journal of Membrane Science. 2011;**375**:55-62. DOI: 10.1016/j.memsci.2011.01.060

[75] Huang L, Bui N-N, Meyering MT, Hamlin TJ, McCutcheon JR. Novel hydrophilic nylon 6,6 microfiltration membrane supported thin film composite membranes for engineered osmosis. Journal of Membrane Science. 2013;**437**:141-149. DOI: 10.1016/j.memsci.2013.01.046

[76] Emadzadeh D, Lau WJ, Matsuura T, Ismail AF, Rahbari-Sisakht M. Synthesis and characterization of thin film nanocomposite forward osmosis membrane with hydrophilic nanocomposite sup- port to reduce internal concentration polarization. Journal of Membrane Science. 2014;**449**:74-85. DOI: 10.1016/j.memsci.2013.08.014

[77] Liang HQ, Hung WS, Yu HH, Hu CC, Lee KR, Lai JY, Xu ZK. Forward osmosis membranes with unprecedented water flux. Journal of Membrane Science. 2017;**529**:47-54. DOI: 10.1016/j.memsci.2017.01.056

[78] Kwon SJ, Park SH, Park MS, Lee JS, Lee JH. Highly permeable and mechanically durable forward osmosis membranes prepared using polyethylene lithium ion battery separators. Journal of Membrane Science. 2017;**544**:213-220. DOI: 10.1016/j.memsci.2017.09.022

[79] Kwon SB, Lee JS, Kwon SJ, Yun ST, Lee S, Lee JH. Molecular layer-by-layer assembled forward osmosis membranes. Journal of Membrane Science. 2015;**488**:111-120. DOI: doi.org/10.1016/j.memsci.2015.04.015

[80] Widjojo M, Chung TS, Weber M, Maletzko C, Warzelhan V. The role of sulphonated polymer and macrovoid-free structure in the support layer for thin-film composite (TFC) forward osmosis (FO) membranes. Journal of Membrane Science. 2011;**383**: 214-223. DOI: 10.1016/j.memsci.2011.08.041

[81] Duong PHH, Chisca S, Hong PY, Cheng H, Nunes SP, Chung TS. Hydroxyl functionalized polytriazole-co-polyoxadiazole as substrates for forward osmosis membranes. ACS Applied Materials & Interfaces. 2015;**7**:3960-3973. DOI: 10.1021/am508387d

[82] Zhou Z, Lee JY, Chung TS. Thin film composite forward-osmosis membranes with enhanced internal osmotic pressure for internal concentration polarization reduction. Chemical Engineering Journal. 2014;**249**:236-245. DOI: 10.1016/j.cej.2014.03.049

[83] Sahebi S, Phuntsho S, Woo YC, Park MJ, Tijing LD, Hong S, Shon HK. Effect of sulphonated polyethersulfone substrate for thin film composite forward osmosis membrane. Desalination. 2013;**389**:129-136. DOI: 10.1016/j.desal.2015.11.028

[84] Li X, Zhang S, Fu F, Chung TS. Deformation and reinforcement of thin-film composite (TFC) polyamide-imide (PAI) membranes for osmotic power generation. Journal of Membrane Science. 2013;**434**:204-217. DOI: 10.1016/j.memsci.2013.01.049

[85] Liu X, Ng HY. Fabrication of layered silica-polysulfone mixed matrix substrate membrane for enhancing performance of thin-film composite forward osmosis membrane. Journal of Membrane Science. 2015;**481**:148-163. DOI: 10.1016/j.memsci.2015.02.012

[86] Ma N, Wei J, Qi S, Zhao Y, Gao Y, Tang CY. Nanocomposite substrates for controlling internal concentration polarization in forward osmosis membranes. Journal of Membrane Science. 2013;**441**:54-62. DOI: 10.1016/j.memsci.2013.04.004

[87] Lu P, Liang S, Qiu L, Gao Y, Wang Q. Thin film nanocomposite forward osmosis membranes based on layered double hydroxide nanoparticles blended substrates. Journal of Membrane Science. 2016;**504**:196-205. DOI: 10.1016/j.memsci.2015.12.066

[88] Huang L, McCutcheon JR. Hydrophilic nylon 6,6 nanofibers supported thin film composite membranes for engineered osmosis. Journal of Membrane Science. 2014;**457**:162-169. DOI: 10.1016/j.memsci.2014.01.040

[89] Huang L, Arena JT, McCutcheon JR. Surface modified PVDF nanofiber supported thin film composite membranes for forward osmosis. Journal of Membrane Science. 2016;**499**:352-360. DOI: 10.1016/j.memsci.2015.10.030

[90] Zhang H, Jiang W, Cui H. Performance of anaerobic forward osmosis membrane bioreactor coupled with microbial electrolysis cell (AnOMEBR) for energy recovery and membrane fouling alleviation. Chemical Engineering Journal. 2017;**321**:375-383. DOI: 10.1016/j.cej.2017.03.134

[91] Klaysom C, Hermans S, Gahlaut A, Van Craenenbroeck S, Vankelecom IFJ. Polyamide/polyacrylonitrile (PA/PAN) thin film composite osmosis membranes: Film optimization, characterization and performance evaluation. Journal of Membrane Science. 2013;**445**:25-33. DOI: 10.1016/j.memsci.2013.05.037

[92] Jia Q, Han H, Wang L, Liu B, Yang H, Shen J. Effects of CTAC micelles on the molecular structures and separation performance of thin-film composite (TFC) membranes in forward osmosis processes. Desalination. 2014;**340**:30-41. DOI: 10.1016/j.desal.2014.02.017

[93] Lu X, Arias Chavez LH, Romero-Vargas Castrillón S, Ma J, Elimelech M. Influence of active layer and support layer surface structures on organic fouling propensity of thin-film composite forward osmosis membranes. Environmental Science and Technology. 2015;**49**:1436-1444. DOI: 10.1021/es5044062

[94] Tiraferri A, Kang Y, Giannelis EP, Elimelech M. Highly hydrophilic thin-film composite forward osmosis membranes functionalized with surface-tailored nanoparticles. ACS Applied Materials & Interfaces. 2012;**4**:5044-5053. DOI: 10.1021/am301532g

[95] You S, Lu J, Tang CY, Wang X. Rejection of heavy metals in acidic wastewater by a novel thin-film inorganic forward osmosis membrane. Chemical Engineering Journal. 2017;**320**:532-538. DOI: 10.1016/j.cej.2017.03.064

[96] Salehi H, Rastgar M, Shakeri A. Anti-fouling and high water permeable forward osmosis membrane fabricated via layer by layer assembly of chitosan/graphene oxide. Applied Surface Science. 2017;**413**:99-108. DOI: 10.1016/j.apsusc.2017.03.271

[97] Fang W, Wang R, Chou S, Setiawan L, Fane AG. Composite forward osmosis hollow fiber membranes: Integration of RO- and NF-like selective layers to enhance membrane properties of anti- scaling and anti-internal concentration polarization. Journal of Membrane Science. 2012;**394-395**:140-150. DOI: 10.1016/j.memsci.2011.12.034

[98] Fang W, Liu C, Shi L, Wang R. Composite forward osmosis hollow fiber membranes: Integration of RO- and NF-like selective layers for enhanced organic fouling resistance. Journal of Membrane Science. 2015;**492**:147-155. DOI: 10.1016/j.memsci.2015.05.045

Chapter 3

Forward Osmosis Membranes – A Review: Part II

Murat Eyvaz, Serkan Arslan, Derya İmer,
Ebubekir Yüksel and İsmail Koyuncu

Additional information is available at the end of the chapter

http://dx.doi.org/ 10.5772/intechopen.74659

Abstract

Forward osmosis (FO) is a technical term describing the natural phenomenon of osmosis: the transport of water molecules across a semipermeable membrane by osmotic pressure from a feed solution (FS) to a draw solution (DS). The diluted DS is then reconcentrated to recycle the draw solutes as well as to produce purified water. As the driving force is only the osmotic pressure difference between two solutions, meaning that there is no need to apply an external energy, this results in low fouling propensity of membrane and minimization of irreversible cake forming, which are the main problems controverted by membrane applications, especially in biological treatment systems (e.g., FO membrane bioreactor (FO-MBR)). The purpose of the book chapter is to bring an overview on the FO membrane manufacturing, characterizing and application area at laboratory or full scales. This book chapter is published in two parts. In the second part, which appears here, characterization of mass transport in FO membranes, fouling mechanisms and foulants on FO membranes in naturally asymmetric structure and application areas of FO membranes in the literature are mentioned. Cutting-edge technologies on FO studies are comprehensively reviewed and following major and minor titles are stated truly on the new technologies.

Keywords: forward osmosis, characterization, structural parameter, membrane fouling, concentration polarization, water/wastewater treatment, desalination, hybrid processes, membrane bioreactor

1. Introduction

FO membranes are preferred over the last few years due to the high rejection of a wide range of contaminants and the lack of hydraulic pressure, resulting in less irreversible fouling on the membrane surface compared to pressure-driven membranes. However, due to the asymmetric structure of the FO membrane, concentration polarization (CP) becomes more important, which motivates many researchers to focus on the selection and/or development of new membrane materials for both active and support layers to decrease CP.

In this second part of the chapter, characterization of FO membranes, such as determining rejection capabilities of membrane layers by analytical approaches and experimental procedures, is thoroughly stated by considering both review and research articles in the available literature. In the following section, fouling phenomena in FO membranes are referred by considering membrane orientation, and before Conclusion, application areas of FO process are presented. Since the permeate (diluting DS) of the FO membrane is not actually a product water, this filtrate (diluted DS) needs to be treated again. For this reason, the FO process needs an additional process to recover the water from the diluted draw solution. In this context, hybrid FO processes are also included in this section. Finally, the general summary of the research is evaluated and the future prospects for FO membranes and applications are introduced.

2. Characterization of FO membranes

Although the model development on characterization for FO membranes is described in some literature [1], more general information from some is given here. The membrane in separation process using osmotic pressure as driving force must be capable of rejecting both the FS and the DS. When there is no solute retention in membrane, the FS and DS are easily diffused from the membrane, and osmosis does not occur. All existing membranes that can be used for this purpose are asymmetric. Many of the problems in the FO process resulted from this asymmetric structure. As with all membrane processes, mass transfer boundary layers form near the selective interface. On the FO membrane, these boundary layers occur on both sides of the selective layer interface. However, in an asymmetric membrane, one of these interfaces is embedded in the support layer. Therefore, the support layer significantly reduces the mixing and prevents the mass transfer [2]. The support layers in the TFC RO membranes are relatively thick on the FO membranes and have 25–45% porosity [3]. Solutes must be transported by support layer to reach to the selective layer on which diffusion or rejection is performed. If the mass transfer in these layers is weak, the situation called ICP occurs. Similar to conventional CP, ICP reduces the osmotic driving force. In an FO membrane where there is an asymmetric support layer in which no mixing occurs, the osmotic driving forces can be severely reduced, resulting in no water flux from the membrane [4]. The severity of ICP is greatly influenced by the support layer. This structure is often referred to a metric known as the structural parameter, S

$$S = \frac{t\tau}{\varepsilon} \quad (1)$$

where t is the thickness, τ is the tortuosity, and ε is the porosity of the support layer. In the FO process, membranes with lower S values are preferred to reduce ICP severity. To this end, a number of studies have been conducted on the production and modification of new FO membranes with low S values since 1990. Tiraferri et al. [5] conducted studies on the effects of solvent quality, dope polymer concentration, backing layer wetting, and casting blade gate on support layer production on one of the first TFC membranes designed for the FO membrane. The pore morphology of the support layer was characterized with the aid of cross-sectional SEM images and reported that the optimum FO membrane must be formed from a mixed structured backing layer and that the upper part of the thin sponge-like layer should be placed on high porosity macrovoids. Shi [6] investigated UF-type phase inversion cast supports for hollow fiber FO membranes and reported that substrates with 300 kDa (molecular weight cut-off (MWCO)) should be preferred to obtain a "good" semipermeable skin.

It has also been claimed that, considering the suitability of the substrate for IP, taking into account the MWCO parameter is more appropriate than the mean pore size. It is estimated that membrane thickness is more important than porosity and tortuosity in recent studies with nanofiber membranes [7]. Moreover, the support layer pore diameter, which is thought to be very effective only in the formation of the selective layer, has also been shown to influence ICP [8]. The influence of the support layer structure on transport is typically expressed using the structural parameter concept. To calculate S, the membrane thickness (can be measured by SEM and relatively easily), porosity, and tortuosity should be measured independently. However, it is quite difficult to measure these last two, especially tortuosity, accurately and reliably. The reason for this is that the characterization of the pore structure of soft materials is an area where work is still developing and there is no standardization for the comprehensive and accurate characterization of 3D structures. Hence, researchers on FO use and develop numerical models more commonly than calculating S parameters with Eq. (1).

Experimental measurements are used when the S parameter is calculated, and therefore, the experimental conditions as a factor are emerging from the structural properties of the membrane. This means that changes in experimental conditions will directly affect the estimated S value. Therefore, no significant comparison can be made between these support layers unless the same experimental conditions are used to test different membranes. In a study by Cath et al., this limitation of the semiempirical method is clearly emphasized [9]. In this study, researchers from 7 different laboratory groups tested 2 different membranes from the same production line under the same experimental conditions but on different systems. One was an HTI-CTA membrane commercially available from HTI, and the other was a TFC membrane from Oasys Water. Significant deviations could be observed between the effective S values obtained by different groups as shown in **Figure 1**. Therefore, researchers report that the experimental conditions are the main factors in the calculation of the effective S parameter in semiempirical calculation method [1].

More recently, a simple characterization method based on a combination of a single FO test and a statistical approach has been developed to avoid pressure RO testing, which can damage the FO membrane or misread membrane properties in the characterization of FO membranes [10]. In this single test, the membrane is operated in AL-FS mode to measure water and reverse salt flux using deionized water (DI) as feed and NaCl as the DS. The statistical approach uses

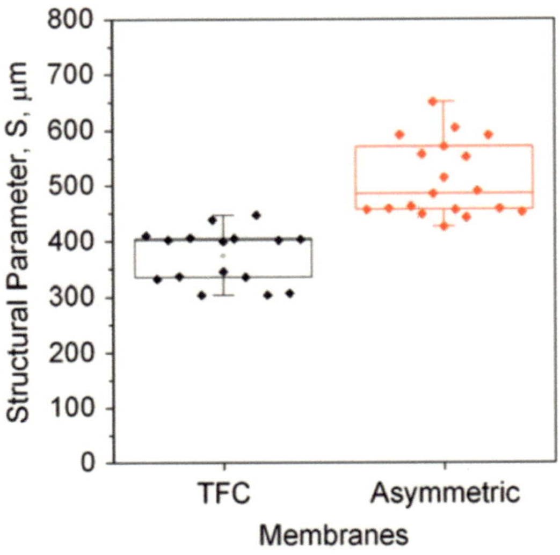

Figure 1. Structural parameters of TFC and asymmetric FO membranes [9].

both ICP and external concentration polarization (ECP) models to predict Jw and Js on the tested membrane and finds the most appropriate water and salt permeability (A and B) and salt diffusion resistance in the support layer. Verifications using various experimental results in this study and other literature have shown that this new FO membrane characterization method sets parameters (A, B, and KICP) more reliably than the conventional characterization method based on the pressure-RO experiment to estimate the experimental Jw and Js. The consideration of ECP helps to determine more accurate FO membrane parameters (especially KICP), but it is difficult to accurately model the ECP for the FO membrane channel tested.

The evaluation of porosity and tortuosity has been carried out with traditional characterization techniques such as SEM and porosimetry as well as newer tools such as x-ray computed tomography (XCT). While none of these techniques comply with all of the difficulties listed above, some are more suitable than others according to the type of the membrane material being tested. Imaging approaches provide good visuals for evaluating the qualities of porous membranes. However, expensive and time-consuming techniques are required to obtain this information from images. It also requires usage expertise. But all of these, as well as resolution and field-of-view (FOV) limitations, are disadvantages that reduce the quantitative value of these images.

Membrane pore structure analysis can also be done without relying on the images. There are a number of analytical techniques that can examine the pore structure by means of probing. While these approaches do not reintroduce visual presentation of membranes, they can provide critical characterization information about FO membrane, including porosity and tortuosity, by using basic models.

Compared with imaging techniques, analytical techniques allow for greater comparisons between different FO membrane structures by easily analyzing a larger sample volume. However, the

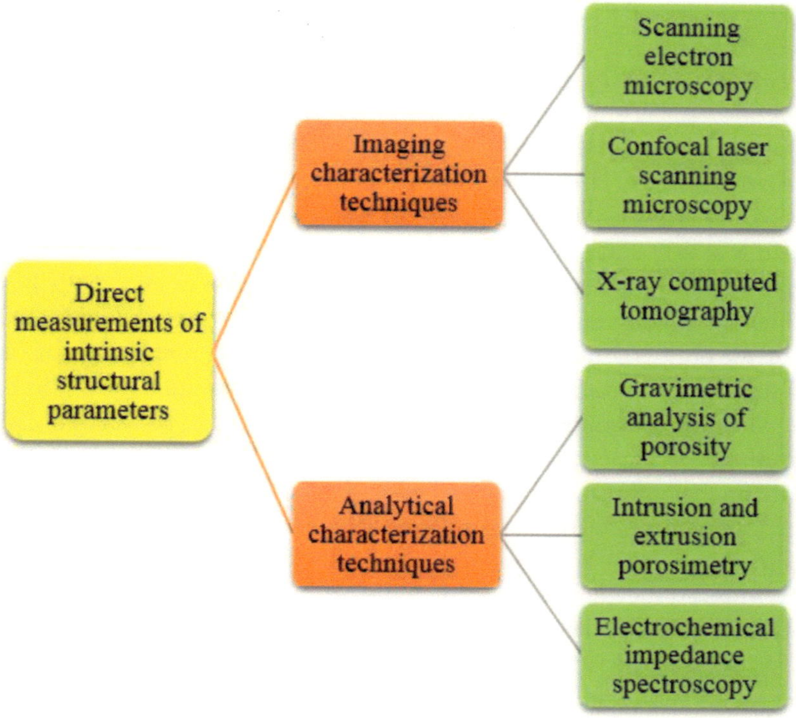

Figure 2. Direct measurement techniques of intrinsic structural parameters (adapted from [1]).

assumptions used to derive the models calculating the porosity and tortuosity must be carefully considered before adapting to the sample being analyzed. Similarly, when analyzing data from them, the biases of different analytical techniques should be considered [1]. Direct measurement techniques of intrinsic structural parameters are presented together in **Figure 2**.

Designers of membranes for osmotic processes need to be able to better calculate the mass transfer resistance of the membrane to overcome the difficulties in nature of osmotic systems. Unless major structural parameters such as porosity and tortuosity are known, wrong areas of designs may be focused on. In order to overcome these difficulties, the above-mentioned methods for membrane characterization need to be further developed [1].

3. Fouling in FO membranes

Today, the greatest challenges of FO technology can be summarized into three main classes: the difficulty of developing a correct and an effective FO membrane, the lack of recyclable and economical DS, and the limited availability of information on membrane fouling [11]. Although fouling of FO membranes is more reversible than RO membranes, removal of contaminants may become more difficult when the feed stream in the FO membrane contacts the support layer [12, 13].

She et al. [14] investigated the membrane fouling in osmotically driven membrane processes and concluded that fouling in pressure-driven membranes can occur at different locations of the membrane [15, 16]. As shown in **Figure 3(a)**, the foulants in the FS are transported to the active layer surface of the membrane in the AL-FS mode, resulting in a cake layer similar to fouling of the RO membranes. This type of pollution is called external pollution. Fouling occurring in the FO membrane in AL-DS mode is more complicated. **Figure 3(b)** shows possible fouling scenarios in AL-DS orientation. If the contaminant has a relatively small size and is able to enter the porous support layer by convection of the FS, it will either be adsorbed through the walls of the pores of the support layer or eventually be retained by the active layer and accumulate on the back surface of the active layer. Subsequently, the foulants entering the porous support layer will adhere to the contaminants that are adsorbed on the walls of the support layer pores or to the accumulated contaminants on the back surface of the active layer, thus leading to "pore clogging." This form of pollution is called internal fouling (scenario (1) in **Figure 3(b)**). In severe fouling conditions, contaminants will continue to accumulate on the outer surface of the porous support layer, as well as internal pore clogging. This type of membrane fouling is referred to as combined internal and external fouling (scheme (2) in **Figure 3(b)**). If the foulants have relatively large sizes and cannot enter the porous support layer, they may only accumulate on the outer surface of the porous support layer. In this case, only external fouling occurs (scenario (3) in **Figure 3(b)**). If contaminants are present in the feedwater in different sizes, both external fouling and internal fouling may occur (scenarios (4) and (5) in **Figure 3(b)**).

According to She et al. [15], compared to internal fouling, it is easier to remove the external fouling from the membrane surface by optimizing the hydrodynamic conditions of the feed stream (such as by increasing the cross-flow rate, applying pulsed flow [17] and employing air scouring [18]). For this reason, most researchers suggest AL-FS orientation in the FO process to prevent undesired internal fouling, even though the ICP in AL-FS is more severe than in AL-DS mode [13, 19]. However, external fouling is more reversible in FO membranes, as there is no such matter as compaction of pollution due to hydraulic pressure in the RO membrane [20]. On the other hand, the internal fouling within the porous support layer functions as an unmixed layer. Internal pollution is less reversible than external pollution, as it is more difficult to control the optimization of hydrodynamic conditions [21]. Internal fouling usually occurs in PRO membranes operating in AL-DS mode [22]. Although the osmotic backwash method has been developed to clean contaminants in the support layer [21], the development

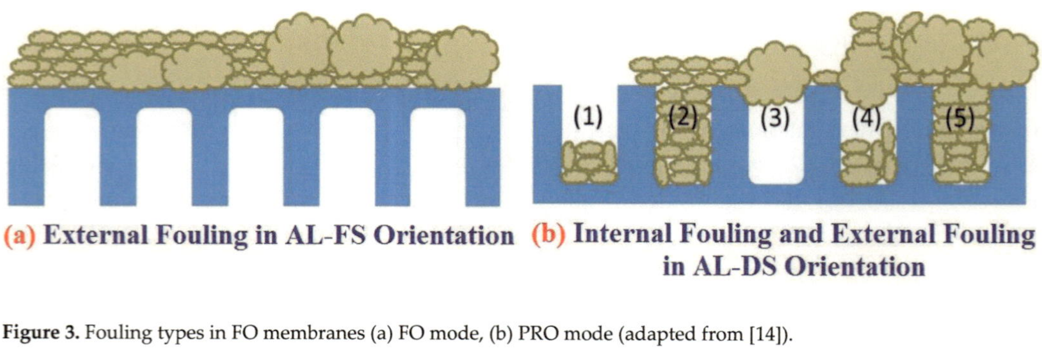

Figure 3. Fouling types in FO membranes (a) FO mode, (b) PRO mode (adapted from [14]).

of more effective strategies for internal pollution control will still be an important research topic in the future.

Classification and interaction of foulants in osmotic pressure-driven membrane processes [22] can be grouped into four main classes: (i) colloidal fouling by accumulation of colloidal particles on the membrane, (ii) organic fouling by deposition and adsorption of the macromolecular organic compounds on the membrane surface, (iii) inorganic scaling by precipitation or crystallization of inorganic compounds that are poorly soluble on the membrane surface, and (iv) biofouling by adhesion and accumulation of the microorganisms to the membrane surface and eventually biofilm development. The specific pollutants in the different groups are closely related to the characteristics of the feedwater. Contaminants specifically present in raw and treated wastewaters are particles, colloids, and organic macromolecules such as polysaccharides, humic substances, and proteins [23]. In addition, these substances are also commonly found in natural waters such as rivers, seawater, and ground waters [13]. Zhou et al. [23] used gas chromatography-mass spectrometry (GC–MS) to identify soluble microbial products (SMPs) containing a large portion of polysaccharides, proteins and humic substances in raw and wastewater. Recently, organic carbon detection-organic nitrogen detection (LC-OCD-OND) has become increasingly popular for the identification of these pollutants [24]. Organic contaminants deposited on the membrane can be identified by Fourier transform infrared (FTIR) spectroscopy, solid-state 13C-nuclear magnetic resonance (NMR) spectroscopy, and high performance size exclusion chromatography (HP-SEC) [25]. Total organic carbon (TOC) measurement and UV analysis were also performed to determine the density of organic foulant deposition on the membrane [26]. Transparent exopolymer particles (TEPs) are another important organic pollutant typically found in natural waters. TEP in the feedwater is determined by two methods: microscopic counting and colorimetric detection [27].

Silica is a major inorganic foulant and is usually present in dissolved form or as colloidal particles in sea water, brackish water, and wastewater [24]. In addition, other inorganic contaminants are dissolved salts such as calcium carbonate, calcium sulfate, and calcium phosphate [28]. These inorganic contaminants deposited on the membrane surface can be extensively characterized by scanning electron microscopy-energy dispersive X-ray diffraction (SEM-EDX) [28] and X-ray diffraction (XRD) [29]. Microorganisms are mainly found in activated sludge in membrane bioreactors (MBR) as biofoulants [28]. These microorganisms can also be found in natural waters and cause biofouling in seawater and brackish water desalination [24]. Microbial populations within the biofilm can be characterized by analysis of DNA extracted from living cells using microbiological methods such as polymerase chain reaction denaturing gradient gel electrophoresis (PCR-DGGE) and fluorescent in situ hybridization (FISH) [23].

She et al. [15] indicated that membranes in osmotic pressure-driven membrane processes are contaminated by natural or industrial waters and wastewaters, and membrane fouling involves the combination of the four fouling categories above [24]. The understanding of mixed pollution mechanisms is difficult because of the various and numerous pollutants. Many studies to understand these fouling mechanisms are generally based on the consideration of a single foulant and the use of a synthetic FS [13, 19]. Meanwhile, the number of studies on fouling of FO membranes is also increasing. In particular, osmotic MBR studies have the potential to conduct research with more complex wastewaters [28, 30]. Working with a single model of foulant is more advantageous in terms of easier control of the selected foulant and understanding of the foulant-foulant or foulant-membrane interactions. The physicochemical properties are also important factors affecting the stability of contaminants in the FS, as well as information on the tendency to contaminate the membrane [31]. With the understanding of the fouling mechanisms in a single foulant system, future studies may focus on the study of the fouling mechanisms for mixed foulant systems, which may lead to a better understanding of the membrane fouling mechanisms.

Colloidal and organic fouling with highly complex mechanisms in FO membranes is affected by a number of physical and chemical factors, and in general, these factors can be divided into five groups: (i) operating conditions such as initial water flow, cross-flow rate, spacer features, ventilation, and temperature; (ii) feedwater characteristics such as foulant type, concentration, pH, temperature, ionic strength, and ionic composition; (iii) DS properties such as solute type and concentration; (iv) Membrane properties such as structural and surface characteristics; (iv) membrane orientation as AL-FS and AL-DS [31].

The composition of the FS is one of the most important factors affecting membrane fouling. The effect of the feedwater composition on FO membrane fouling is similar to that of pressure-based membrane processes, and recently some investigations have been conducted on this topic [13, 32]. Generally, the degree and rate of fouling are strongly dependent on the properties and concentration of pollutants in the feedwater. In addition, since the FS chemistry significantly affects the physico-chemical properties of the contaminant [22, 33], it will also play a role in foulant-foulant and foulant-membrane interactions and determine the membrane's fouling behavior.

The composition and concentration of the DS, the main source of osmotic pressure in the FO process, not only affects water and salt flux but also plays a role on the membrane fouling. In general, as the DS concentration increases, the initial water flux increases and exacerbates membrane fouling. Studies in the literature have reported membrane fouling increases with increasing DS concentration [13, 19, 34]. The high hydraulic drag force caused by the high flux also leads to the accumulation of foulant on the surface of the membrane. In this context, the change in DS concentration leads mainly to changes in hydrodynamic conditions. For this reason, pollution behavior due to DS concentration can be well explained by the flux-dependent fouling mechanism in which hydrodynamic conditions play a dominant role.

Membrane material and properties may also affect membrane fouling behavior. Membranes used in osmotic processes are generally originated from a nonporous active layer formed on a porous support layer [35–37]. The intrinsic separation properties of the active layer and the structural properties of the support layer govern the transport of water and solutes, which may affect membrane fouling behavior. Membranes with superior separation properties and structural properties (i.e., more water permeability, high selectivity, and membranes with smaller structural parameters) can provide higher water flux. However, the increased hydrodynamic drag force due to increased water flow will also increase the membrane fouling potential. On the other hand, membranes with low separation and selectivity properties may increase the risk of membrane fouling as there may be more solute transfer between DS and

FS. When designing or selecting membranes for FO applications in the future, the separation and structural properties of the membranes should be considered not only in terms of water flow performance but also in terms of the fouling behavior [15].

Membrane fouling and CP behave differently in different orientations of membrane (AL-FS or AL-DS) in osmotic pressure-driven membranes (**Figure 4**). Therefore, fouling and CP are defined as cake-enhanced external concentration polarization (CE-ECP) in the AL-FS mode [38], while in the AL-DS mode, it is defined as pore clogging-enhanced internal concentration polarization (PCE-ICP) [19]. It is reported that the main factor that dominates water flux in osmotic pressure-driven membranes is ICP and PCE-ICP presumably plays a leading role in the flux declining. Furthermore, while CE-ECP is very effective in AL-FS mode membrane fouling, a strong ICP effect can moderate flux decline rate. On the other hand, PCE-ICP can cause much more severe flux declines. However, systematic studies are still needed to explore the effects of CE-ECP and PCE-ICP on membrane clogging in osmotic pressure-driven membranes.

As shown in **Figure 5**, membrane fouling, CP (both ICP and ECP), and RSD are closely interrelated and can be modeled using the osmotic-resistance filtration model. Factors and mechanisms affecting FO membrane fouling such as hydrodynamic conditions, feedwater composition, membrane properties, and cake-enhanced concentration polarization (CE-CP) are also applicable for NF/RO processes. Osmotic pressure is the indispensable parameter for osmotically driven membrane processes. The composition and concentration of this solution may also affect other factors by means of membrane fouling. This is the point where osmotically driven

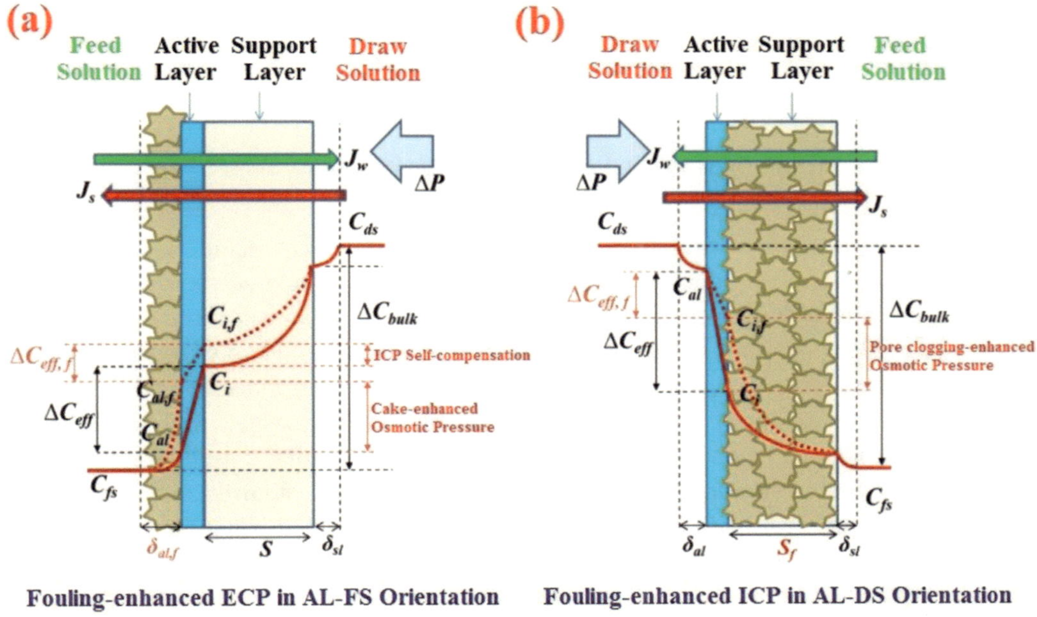

Figure 4. Schematic illustration of concentration profile across the membrane due to fouling-enhanced concentration polarization (a) fouling-enhanced ECP in AL-FS orientation. (b) Fouling-enhanced ICP in AL-DS orientation (adapted from [14]).

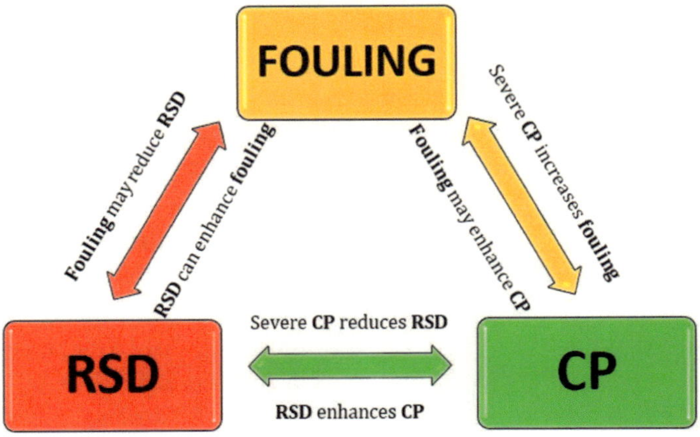

Figure 5. The intrinsic interrelationship among membrane fouling: CP (concentration polarization) and RSD (reverse salt diffusion) (adapted from [14]).

membranes are separated mainly from hydraulic pressure-driven membranes in terms of the fouling mechanism. Membrane orientation (AL-FS/AL-DS or FO/PRO) is another factor affecting membrane fouling, and FO mode is more preferred as it is less prone to fouling and provides a more stable water flux. However, PRO mode can also be preferred for strong membrane stability under high pressure and fewer ICPs. However, this mode has a tendency for internal fouling, which is less reversible. Both the size exclusion mechanism and CE-CP can affect membrane fouling, which can increase or decrease the rejection of contaminants. Modification of the membrane may be a strategy to reduce the fouling of the membrane and to increase reversibility of membrane fouling, which facilitates membrane cleaning [14].

4. Application areas of forward osmosis membranes

FO can be applied for the treatment of various kind of wastewaters including strong industrial effluents, i.e., from textile processes, oil and gas well fracturing waste streams, landfill leachates, nutrient-rich liquids, activated sludge, municipal wastewater, and even nuclear-origin wastewaters have been mentioned [39]. The applications of FO process can be classified as in **Figure 6**. The FO membrane rejects particles, pathogens, and emerging substances with an average porosity of 0.25–0.37 nm [40]. FO is also able to reject high levels of salt that cannot be achieved by normal treatment systems, and the total dissolved solids (TDS) from complex water can effectively be removed [41]. FO is no required for pretreatment of feedwaters (wastewaters) with complex contents. Conversely, RO and NF processes are more susceptible to fouling. Pretreatment is required to increase membrane lifetime and reduce costs [42]. FO can also be used for dewatering applications [43], useful for an efficient anaerobic digestion of wastewater, and is simpler and more environmentally friendly than classical dewatering processes [11]. High saline currents can be processed by the FO, not by the RO [44].

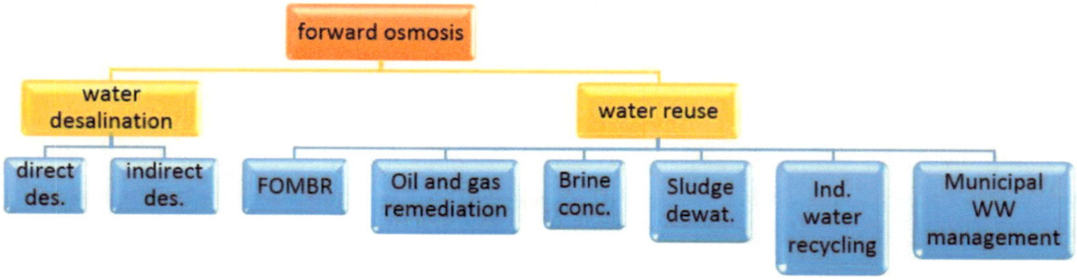

Figure 6. Forward osmosis applications (adapted from [45]).

4.1. Water/wastewater treatment

According to FO literature in the last 10 years, about 7% of the studies have used complex water. However, the number of studies on wastewater is also increasing. The advantages of FO observed in these studies encourage they prefer FO instead of current technologies in future studies [45].

In municipal wastewater treatment processes, integrated FO-membrane distillation (MD) system is applied for sewer mining. In a continuous operating period, a stable water flux has been achieved at a recovery rate of up to 80% [46]. FO rejects most organic pollutants at a moderate level, whereas MD rejects almost the entire residue. Recovery of clean water from secondary wastewater was performed by FO electro dialysis (FOeED)-integrated system powered by photovoltaic energy source. This process removed total organic carbon from wastewater and produced fresh water [47]. Utilizing natural energies (osmotic pressure and solar energy), this hybrid system is a convenient process for potable water supply in isolated areas, remote areas, and islands.

MBR, which contains both activated sludge process and membrane filtration, has become one of the most widely applied technologies in wastewater treatment. The integration of the biological system with the FO membrane (FO-MBR or OMBR) can reduce energy consumption in conventional MBR. In recent years, studies on FO-MBR have been increasing [48–50]. This process not only reduces the cost of MBRs used by UF or MF but also provides fouling control through air cleaning in conventional MBR; at the same time, a more stable flux is obtained. Thus, with the help of the FO membrane in the MBR, more efficient removal efficiency is obtained with less fouling tendency without the need for hydraulic pressure [45].

FO was tested for dewatering of the nutrient-rich anaerobic digester concentration [51] in which organic compounds are rejected by FO membrane, and an RO membrane can be used to recover fresh water from a clean and diluted DS. The FO membrane was also used for activated sludge dewatering [52]. The EDTA sodium salt has been tested as DS for dewatering of activated sludge with high nutrient content. The nutrients in the sludge were successfully removed by means of FO membrane. The macromolecular DS can be posttreated with an NF process for the recovery of freshwater. Alternatively, the concentration of the RO membrane was used as DS in Zhu's investigation and an effective sludge thickening was obtained. Thus, RO concentration is also osmotically diluted and safe disposal is possible while the volume of sludge is reduced by that study.

Another important source of pollution for wastewater treatment plants is industrial wastewater. In the US, a company has installed a pilot FO plant for the recycling of dye containing wastewater from textile and carpet mill processes [53]. In another study, the FO process was used to recover heavy metals from industrial wastewaters [54]. The effects of hydrodynamic conditions, organic pollution, temperature, and FS and DS properties on the separation efficiency were investigated. It has been reported that almost all metals such as Pb, Zn, Cu, and Cd have been removed in the study and that the FO process has the potential to be an effective and economical process for the treatment of industrial wastewater.

Linares et al. [45] expressed that, today, most FO applications for industrial wastewater treatment are devoted to the treatment and recovery of wastewater from the oil and gas (O&G) industry. In these applications, capacity for the treatment of emulsifier oil waters with FO has been stated [55]. Fresh water was recovered from wastewater by FO membrane containing up to 200,000 ppm of oil and a reasonable water flux value about 12 LMH was obtained. Many studies at the laboratory or commercial scale have been directly applied to the real wastewater of the O&G industry. Combined with RO in a closed loop, FO was used for drilling wastewater treatment from the gas exploration process [56]. The wastewater recovery capacity of the plant is 242,000 gallons of water per day, reducing the need for additional fresh water. Similar studies and applications have been performed by different companies and research groups using different membrane materials, modules, DS, and process configurations [57–59]. In these studies, it was reported that the volume of wastewater was greatly reduced, the need for fresh water was reduced, and a well-designed FO process could be a much more advantageous option than RO [60].

4.2. Desalination

Conventional desalination technologies include membrane-based separation processes such as RO, NF, and electrodialysis and thermal desalination technologies such as multieffect distillation (MED), multistage flash (MSF), and mechanical vapor compression (MVC). Pretreatment of feedwater has critical precaution to prevent the physical equipment of conventional processes from being damaged by wastewater components and to facilitate their performance by maintaining the consistent quality of the pretreated feedwater. Today, pretreatment technologies for desalination are designed to reduce the potential for contamination of feedwater by removing natural organic matter and suspended solids. However, pretreatment technologies are typically not designed to remove dissolved solids [61]. Inorganic scaling in membrane and thermal desalination processes caused by low solubility dissolved salts in food water limits operating conditions and system performance. In MED and MSF, scaling reduces heat transfer efficiency and system recovery rates and limits operating temperatures [62–65].

Shaffer et al. [65] notified that to prevent the harmful effects of the scaling, the FO pretreatment can act to remove dissolved organic material and dissolved inorganic scalants in addition to suspended solids from the FS. When the FO process is used for pretreatment, the traditional desalination process used for recovery of the DS is only affected by NaCl solution or an ammonia-carbon dioxide solution with negligible fouling and scaling potential of these engineered DS. The reversibility of the FO fouling shows that it can maintain the flow and performance of the FO membranes when they come in contact with raw feedwater with high fouling potential, under proper hydrodynamic conditions. A schematic view of the FO process applied for pretreatment prior to a classical membrane or thermal desalination process is presented in **Figure 7**.

The use of the FO process as pretreatment can improve the performance of conventional desalination processes by removing the small amounts of scalants present in the feedwater. The combined desalination processes can be operated at higher pressures or temperatures without the risk of scaling, resulting in higher system recovery. Testing the process modeling of an FO-RO system [66] and testing both bench-scale FO-RO [67] and FO-NF [68] systems proved the feasibility of pretreatment of the FO process. Furthermore, when FO is used instead of processes such as ion exchange and NF in the pretreatment, there is also the advantage that not only specific cations or anions but also all ions in the feedwater can be removed, in addition to the low membrane fouling tendency [45].

Linares et al. [45] notified that the direct use of FO for desalination is similar to the use of RO and NF processes conventionally used to obtain fresh water from sea water directly. This process uses seawater as FS, while nonvolatile NaCl or volatile ammonia-carbon dioxide is used as the DS [69]. However, in this process, an additional operation is required to recover the DS

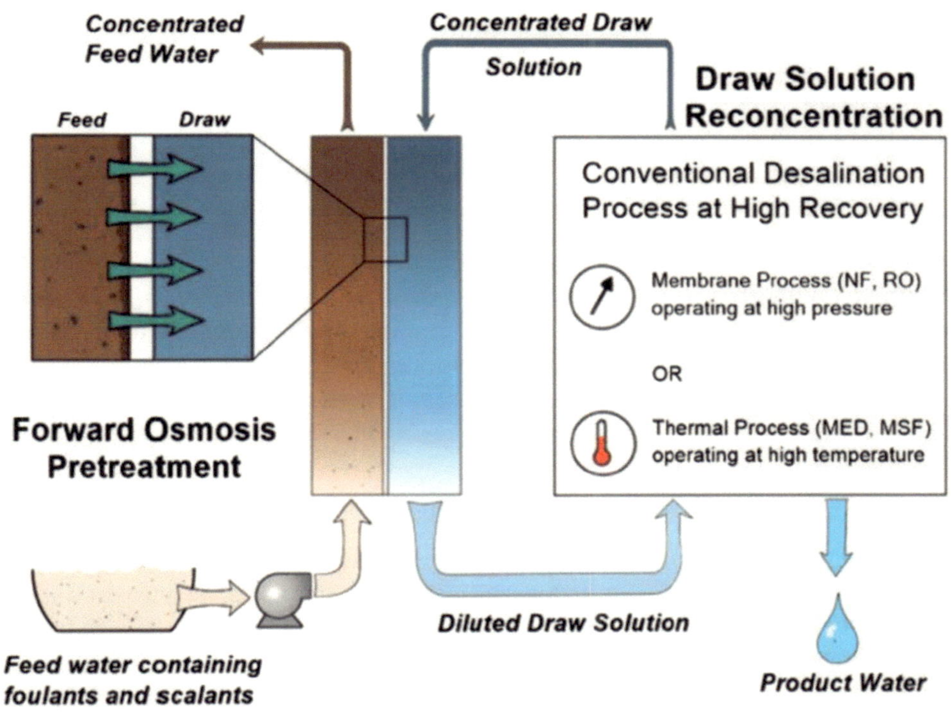

Figure 7. Schematic of FO pretreatment for a conventional membrane desalination process [65].

from the diluted DS solution to obtain fresh water [70]. One of the most common desalination studies is the use of ammonia-carbon dioxide solution as a DS and recovering fresh water with a thermal process and regenerating the osmotic agent [2, 71]. In another study, it was reported that the total equivalent work requirement of this process was less than the conventional desalination process, such as RO, and achieved energy savings of up to 85% when expressed in terms of energy [72]. Researchers have investigated the CP in the direct desalination FO process in which ammonium bicarbonate is used as a DS and have concluded that recovery of fresh water from saline water by FO is a fairly feasible method [73]. There are also different and new DS solution searches to perform an easier and more sustainable DS regeneration in direct FO desalination studies. Generally, an ideal DS should be easy to recover and reusable with high osmotic pressure and high resolution, not toxic, easily available, and inexpensive [70]. In a study where hydrophilic nanoparticles were used as FS for DS and synthetic seawater, about 93% of salt recovery was obtained with flux and UF at around 6 LMH [74]. In a study where divalent salts such as Na_2S0_4 were used as DS and brackish water as FS, 98% of the DS was rejected using NF, while 8–10 LMH flux was obtained [68]. Most DSs investigated for direct FO desalination were not commercially viable due to their high cost, limited maximum water flux they could produce, or low recovery of DS efficiencies. The world's only commercial FO facility for direct sea water treatment, was established in Al Najdah, Oman. This plant is still in operation and has reduced chemical consumption and provides longer membrane life and lower carbon footprint [75] compared to competing technologies such as traditional high-pressure RO membrane systems, saving significant operational and capital costs. These advantages have been associated with the reduction of RO membrane fouling due to the use of FO as a pretreatment step. In the direct FO desalination, similar to the RO desalination, a pretreatment process may be required. Currently, there are very few studies using natural seawater in direct FO desalination. For this reason, the fouling tendency of the FO membrane in these conditions has not been adequately investigated. However, Li et al. [24] reported that a foulant matrix containing natural organic matter and polymerized silica was formed on the membrane when natural seawater was used as feedwater.

In the indirect FO desalination, there is a degraded matrix, such as wastewater or urban stormwater runoff, on the FS side, while DS is using high salinity solution [54, 76]. Potential DS in indirect FO desalination is seawater and brackish water. In addition to being free of charge DS, the main attraction is fresh water recovery through free osmotic energy from the FS, and then a partial dewatered water (diluted DS) that can be desalinated by a low-pressure RO [77]. Thus, the cost of the entire desalination process is also reduced. These studies show that FO desalination integrates fresh water treatment operations from wastewater treatment and seawater, providing a water-energy nexus for coastal cities and a promising process [54, 76].

These studies, in particular the use of primary wastewater as FS for FO, have introduced a concept of the feasibility of FO membrane, which can avoid high-cost treatment of wastewater by conventional treatment processes. For example, an anaerobic process that can be used to treat concentrated primary wastewater (concentrated FS) will provide both biogas production and reduced wastewater treatment costs [78]. Indirect desalination experiments have demonstrated the ability of FO membranes to reject waste water nutrients, especially COD and phosphate and moderately nitrogen. In addition, Linares et al. [76] could adapt the system to the primary clarifier tank using a submerged membrane module, in partial desalination of seawater. This study also showed that FO membranes could reject up to 98% of heavy metals in wastewater. Direct and indirect layouts of desalination systems employing FO membrane are shown in **Figure 8**.

According to a fractional organic carbon analysis carried out in the fouling layer of the FO membrane, it has been reported that this fouling is mainly composed of biopolymers and protein-like substances. A similar result was observed in the FO membrane in the osmotic MBR that was used for municipal wastewater treatment [28]. When the FO system is combined with a low-pressure RO system, this hybrid process has been found to function as a double barrier against selected microcontaminants including pharmaceutically active compounds, hormones, and other organic micropollutants [79]. In practice, most of the micropollutants are rejected by FO membrane using secondary municipal wastewater as FS and sea water as DS, and removal rates were 44–95% for hydrophilic neutral compounds, 48–92% for hydrophilic neutral contaminants, and 96–99% for hydrophilic ionic microcontaminants.

In the FO process coupled with low-pressure RO, the removal of low molecular weight hydrophilic neutral micropollutants was effective (>89%) and the removal of the remaining compounds was over 99% [80]. A membrane cleaning protocol was investigated in the FO application in which municipal secondary wastewater was used as FS and sea water was employed as DS for removing of NOM-fouling through the active layer and removing of transparent exopolymeric particles from the support layer by reporting many cleaning procedures. Osmotic backwashing did not seem to help the recovery of water flux. However, when air was scoured in concentrated wastewater for 15 minutes as a cleaning technique, 89.5% flux recovery was achieved. Cleaning of the active layer with Alconox and EDTA chemistry slightly increased pollution reversibility (93.6%). The chemical cleaning of the support layer removed the reversible pollution of SL up to 94.5%. The irreversible pollution rate in these experiments was 5.5% and it was attributed to biopolymers and trace TEP that cannot be removed from the

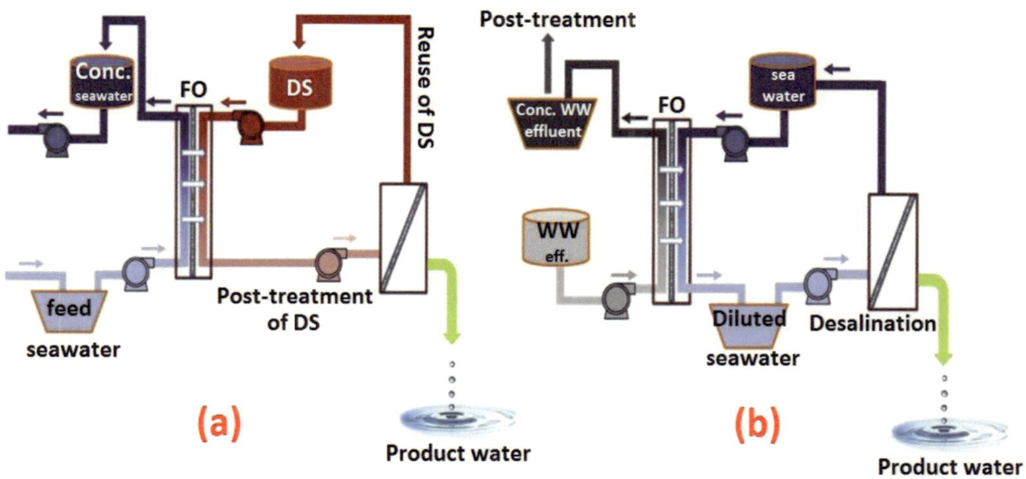

Figure 8. Layout of the two FO processes for desalination (a) direct, (b) indirect (adapted from [45]).

membrane surface [18]. It has been reported that the source of irreversible contaminants on the membrane surface after chemical cleaning and at negligible level is the minimal compaction and of the nature of the FO membrane [45, 79].

In some FO applications, the saline water is used as a major DS rather than the FS. The simplest of these applications is FO, which is used as a pretreatment for RO. In this case, seawater is used as a DS and freshwater is used as an FS for more favorable RO conditions by pressurizing and diluting sea water. Thanks to this pretreatment, the energy required for desalination of the water is greatly reduced. A similar process is pretreatment of RO water using wastewater as FS. The benefit of using such water is that the RO feedwater is diluted to more favorable operating conditions; thus, concentrated feedwater is more appropriate for effective handling. Similarly, a new procedure using ocean water to dewater an algae/nutrient solution for the production of algae biofuels is being investigated [45, 81].

In a recent analysis, McGovern and Lienhard [82] compared the specific energy consumption of a two-pass RO system with FO for desalination of seawater. At 50% recovery, for desalination of seawater containing 35,000 mg/L TDS, the two-pass RO energy consumption has been 3.0 kWh/m^3 including UF (for pretreatment), first- and second- pass RO. The energy consumption for the FO process with the dilution and regeneration process of DS consuming 0.10 and 3.48 kWh/m^3, respectively, for the same conditions was calculated as 3.58 kWh/m^3. Therefore, in order for the FO to be able to compete with the RO in terms of energy consumption, the regeneration process must be significantly more efficient than RO. However, the FO process has the advantage of having less tendency to membrane fouling compared to RO due to the lack of a hydraulic driving pressure. The FO process is also suitable for niche applications where the salinity levels of the water to be treated are higher than the salinity that can be treated by RO process [83].

4.3. Novel/hybrid processes

In their review on emerging desalination technologies, Subramani and Jacangelo [83] reported that the combination of the two technologies (hybrid) has shown that a hybrid technology is more effective than single use. Different hybrid configurations are being evaluated for the treatment of the hard waste waters of various industrial sources. All these industrial sectors require drinking water for various operations and applications. Emerging desalination technologies not only purify these complex wastewaters but also provide water recovery with low operating and maintenance costs and reduce the cost of electricity consumption and membrane cleaning chemicals.

Two hybrid configurations that can be used for the purification of various industrial wastewaters are shown in **Figure 9**. An FO system in **Figure 9a** is combined with an RO system for the treatment of highly contaminated wastewaters [59, 83]. Since hydraulic pressure is not present in FO, the accumulation of contaminants in the membrane is lower and the pretreatment need is eliminated. Again due to the lack of applied pressure, osmotic cleaning using a low salinity solution on the DS side will cause water transport from the DS to the FS [11]. This transport will remove loose deposits of foulants from the membrane surface and lead to more effective cleaning. The concentration of the DS is carried out using a known RO system. Because of the maximum feed pressure limit in RO, the hybrid configuration of FO and RO

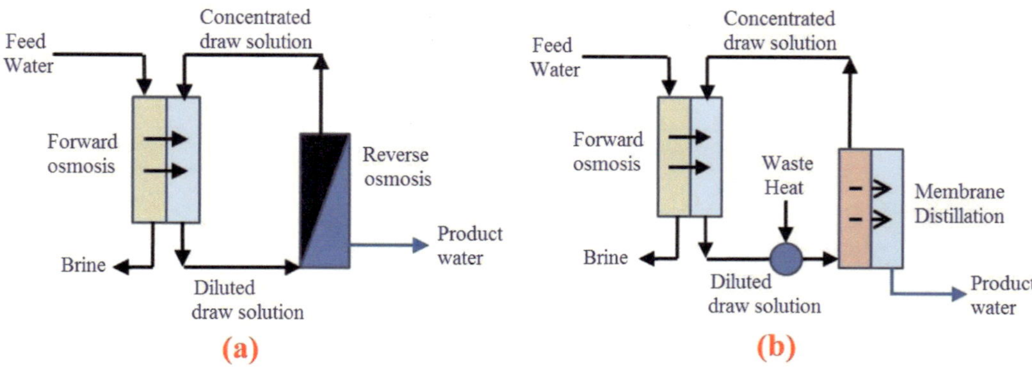

Figure 9. Two hybrid FO applications for wastewater treatment (a) FO-RO, (b) FO-MD [83].

can only be used for the treatment of feedwater streams with low salinity. For feedwater streams with a TDS > 40,000 mg/L, DS recovery can be achieved using a gaseous NH3/CO2 mixture. In this case, additional energy requirements must be taken into account in order to recover the DS using heat or other thermal methods. This configuration is particularly suitable for the refining of reflux water in the petroleum and gas industry when reuse of the water is desired. Purified water can be reused as feedwater for boilers or irrigation [83].

In **Figure 9b**, an FO system is combined with an MD system. The MD system is used for the concentration of the DS [46]. Depending on the salinity of the feed water, various DSs can be used. Since salinity is not a limiting factor for the performance of the MD system, this hybrid configuration can be used to treat wastewater with high salinity. A typical application involves flowback or processing of produced water in the oil and gas industry [84]. This hybrid configuration guarantees a minimum energy requirement when a waste heat source is available to heat the drawing solution and to reconcentrate it using MD [83].

Holloway et al. [51] suggested a hybrid FO-RO system for anaerobic digester concentration. The high energy consumption of the RO (~ 4 kWh/m^3) has been a major limiting factor for the process, although water recovery has been achieved up to 75% with a high concentration of DS (70 g/L NaCl). In a further study [85], seawater was used as a DS solution in a two-stage FO process for sludge concentration to be used as fertilizer. However, high reverse salt flux and membrane fouling due to cake layer formation have been reported as serious problems of the system. Hau et al. [52] suggested a hybrid FO-NF system for a sludge dewatering application. The results showed that the FO performance was better in terms of water flux and reverse salt flux when EDTA was used as DS instead of conventional NaCl or sea water. In addition, FO has successfully rejected more than 90% of the nutrients released from the feed sludge. They also indicated that the NF recovery of EDTA sodium salts exhibiting high charged compounds performed well and had a high salt rejection of 93%. While the water flux was constant during the first hours of operation, the FO membrane was then rapidly reduced due to the increased buildup of the sludge cake layer in the concentrated feed and diluted DS.

Oasys Water Inc. has operated a pilot scale thermal-based hybrid FO system for water with high salinity (>70,000 ppm TDS), which is a product of shale gas industry [41, 60]. The results

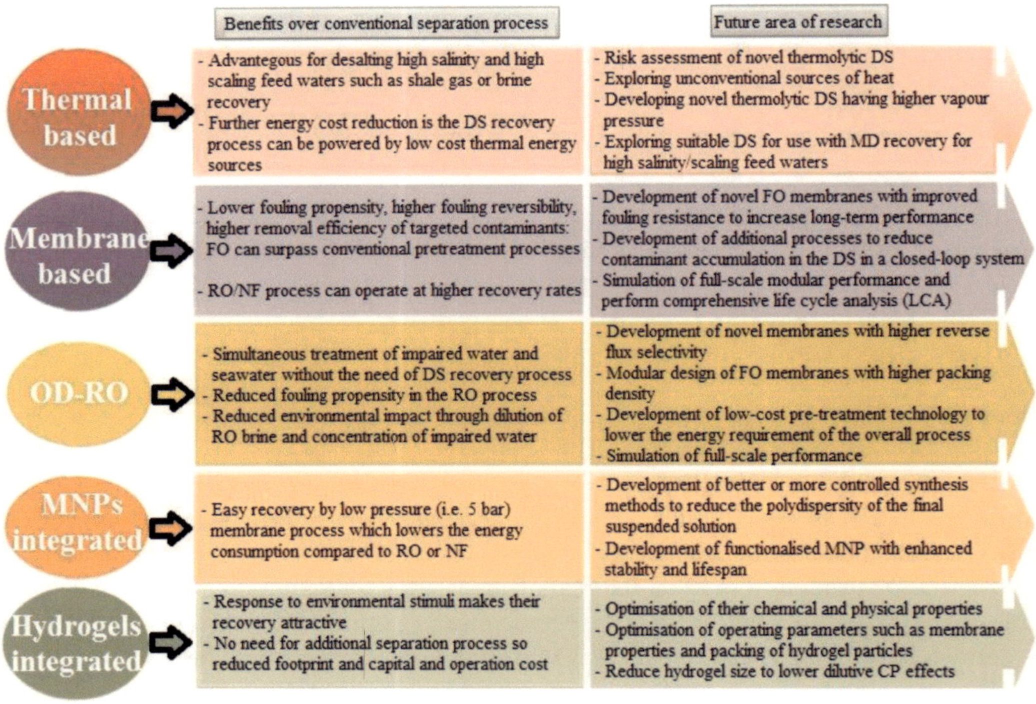

Figure 10. Summary of the benefits of current hybrid FO systems and direction for future research (adapted from [86]).

show that this hybrid system can exhibit feedwater recovery performance (60%) similar to evaporative saline concentration technologies and that the final product water meets surface water discharge criteria in terms of TDS, chlorides, barium, and strontium. However, although the RO required less specific energy when desalinating waters with lower salinity, it was found that this study was not sufficient to purify the challenging feedwater. The hybrid FO distillation system can be integrated to provide a zero-liquid discharge (ZLD) facility designed as a membrane brine concentrator (MBC). The MBC system is ideal for the oil and gas industry and provides up to 85% water recovery while discharging brine with salt concentration up to 25%. A summary of the benefits of current hybrid FO systems and direction of future research are schematized in **Figure 10** [86].

5. Conclusions

FO process has a big potential to be an alternative solution for water/wastewater treatment and desalination purposes over conventional membrane processes. To benefit from this potential at maximum, ICP and low flux challenges should be completely solved or minimized by changing operational parameters. Changing membrane orientation (to increase water flux), utilizing various DSs (to increase osmotic pressure), and changing sludge retention time (i.e., to hinder salt accumulation in FO-MBR) are some of the basic procedures used since former FO studies.

The use of hybrid systems such as FO-RO and FO-MD even with seawater desalination and optimization energy consumption could be more feasible and better alternative than the performance exhibited by the FO process alone for wastewater recovery. However, the indispensable factor affecting the process performance is FO membrane. According to the current studies, utilizing novel nanomaterials, substrates, and layer-by-layer assumptions in manufacturing of FO membrane undoubtedly enhance the water flux and rejection of the pollutants and minimize the membrane fouling but using synthetic wastewater-generally, containing one model foulant or DI water as feed solution makes it difficult to predict how FO membranes will act in real wastewaters or harsh environmental conditions. Therefore, working with complex foulants and real wastewaters to better understand membrane behaviors and using modeling tools for fouling prediction and new cleaning strategies are essential to mitigate intrinsic challenges of the FO membranes.

In ongoing researches, the developed new support layers appear to continue increasing water flux slightly; however, lower water flux remains as a main challenge of the process when compared to the conventional membrane systems. It is also a fact that the diffusion provided by draw solution in the process is not effective alone to increase product water volume; therefore, some promotive factors such as rehabilitated hydrodynamic behaviors or simultaneous filtration could be provided together with diffusion phenomena in further researches.

Acknowledgements

This work was supported by the Scientific and Technological Research Council of Turkey (TUBITAK), grant number: CAYDAG-113Y340.

Author details

Murat Eyvaz[1]*, Serkan Arslan[1], Derya İmer[2], Ebubekir Yüksel[1] and İsmail Koyuncu[2]

*Address all correspondence to: meyvaz@gtu.edu.tr

1 Environmental Engineering Department, Gebze Technical University, Gebze-Kocaeli, Turkey

2 Environmental Engineering Department, İstanbul Technical University, Maslak-İstanbul, Turkey

References

[1] Manickam SS, McCutcheon JR. Understanding mass transfer through asymmetric membranes during forward osmosis: A historical perspective and critical review on measuring structural parameter with semi-empirical models and characterization approaches. Desalination. 2017;**421**:110-126. DOI: 10.1016/j.desal.2016.12.016

[2] Cath TY, Childress AE, Elimelech M. Forward osmosis: Principles, applications, and recent developments. Journal of Membrane Science. 2006;**281**:70-87. DOI: 10.1016/j.memsci.2006.05.048

[3] Manickam S, McCutcheon JR. Characterization of polymeric nonwovens using porosimetry, porometry and X-ray computed tomography. Journal of Membrane Science. 2012;**407**:108-115. DOI: 10.1016/j.memsci.2012.03.022

[4] McCutcheon JR, McGinnis RL, Elimelech M. A novel ammonia-carbon dioxide forward (direct) osmosis desalination process. Desalination. 2005;**174**(1):1-11. DOI: 10.1016/j.desal.2004.11.002

[5] Tiraferri A, Yip NY, Phillip WA, Schiffman JD, Elimelech M. Relating performance of thin-film composite forward osmosis membranes to support layer formation and structure. Journal of Membrane Science. 2011;**367**:340-352. DOI: 10.1016/j.memsci.2010.11.014

[6] Shi L, Chou SR, Wang R, Fang WX, Tang CY, Fane AG. Effect of substrate structure on the performance of thin-film composite forward osmosis hollow fiber membranes. Journal of Membrane Science. 2011;**382**(1):116-123. DOI: 10.1016/j.memsci.2011.07.045

[7] Bui NN. Engineered osmosis for sustainable water and energy: Novel nanofiber supported thin-film composite membrane design & updated flux model proposal [thesis]. Storrs: University of Connecticut; 2013

[8] Huang L, McCutcheon JR. Impact of support layer pore size on performance of thin film composite membranes for forward osmosis. Journal of Membrane Science. 2015;**483**:25-33. DOI: 10.1016/j.memsci.2015.01.025

[9] Cath TY, Elimelech M, McCutcheon JR, McGinnis RL, Achilli A, Anastasio D, Brady AR, Childress AE, Farr IV, Hancock NT. Standard methodology for evaluating membrane performance in osmotically driven membrane processes. Desalination 2012;**312**:31-38. DOI: 10.1016/j.desal.2012.07.005

[10] Lee J, Choi JY, Choi JS, Chu KH, Yoon Y, Kim S. A statistics-based forward osmosismembrane characterization method without pressurized reverse osmosis experiment. Desalination. 2017;**403**:36-45. DOI: 10.1016/j.desal.2016.04.023

[11] Chung TS, Zhang S, Wang KY, Su J, Ling MM. Forward osmosis processes: Yesterday, today and tomorrow. Desalination. 2012;**287**:78-81. DOI: 10.1016/j.desal.2010.12.019

[12] Chung TS, Li X, Ong RC, Ge QC, Wang HL, Han G. Emerging forward osmosis (FO) technologies and challenges ahead for clean water and clean energy applications. Current Opinion in Chemical Engineering. 2012;**1**:246-257. DOI: 10.1016/j.coche.2012.07.004

[13] Mi B, Elimelech M. Chemical and physical aspects of organic fouling of forward osmosis membranes. Journal of Membrane Science. 2008;**320**:292-302. DOI: 10.1016/j.memsci.2008.04.036

[14] She Q, Wang R, Fane AG, Tang CY. Membrane fouling in osmotically driven membrane processes: A review. Journal of Membrane Science. 2016;**499**:201-233. DOI: 10.1016/j.memsci.2015.10.040

[15] Zhang M, Hou D, She Q, Tang CY. Gypsum scaling in pressure retarded osmosis: Experiments, mechanisms and implications. Water Research. 2014;**48**:387-395. DOI: 10.1016/j.watres.2013.09.051

[16] Yip NY, Elimelech M. Influence of natural organic matter fouling and osmotic backwash on pressure retarded osmosis energy production from natural salinity gradients. Environmental Science and Technology. 2013;**47**:12607-12616. DOI: 10.1021/es403207m

[17] Boo C, Elimelech M, Hong S. Fouling control in a forward osmosis process integrating seawater desalination and wastewater reclamation. Journal of Membrane Science. 2013;**444**:148-156. DOI: 10.1016/j.memsci.2013.05.004

[18] Valladares-Linares R, Li Z, Yangali-Quintanilla V, Li Q, Amy G. Cleaning protocol for a FO membrane fouled in wastewater reuse. Desalination and Water Treatment. 2013; **51**:4821-4824. DOI: 10.1080/19443994.2013.795345

[19] Tang CY, She Q, Lay WCL, Wang R, Fane AG. Coupled effects of internal concentration polarization and fouling on flux behavior of forward osmosis membranes during humic acid filtration. Journal of Membrane Science. 2010;**354**:123-133. DOI: 10.1016/j.memsci.2010.02.059

[20] Mi B, Elimelech M. Organic fouling of forward osmosis membranes: Fouling reversibility and cleaning without chemical reagents. Journal of Membrane Science. 2010;**348**:337-345

[21] Arkhangelsky E, Wicaksana F, Chou S, Al-Rabiah AA, Al-Zahrani SM, Wang R. Effects of scaling and cleaning on the performance of forward osmosis hollow fiber membranes. Journal of Membrane Science. 2012;**415-416**:101-108. DOI: 10.1016/j.memsci.2012.04.041

[22] Fritzmann C, Löwenberg J, Wintgens T, Melin T. State-of-the-art of reverse osmosis desalination. Desalination. 2007;**216**:1-76. DOI: 10.1016/j.desal.2006.12.009

[23] Zhou W, Wu B, She Q, Chi L, Zhang Z. Investigation of soluble microbial products in a full-scale UASB reactor running at low organic loading rate. Bioresource Technology. 2009;**100**:3471-3476. DOI: 10.1016/j.biortech.2009.03.006

[24] Li ZY, Yangali-Quintanilla V, Valladares-Linares R, Li Q, Zhan T, Amy G. Flux patterns and membrane fouling propensity during desalination of seawater by forward osmosis. Water Research. 2012;**46**:195-204. DOI: 10.1016/j.watres.2011.10.051

[25] Meng F, Chae SR, Drews A, Kraume M, Shin HS, Yang F. Recent advances in membrane bioreactors (MBRs): Membrane fouling and membrane material. Water Research. 2009; **43**:1489-1512. DOI: 10.1016/j.watres.2008.12.044

[26] Wang YN, Tang CY. Fouling of nanofiltration, reverse osmosis, and ultrafiltration membranes by protein mixtures: The role of inter-foulant-species interaction. Environmental Science and Technology. 2011;**45**:6373-6379. DOI: 10.1021/es2013177

[27] Berman T, Mizrahi R, Dosoretz CG. Transparent exopolymer particles (TEP): A critical factor in aquatic biofilm initiation and fouling on filtration membranes. Desalination. 2011;**276**:184-190. DOI: 10.1016/j.desal.2011.03.046

[28] Zhang J, Loong WLC, Chou S, Tang C, Wang R, Fane AG. Membrane biofouling and scaling in forward osmosis membrane bioreactor. Journal of Membrane Science. 2012;**403-404**:8-14. DOI: 10.1016/j.memsci.2012.01.032

[29] Kang NW, Lee S, Kim D, Hong S, Kweon JH. Analyses of calcium carbonate scale deposition on four RO membranes under a seawater desalination condition. Water Science and Technology. 2011;**64**:1573-1580. DOI: 10.2166/wst.20U.671

[30] Zhang H, Ma Y, Jiang T, Zhang G, Yang F. Influence of activated sludge properties on flux behavior in osmosis membrane bioreactor (OMBR). Journal of Membrane Science. 2012;**390-391**:270-276. DOI: 10.1016/j.memsci.2011.11.048

[31] Tang CY, Chong TH, Fane AG. Colloidal interactions and fouling of NF and RO membranes: A review. Advances in Colloid and Interface Science. 2011;**164**:126-143. DOI: 10.1016/j.cis.2010.10.007

[32] Kim Y, Elimelech M, Shon HK, Hong S. Combined organic and colloidal fouling in forward osmosis: Fouling reversibility and the role of applied pressure. Journal of Membrane Science. 2014;**460**:206-212. DOI: 10.1016/j.memsci.2014.02.038

[33] Palecek SP, Zydney AL. Intermolecular electrostatic interactions and their effect on flux and protein deposition during protein filtration. Biotechnology Progress. 1994;**10**:207-213. DOI: 10.1021/bp00026a010

[34] Zou S, Wang YN, Wicaksana F, Aung T, Wong PCY, Fane AG, Tang CY. Direct microscopic observation of forward osmosis membrane fouling by microalgae: Critical flux and the role of operational conditions. Journal of Membrane Science. 2013;**436**:174-185. DOI: 10.1016/j.memsci.2013.02.030

[35] Song X, Liu Z, Sun DD. Nano gives the answer: Breaking the bottleneck of internal concentration polarization with a nanofiber composite forward osmosis membrane for a high water production rate. Advanced Materials. 2011;**23**:3256-3260. DOI: 10.1002/adma.201100510

[36] Chou S, Wang R, Shi L, She Q, Tang C, Fane AG. Thin-film composite hollow fiber membranes for pressure retarded osmosis (PRO) process with high power density. Journal of Membrane Science. 2012;**389**:25-33. DOI: 10.1016/j.memsci.2011.10.002

[37] Han G, Chung TS. Robust and high performance pressure retarded osmosis hollow fiber membranes for osmotic power generation. AICHE Journal. 2014;**60**:1107-1119. DOI: 10.1002/aic.14342

[38] Boo C, Lee S, Elimelech M, Meng Z, Hong S. Colloidal fouling in forward osmosis: Role of reverse salt diffusion. Journal of Membrane Science. 2012;**390-391**:277-284. DOI: 10.1016/j.memsci.2011.12.001

[39] Lutchmiah K, Verliefde ARD, Roest K, Rietveld LC, Cornelissen ER. Forward osmosis for application in wastewater treatment: A review. Water Research. 2014;**58**:179-197. DOI: 10.1016/j.watres.2014.03.045

[40] Fang Y, Bian L, Bi Q, Li Q, Wang X. Evaluation of the pore size distribution of a forward osmosis membrane in three different ways. Journal of Membrane Science. 2014; **454**(0):390-397. DOI: 10.1016/j.memsci.2013.12.046

[41] Coday BD, Xu P, Beaudry EG, Herron J, Lampi K, Hancock NT, Cath TY. The sweet spot of forward osmosis: Treatment of produced water, drilling wastewater, and other complex and difficult liquid streams. Desalination. 2014;**333**(1):23-35. DOI: 10.1016/j.desal.2013.11.014

[42] Kim ES, Liu Y, El-Din MG. The effects of pretreatment on nanofiltration and reverse osmosis membrane filtration for desalination of oil sands process-affected water. Separation and Purification Technology. 2011;**81**(3):418-428. DOI: 10.1016/j.seppur.2011.08.016

[43] Zhu H, Zhang L, Wen X, Huang X. Feasibility of applying forward osmosis to the simultaneous thickening, digestion, and direct dewatering of waste activated sludge. Bioresource Technology. 2012;**113**(0):207-213. DOI: 10.1016/j.biortech.2011.12.064

[44] Hydranautics. Element Spec Sheets. Hydranautics a Nitto Group Company. 2014. Available from: http://membranes.com/index.php?pagename=spec_sheets [Accessed: August 11, 2017]

[45] Linares RV, Li Z, Sarp S, Bucs S, Amy G, Vrouwenvelder JS. Forward osmosis niches in seawater desalination and wastewater reuse. Water Research. 2014;**66**:122-139. DOI: 10.1016/j.watres.2014.08.021

[46] Xie M, Nghiem LD, Price WE, Elimelech M. A forward osmosis membrane distillation hybrid process for direct sewer mining: System performance and limitations. Environmental Science and Technology. 2013;**47**(23):13486-13493. DOI: 10.1021/es404056e

[47] Zhang Y, Pinoy L, Meesschaert B, Van der Bruggen B. A natural driven membrane process for brackish and wastewater treatment: Photovoltaic powered ED and FO hybrid system. Environmental Science and Technology. 2013;**47**(18):10548-10555. DOI: 10.1021/es402534m

[48] Li ZY, Valladares-Linares R, Yangali-Quintanilla V, Amy G. A sequential batch reactor forward osmosis system for water reuse. In: Proceedings of American Membrane Technology Association Membrane Technology Conference and Exhibition. Phoenix: AWWA/AMTA Publishing; February 27–March 1, 2012

[49] Qiu G, Zhang S, Raghavan DSSS, Das S, Ting YP. The potential of hybrid forward osmosis membrane bioreactor (FOMBR) processes in achieving high throughput treatment of municipal wastewater with enhanced phosphorus recovery. Water Research. 2016;**105**:370-382. DOI: 10.1016/j.watres.2016.09.017

[50] Zhang S, Liu P, Chen Y, Jin J, Hu L, Jian X. Preparation of thermally stable composite forward osmosis hollow fiber membranes based on copoly(phthalazinone biphenyl ether sulfone) substrates. Chemical Engineering Science. 2017;**166**:91-100. DOI: 10.1016/j.ces.2017.03.026

[51] Holloway RW, Childress AE, Dennett KE, Cath TY. Forward osmosis for concentration of anaerobic digester centrate. Water Research. 2007;**41**(17):4005-4014. DOI: 10.1016/j.watres.2007.05.054

[52] Hau NT, Chen SS, Nguyen NC, Huang KZ, Ngo HH, Guo W. Exploration of EDTA sodium salt as novel draw solution in forward osmosis process for dewatering of high nutrient sludge. Journal of Membrane Science. 2014;**455**:305-311. DOI: 10.1016/j.memsci.2013.12.068

[53] Catalyx. Forward osmosis for recycling dye wastewater. Filtration and Separation. 2009;**46**(3):14. DOI: 10.1016/S0015-1882(09)70120-X

[54] Li Z, Valladares Linares R, Abu-Ghdaib M, Zhan T, Yangali- Quintanilla V, Amy G. Osmotically driven membrane process for the management of urban runoff in coastal regions. Water Research. 2004;**48**:200-209. DOI: 10.1016/j.watres.2013.09.028

[55] Duong PHH, Chung TS. Application of thin film composite membranes with forward osmosis technology for the separation of emulsified oil-water. Journal of Membrane Science. 2014;**452**:117-126. DOI: 10.1016/j.memsci.2013.10.030

[56] HTI. Oil Wastewater Treatment & Gas Wastewater Treatment: Lead Story [Internet]. 2011. Available from: http://www.htiwater.com/divisions/oil-gas/lead_story.html [Accessed: Sep 11, 2017]

[57] Nelson CE, Ghosh AK. Oil & Natural Gas Technology-Membrane Technology for Produced Water in Lea County. Lea County Government and New Mexico Institute of Mining and Technology [Internet]. 2011. Available from: https://www.netl.doe.gov/File%20Library/Research/Oil-Gas/nt0005227-final-report.pdf [Accessed: Sep 15, 2017]

[58] Abousnina RM. Oily wastewater treatment: Removal of dissolved organic components by forward osmosis [thesis]. University of Wollongong; 2012

[59] Hickenbottom KL, Hancock NT, Hutchings NR, Appleton EW, Beaudry EG, Xu P, Cath TY. Forward osmosis treatment of drilling mud and fracturing wastewater from oil and gas operations. Desalination. 2013;**312**:60-66. DOI: 10.1016/j.desal.2012.05.037

[60] McGinnis RL, Hancock NT, Nowosielski-Slepowron MS, McGurgan GD. Pilot demonstration of the NH_3/CO_2 forward osmosis desalination process on high salinity brines. Desalination. 2013;**312**:67-74. DOI: 10.1016/j.desal.2012.11.032

[61] Greenlee LF, Lawler DF, Freeman BD, Marrot B, Moulin P. Reverse osmosis desalination: Water sources, technology, and today's challenges. Water Research. 2009;**43**:2317-2348. DOI: 10.1016/j.watres.2009.03.010

[62] El-Dessouky HT, Ettouney HM. Fundamentals of Salt Water Desalination. 1st ed. Amsterdam: Elsevier Science Ltd; 2002. p. 690. DOI: 10.1016/B978-0-444-50810-2.50018-X

[63] Van der Bruggen B, Vandecasteele C. Distillation vs. membrane filtration: Overview of process evolutions in seawater desalination. Desalination. 2002;**143**:207-218. DOI: 10.1016/S0011-9164(02)00259-X

[64] Kim HI, Kim SS. Plasma treatment of polypropylene and polysulfone supports for thin film composite reverse osmosis membrane. Journal of Membrane Science. 2006;**286**:193-201. DOI: 10.1016/j.memsci.2006.09.037

[65] Shaffer DL, Werber JR, Jaramillo H, Lin S, Elimelech M. Forward osmosis: Where are we now? Desalination. 2015;**356**:271-284. DOI: 10.1016/j.desal.2014.10.031

[66] Zaviska F, Zou L. Using modelling approach to validate a bench scale forward osmosis pre-treatment process for desalination. Desalination. 2014;**350**:1-13. DOI: 10.1016/j.desal.2014.07.005

[67] Bamaga OA, Yokochi A, Beaudry EG. Application of forward osmosis in pretreatment of seawater for small reverse osmosis desalination units. Desalination and Water Treatment. 2009;**5**:183-191. DOI: 10.5004/dwt.2009.574

[68] Zhao S, Zou L, Mulcahy D. Brackishwater desalination by a hybrid forward osmosis–nanofiltration system using divalent draw solute. Desalination. 2012;**284**:175-181. DOI: 10.1016/j.desal.2011.08.053

[69] Chekli L, Phuntsho S, Shon HK, Vigneswaran S, Kandasamy J, Chanan A. A review of draw solutes in forward osmosis process and their use in modern applications. Desalination and Water Treatment. 2012;**43**(1-3):167-184. DOI: 10.1080/19443994.2012.672168

[70] Li D, Zhang X, Simon GP, Wang H. Forward osmosis desalination using polymer hydrogels as a draw agent: Influence of draw agent, feed solution and membrane on process performance. Water Research. 2013;**47**(1):209-215. DOI: 10.1016/j.watres.2012.09.049

[71] Gray GT, McCutcheon JR, Elimelech M. Internal concentration polarization in forward osmosis: Role of membrane orientation. Desalination. 2006;**197**(1-3):1-8. DOI: 10.1016/j.desal.2006.02.003

[72] McGinnis RL, Elimelech M. Energy requirements of ammonia-carbon dioxide forward osmosis desalination. Desalination. 2007;**207**(1-3):370-382. DOI: 10.1016/j.desal.2006.08.012

[73] Chanukya BS, Patil S, Rastogi NK. Influence of concentration polarization on flux behavior in forward osmosis during desalination using ammonium bicarbonate. Desalination. 2013;**312**:39-44. DOI: 10.1016/j.desal.2012.05.018

[74] Ling MM, Chung TS. Desalination process using super hydrophilic nanoparticles via forward osmosis integrated with ultrafiltration regeneration. Desalination. 2011;**278**(1-3):194-202. DOI: 10.1016/j.desal.2011.05.019

[75] Modern Water. Membrane Processes Forward Osmosis: Desalination [Internet]. 2013. Available from: https://www.modernwater.com/pdf/MW_Factsheet_Membrane_HIGHRES.pdf [Accessed: Sep 14, 2017]

[76] Valladares Linares R, Li Z, Abu-Ghdaib M, Wei CH, Amy G, Vrouwenvelder JS. Water harvesting from municipal wastewater via osmotic gradient: An evaluation of process performance. Journal of Membrane Science. 2013;**447**:50-56. DOI: 10.1016/j.memsci.2013.07.018

[77] Cath TY, Hancock NT, Lundin CD, Hoppe-Jones C, Drewes JE. A multi-barrier osmotic dilution process for simultaneous desalination and purification of impaired water. Journal of Membrane Science. 2010;**362**(1-2):417-426. DOI: 10.1016/j.memsci.2010.06.056

[78] McCarty PL, Bae J, Kim J. Domestic wastewater treatment as a net energy producer – Can this be achieved? Environmental Science and Technology. 2011;**45**(17):7100-7106. DOI: 10.1021/es2014264

[79] Cath TY, Drewes JE, Lundin CD. A novel hybrid forward osmosis process for drinking water augmentation using impaired water and Saline water sources. In: Proceedings of the 24th Annual WateReuse Symposium. Seattle: Water Research Foundation; Sep 13-16, 2009. Available from: http://inside.mines.edu/~tcath/research/projects/Cath_WRS_2009_AwwaRF4150.pdf [Accessed: October 23, 2017]

[80] Valladares Linares R, Yangali-Quintanilla V, Li Z, Amy G. Rejection of micropollutants by clean and fouled forward osmosis membrane. Water Research. 2011;**45**(20):6737-6744. DOI: 10.1016/j.watres.2011.10.037

[81] Hoover LA, Phillip WA, Tiraferri A, Yip NY, Elimelech M. Forward with osmosis: Emerging applications for greater sustainability. Environmental Science and Technology. 2011;**45**(23):9824-9830. DOI: 10.1021/es202576h

[82] McGovern RK, Lienhard V. On the potential of forward osmosis to energetically outperform reverse osmosis desalination. Journal of Membrane Science. 2014;**469**:245-250. DOI: 10.1016/j.memsci.2014.05.061

[83] Subramani A, Jacangelo JG. Emerging desalination technologies for water treatment: A critical review. Water Research. 2015;**75**:164-187. DOI: 10.1016/j.watres.2015.02.032

[84] Department of Energy (USDOE). Advanced, Energy-efficient Hybrid Membrane System for Industrial Water Reuse [Internet]. Available from: https://energy.gov/sites/prod/files/2016/12/f34/0877-Hybrid%20Membrane%20System-090716_compliant.pdf, [Accessed: Sep 10, 2017]

[85] Nguyen NC, Chen SS, Yang HY, Hau NT. Application of forward osmosis on dewatering of high nutrient sludge. Bioresource Technology. 2013;**132**:224-229. DOI: 10.1016/j.biortech.2013.01.028

[86] Tsai JH, Macedonio F, Drioli E, Giorno L, Chou CY, Hu FC, Li CL, Chuang CJ, Tung KL. Membrane-based zero liquid discharge: Myth or reality? Journal of the Taiwan Institute of Chemical Engineers. 2017;**80**:192-202. DOI: 10.1016/j.jtice.2017.06.050

Chapter 4

Effect of Internal and External Concentration Polarizations on the Performance of Forward Osmosis Process

Amrit Bhinder, Simin Shabani and
Mohtada Sadrzadeh

Additional information is available at the end of the chapter

http://dx.doi.org/10.5772/intechopen.71343

Abstract

Forward osmosis (FO) as an osmotically driven membrane process is severely affected by the concentration polarization phenomenon on both sides of the membrane as well as inside the support layer. Though the effect of internal concentration polarization (ICP) in the porous support on the draw solution side is far more pronounced than that of the external concentration polarization (ECP), still the importance of ECP cannot be neglected. The ECP becomes particularly important when the feed flow rate is enhanced to increase the permeation flux by increasing the agitation and turbulence on the membrane surface. To capture the effect of ECP a suitable value of mass transfer coefficient must be determined. In this chapter, an FO mass transport model that accounts for the presence of both ICP and ECP phenomena is first presented on the basis of solution-diffusion model coupled with diffusion-convection. Then, three methods for the estimation of mass transfer coefficient based on empirical Sherwood (Sh) number correlations, pressure-driven reverse osmosis (RO), and osmosis-driven pressure retarded osmosis (PRO) are proposed. Finally, a methodology for the prediction of water flux through FO membranes using the theoretical model and calculated/measured parameters (hydraulic permeability, salt resistivity of the support layer, and mass transfer coefficient) is presented.

Keywords: forward osmosis, concentration polarization, mass transfer coefficient, reverse osmosis, pressure retarded osmosis

1. Introduction

With the increasing application of membrane-based separation processes in desalination and wastewater treatment, vast efforts have been devoted to making them more energy efficient.

In the hunt of more economical and efficient method, forward osmosis (FO) has been developed as an alternative to the conventional pressure-driven separation processes like reverse osmosis (RO) and nanofiltration (NF) [1–3]. FO is an osmotically driven membrane separation process, where water molecules are transferred from a dilute feed solution to a more concentrated draw solution through a semi-permeable membrane which selectively rejects a broad range of dissolved contaminants in the wastewater [3]. The driving force for water transport is the chemical potential difference between the draw and feed solutions, thus eliminating the use of hydraulic pressure and consequently enhances energy efficiency [4–6].

Besides being energy efficient, FO process is less prone to fouling as compared to pressure-driven NF and RO processes. However, FO suffers from an enhanced concentration polarization effect inside the support layer known as internal concentration polarization (ICP), where the solvent (commonly water) permeates through the support and dilutes the draw concentration at the inner side of the active layer. The ICP reduces the real driving force for mass transfer, thereby reducing the performance of the FO process, significantly [7, 8]. In addition to ICP, FO suffers from an external concentration polarization (ECP). In fact, in a typical pressure driven process, ECP occurs on one side of the membrane (feed side), whereas in the FO, this phenomenon happens on both sides (feed and draw). The polarization that occurs on the feed side is concentrative and is different in nature from the dilutive polarization on the draw side due to incoming permeate flux. The first polarization is called concentrative ECP and the second one that takes places in the draw side is termed as dilutive ECP. The ICP is not affected by the hydrodynamics of the flow and is more severe than the ECP which makes the theoretical study of transport phenomena in an FO process very challenging.

Early attempts to model the mass transfer through an FO membrane was conducted by Lee et al. [9]. They considered the ICP inside the porous support and developed a model to predict the performance of a pressure retarded osmosis (PRO) process. In the PRO process, which is used for energy generation from an osmotic pressure difference, the membranes are oriented in the exact opposite configuration of FO with the active layer facing the draw solution. Later, Loeb et al. [10] followed the same approach and developed a model for the FO process. McCutcheon et al. [11] coupled the boundary layer film theory to capture the effect of ECP on the active layer as well as the ICP in the porous support for both FO and PRO processes. Suh and Lee [12] fine-tuned this model by considering the dilutive ECP phenomenon on the draw side which was neglected by previous researchers. They suggested that the effect of diluted draw solution on the ECP must be taken into account, particularly for low cross-flow velocities and high water flux. Even though the above models provide a comprehensive framework of relationships for the ICP and ECP on both sides of FO membranes and can predict the flux satisfactorily at a particular flow rate, they are not sufficiently sensitive to a change in the feed flow rate.

The change in water flux with a change in the flow rate is captured by the mass transfer coefficient (k) on either side of the membrane. The mass transfer coefficient is commonly calculated using Sherwood (Sh) number relations which are empirical correlations as a function of Reynolds Schmidt numbers [13]. The Sh number relationships available in the literature were either adapted from the analogy between heat and mass transfer or were derived for flow in non-porous smooth [13–15]. These relations were later modified for the ultrafiltration (UF) experiments [13, 14]. UF is a pressure-driven process with a different flow hydrodynamics from FO process which is driven by the osmotic pressure gradient. Also, the topology of a typical UF membrane is rougher (on a microscopic scale) and porous than an FO membrane that might affect the Sh number which is a linear function of a frictional factor. Hence, the correlations of Sh number derived from UF experiments may not be valid for the FO process.

Although extensive research has been carried out on the derivation of empirical and semi-empirical Sh number correlations for pressure driven membrane processes (at various operating conditions and spacer geometries) [14, 16, 17], no such efforts have been made to better understand the boundary layer phenomena in an FO process. It is worth mentioning that, based on the film theory the severity of the ECP depends upon the value of mass transfer coefficient. Since the concentration profile in the boundary layer is exponential in nature, a small error in the value of mass transfer coefficient may magnify error to a large extent. Hence, to develop a robust model for the FO process, there is crucial need to find an appropriate correlation of mass transfer coefficient for each specific membrane process with certain hydrodynamic properties of channel and membrane characteristics.

In this chapter an attempt has been made to provide (i) the theoretical background of internal and external concentration polarization phenomena, and (ii) different methods that can be used for the estimation of mass transfer coefficient in the FO process. Since the support layer of a thin film composite FO membrane is made from a porous material (e.g., polysulfone, PSf), having a similar structure and porosity as that of a UF membrane, the literature Sh number correlations might be valid on this side of the membrane. But for the selective layer of the membrane, which is smooth and non-porous, these relationships are not necessarily usable. Hence, more practical methods to get an estimate of the value of mass transfer coefficient on the active side of the membrane in an FO process by (i) RO and (ii) PRO experiments are proposed. These mass transfer coefficients can then be used in the theoretical model to predict the water flux with a change in the feed flow velocity.

2. Theory

2.1. Water flux in FO

Water flux in a pressure-driven membrane separation process is directly proportional to the applied pressure (Δp) and the osmotic pressure difference between the two solutions ($\Delta\pi$) [11].

$$J_w = A(\Delta p - \sigma\Delta\pi) \tag{1}$$

where A is the pure water permeability, and σ is the reflection coefficient which describes the fraction of the solutes reflected or rejected by the membrane. For ideal membranes with no solute transport, its value is unity. In an FO process, no pressure is applied ($\Delta p = 0$) and the water flux through the membrane is just due to the difference in the osmotic pressures of the draw and feed solutions, given by:

$$J_w = A(\pi_{D,b} - \pi_{F,b}) \tag{2}$$

where $\pi_{D,b}$ and $\pi_{F,b}$ are the osmotic pressures of the draw and feed solutions, respectively.

2.2. Concentration polarization

In an FO operation, the actual flux is far less than the theoretical flux obtained from Eq. (2) which shows a decline in driving force. On the feed side, where the solvent permeates through the membrane, the solutes are retained by the membrane increasing their concentration on the membrane surface that is referred to as concentrative ECP. The permeate entering the draw side dilutes the draw solution at the membrane surface that is known as diluted ECP. **Figure 1(a)** and **(b)** depict concentrative and dilutive ECP, as well as ICP, occurring in FO and PRO processes. Both these phenomena contribute to a decrease in the net osmotic driving force across the membrane and hence lowering the flux.

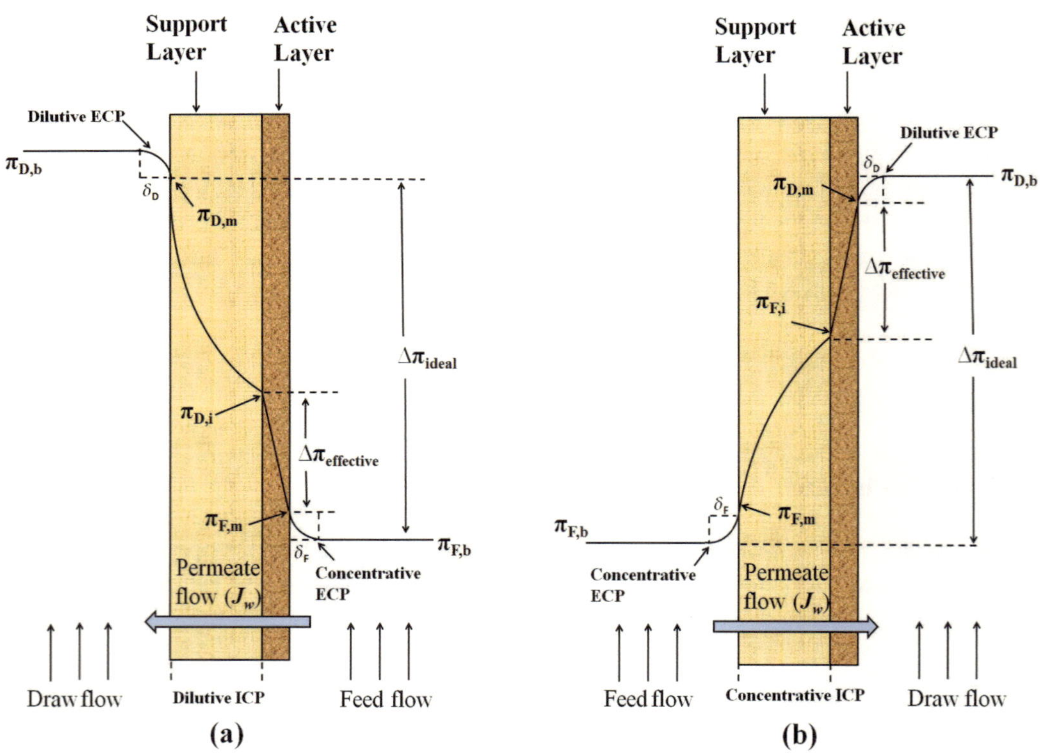

Figure 1. Direction of water flux and the concentration profile developed across the membrane in (a) FO mode and (b) PRO mode.

The ECP can be mitigated by inducing turbulence which enhances the mixing and consequently levels the concentration difference between the bulk and adjacent solution to the membrane surface. However, the concentration polarization in FO is not just limited to ECP. The structure of FO membranes is typically asymmetric, i.e. a thin active layer which governs the molecular transport rate is coated on a porous support that provides mechanical strength. In the FO mode (when active layer and support are facing the feed and draw solutions, respectively), a more severe concentration polarization takes places inside the porous support layer of the membrane, known as ICP. The enhanced dilution of the draw solution inside the porous support contributes to a massive decline in the osmotic pressure difference, thereby decreasing the flux more severely.

2.2.1. *Internal concentration polarization*

At steady state, the salt leakage (J_s) from the active layer (if the membrane is not perfect) originates from the convective flow of solute ($J_w c$) away from the active layer and diffusive flow of solute $D^{"} dc/dx$ toward the active layer due to concentration gradient inside the porous support [18]:

$$J_s = J_w c - D^{"} \frac{dc}{dx} \tag{3}$$

where c is the concentration of solute at any point inside support layer, $D^{"}$ is the salt diffusion coefficient in the porous support and is given by:

$$D^{"} = \frac{D\varepsilon}{\tau} \tag{4}$$

where D is the diffusivity of solute in water, and ε and τ are the porosity and tortuosity of the support, respectively. Appropriate boundary conditions (as shown in **Figure 2**) are represented as:

$$c = c_{\mathrm{D,i}} \text{ at } x = 0$$

$$c = c_{\mathrm{D,m}} \text{ at } x = t$$

Applying these boundary conditions a relation for the concentration of solution inside the porous support near the active layer ($c_{\mathrm{D,i}}$) is derived as follows:

$$c_{\mathrm{D,i}} = \frac{c_{\mathrm{D,m}} + J_s/J_w}{\exp(J_w K)} - \frac{J_s}{J_w} \tag{5}$$

where $c_{\mathrm{D,m}}$ is the concentration of solution on the support layer adjacent to the bulk solution. Here $K = \tau t/(D\varepsilon)$ is defined as the solute resistivity inside the porous support.

Considering a perfect membrane with 100% salt rejection, the value of J_s can be neglected, and Eq. (5) simplifies to:

$$c_{\mathrm{D,i}} = \frac{c_{\mathrm{D,m}}}{\exp(J_w K)} \tag{6}$$

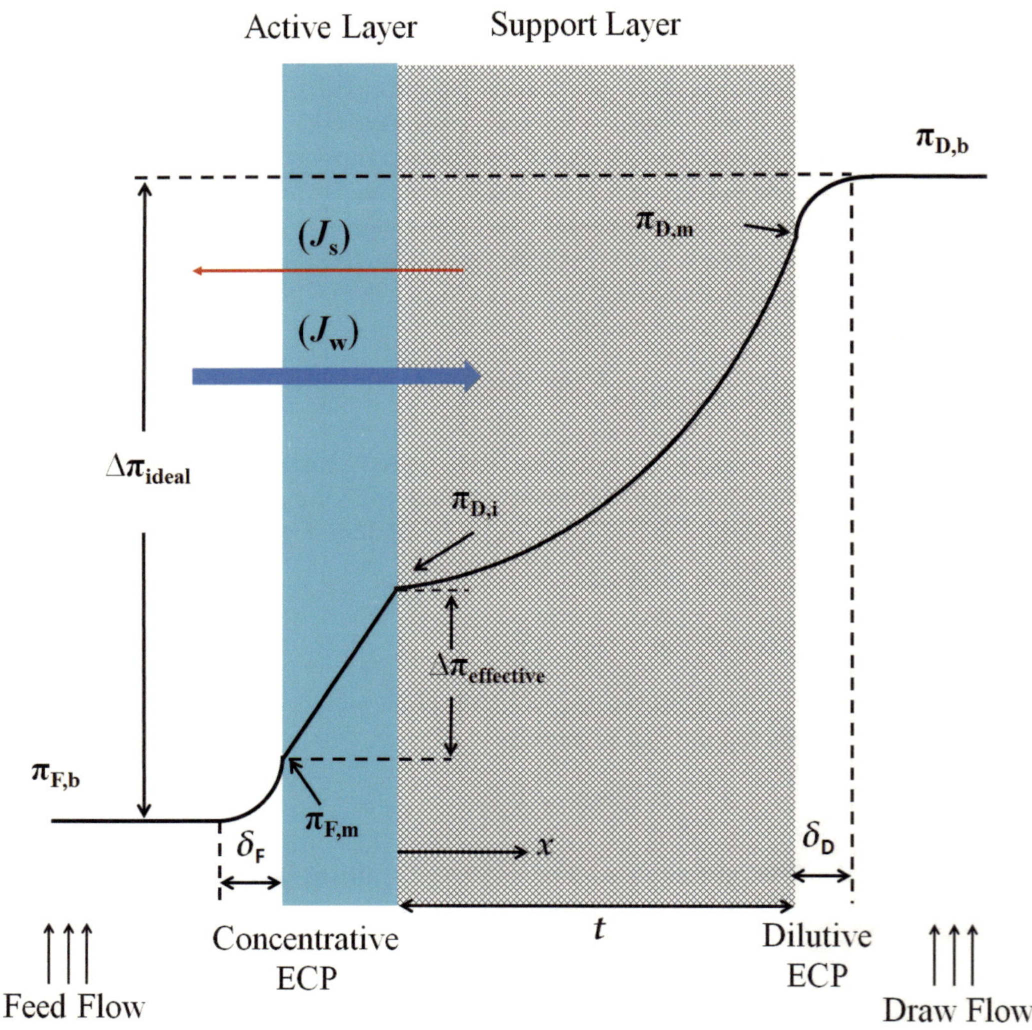

Figure 2. Concentration profiles (considering dilutive ECP) and the concentration boundary thickness developed on both sides of the membrane during an FO osmosis process.

The salt concentration ratio is approximately proportional to the osmotic pressure ratio of the solution, which gives:

$$\pi_{D,i} = \frac{\pi_{D,m}}{\exp(J_w K)} \tag{7}$$

The actual flux is generated by the concentration difference across the active layer of the membrane and is given by [1, 11]:

$$J_w = A(\pi_{D,i} - \pi_{F,m}) \tag{8}$$

Substituting Eq. (7) in Eq. (8), the following equation is obtained for the actual flux through the membrane:

$$J_w = A\left[\pi_{D,m}\exp(-J_w K) - \pi_{F,m}\right] \tag{9}$$

2.2.2. External concentration polarization

The concentrative ECP occurring on the feed side of the membrane can be captured by using the same differential equation and applying appropriate boundary conditions between the membrane surface and bulk solution on the feed side [1, 7]:

$$c = c_{F,b} \text{ at } x = 0$$

$$c = c_{F,m} \text{ at } x = \delta_F$$

where $c_{F,m}$ and $c_{F,b}$ are the concentrations of solute on the membrane surface and in the bulk feed solution, respectively. δ_F is the thickness of concentration boundary layer on the active layer of the membrane. Solving the differential equation and applying the above boundary conditions the following relation is derived for the electrolyte concentration on the membrane surface:

$$c_{F,m} = \left(c_{F,b} + \frac{J_s}{J_w}\right)\exp\left(\frac{J_w\delta_F}{D}\right) - \frac{J_s}{J_w} \tag{10}$$

Again for a high solute rejecting membrane $J_s \approx 0$, hence

$$c_{F,m} = c_{F,b}\exp\left(\frac{J_w\delta_F}{D}\right) \tag{11}$$

D/δ_F in this equation is mass transfer coefficient on the feed side of the membrane (k_F). By replacing the concentrations in Eq. (11) with the corresponding osmotic pressures and substituting this equation into Eq. (9), the following equation for flux is obtained:

$$J_w = A\left[\pi_{D,m}\exp(-J_w K) - \pi_{F,b}\exp\left(\frac{J_w}{k_F}\right)\right] \tag{12}$$

The effect of ICP in the support layer and ECP on the feed side are accounted in Eq. (12). By considering the effect of dilutive ECP on the draw side, a concentration boundary layer forms on the support layer of the membrane and $\pi_{D,m}$ will not be equal to $\pi_{D,b}$. Using appropriate boundary conditions:

$$c = c_{D,m} \text{ at } x = 0$$

$$c = c_{D,b} \text{ at } x = \delta_D$$

The following equation is derived for the concentration of solution on the support layer ($c_{D,m}$).

$$c_{D,m} = \left(c_{D,b} + \frac{J_s}{J_w}\right)\exp\left(-\frac{J_w\delta_D}{D}\right) - \frac{J_s}{J_w} \tag{13}$$

where $c_{D,b}$ is the bulk draw solution concentration and δ_D is the thickness of concentration boundary layer on the porous support. Applying similar assumption of $J_s \approx 0$ and inserting the mass transfer coefficient on the draw side of the membrane as $k_D = D/\delta_D$ we get:

$$c_{D,m} = c_{D,b} \exp\left(-\frac{J_w}{k_D}\right) \tag{14}$$

Finally, the modified flux equation by incorporating the ICP in the support layer and ECP on both sides of the membrane is acquired as follows:

$$J_w = A\left[\pi_{D,b} \exp(-J_w K) \exp\left(-\frac{J_w}{k_D}\right) - \pi_{F,b} \exp\left(\frac{J_w}{k_F}\right)\right] \tag{15}$$

A summary of the main mass transfer equations, boundary conditions, and concentration relations is presented in **Table 1**.

A similar analogy can be applied when the process is operated in the PRO mode. In this mode, the feed and draw solutions face the support and active layers, respectively. Hence, the ECP occurs on the draw side and is dilutive in nature, i.e. the draw solution becomes diluted near the membrane surface by the incoming permeate that leads to a decrease in osmotic driving force. The dilutive ECP phenomenon provides the following relation for the ratio of draw solution concentration on the membrane surface ($c_{D,m}$) and in the bulk solution ($c_{D,b}$) [7]:

$$\frac{c_{D,m}}{c_{D,b}} = \exp\left(-\frac{J_w}{k}\right) \tag{16}$$

where k is the mass transfer coefficient on the draw side of the membrane.

On the feed side of the membrane, the ICP occurs that increases the concentration of salt inside the porous support and makes it concentrative in nature, thus decreasing the driving force. The modulus for concentrative ICP is given by the following relation:

$$\frac{c_{F,i}}{c_{F,b}} = \exp(J_w K) \tag{17}$$

Assumption	Mass transfer equation	Boundary condition	Concentration relations
ICP in the support layer	$J_s = J_w c - D^* \frac{dc}{dx}$	$\begin{cases} x = t, & c = c_{D,m} \\ x = 0, & c = c_{D,i} \end{cases}$	$c_{D,i} = \frac{c_{D,m} + J_s/J_w}{\exp(J_w K)} - \frac{J_s}{J_w}$
ECP on the draw side	$J_s = J_w c - D^* \frac{dc}{dx}$	$\begin{cases} x = 0, & c = c_{D,m} \\ x = \delta_D, & c = c_{D,b} \end{cases}$	$c_{D,m} = \left(c_{D,b} + \frac{J_s}{J_w}\right) \exp\left(-\frac{J_w \delta_D}{D}\right) - \frac{J_s}{J_w}$
ECP on the feed side	$J_s = J_w c - D^* \frac{dc}{dx}$	$\begin{cases} x = 0, & c = c_{F,b} \\ x = \delta_F, & c = c_{F,m} \end{cases}$	$c_{F,m} = \left(c_{F,b} + \frac{J_s}{J_w}\right) \exp\left(\frac{J_w \delta_F}{D}\right) - \frac{J_s}{J_w}$

Table 1. A summary of governing equations and conditions considering both ECP and ICP [12].

http://dx.doi.org/10.5772/intechopen.71343

where $c_{F,i}$ and $c_{F,b}$ are the concentrations of the feed solution inside the porous support close to the active layer and in the bulk solution, respectively. By incorporating the dilutive ECP and concentrative ICP phenomena in the PRO process, an analytical model, analogous to FO, is obtained as follows:

$$J_w = A\left[\pi_{D,b} \exp\left(-\frac{J_w}{k}\right) - \pi_{F,b} \exp(J_w K)\right] \quad (18)$$

3. Standard experiments to use the data analysis model

Draw solutions with various concentrations of a particular salt in deionized water are first prepared. Then, the properties of both feed and draw solutions like viscosity, density, diffusion coefficient, and osmotic pressure are measured or taken from literature [19]. The Osmotic pressure of unknown feed and draw solutions can be found experimentally by using automatic osmometers. This instrument estimates the osmotic pressure by measuring the depression in the freezing point of the solution. The osmotic pressure of at least three solutions is measured, and a linear relationship is obtained between the osmotic pressure and the concentration.

The FO experiments are conducted by a cross-flow filtration setup. The schematic diagram of a typical setup is shown in **Figure 3**. The membrane cell has channels on both sides of the membrane for the flow of feed and draw solutions. The length, width, and depth of the channels should be measured precisely for the calculation of mass transfer coefficient. The effective filtration area of the membrane is measured to calculate the water flux. Feed and permeate spacers are typically used on both draw and feed channels in the cell to provide mechanical

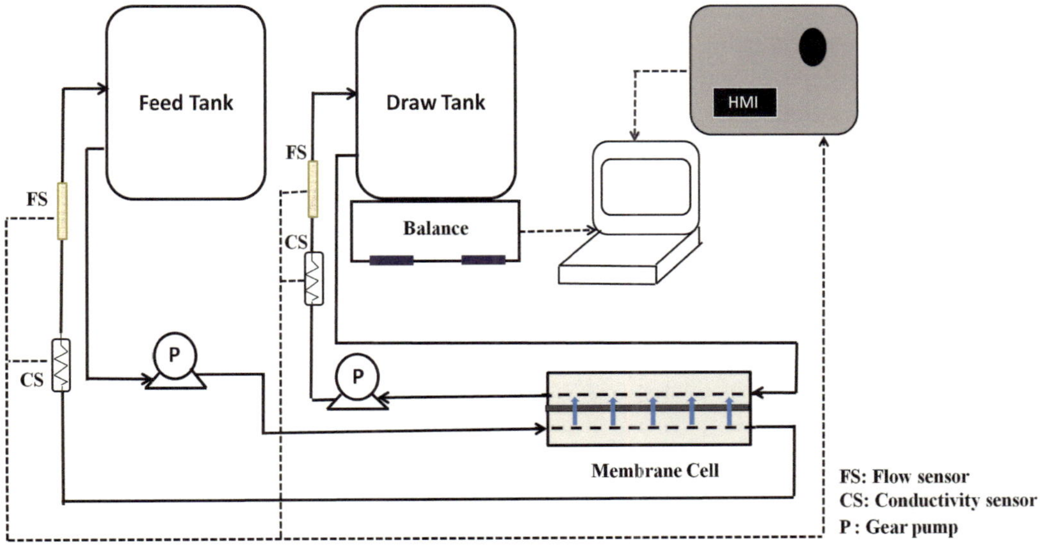

Figure 3. Schematic of a bench scale cross-flow FO setup.

support to the membrane. The feed and draw solution flow rates and the temperature of the experiment are maintained at certain values and are used for the calculation of mass transfer coefficients and osmotic pressure, respectively. The water flux through the membrane can be calculated by recording the change in the weight of the draw solution with time using a weighing scale. During the experiment, the conductivity and temperature of both feed and draw solutions must be monitored.

The same setup but opposite configuration is utilized for the PRO experiments. For both FO and PRO experiments, membranes are needed to be kept in the DI water for about 24 hours before experiments. After mounting the membranes in the module, the flow rates of the feed and the draw solutions are adjusted to desired values. The system is allowed to stabilize, and then the change in weight of the draw solution is recorded over time. Due to the change of draw solution concentration, a certain amount of concentrated draw solution needs to be gradually added to the solution. The conductivity of the draw solution is monitored online, and the addition of concentrated solution stopped when the conductivity of the draw solution reaches to the desired concentration of the solution. A similar procedure needs to be followed to increase the concentration the feed solution.

The pure water permeability of the membrane should be acquired using an RO setup with DI water as feed solution. The water flux is measured at different transmembrane pressures, and the slope is obtained as pure water permeability (A).

4. Estimation of mass transfer coefficient

The value of mass transfer coefficient depends on the hydrodynamics of the flow, applied driving force, water flux through the membrane, characteristics of the membrane (roughness and porosity) and the type of *solute* [14]. In this section, we provide three different methodologies to find the mass transfer coefficient: (i) empirical equations based on Sh number, (ii) pressure-driven method using RO, and (ii) osmotic pressure-driven method using PRO.

4.1. Empirical equations based on Sh number

Mass transfer coefficient is a parameter which describes the ratio between the actual mass (molar) flows of species into or out of a flowing fluid and the driving force that creates that flux. Mass transfer coefficient depends on module configuration, solute diffusion coefficient, viscosity, density, and velocity of feed solution [20]. It is related to the Sh number which shows the ratio of the convective mass transfer to diffusive rate and can be defined as follows [21]:

$$k = \frac{\mathrm{Sh}\,.\,D}{d_h} = a\,\mathrm{Re}^b\,\mathrm{Sc}^c \left(\frac{d_h}{L}\right)^d \tag{19}$$

where a, b, c, and d are constants, D is the diffusion coefficient of solute in draw or feed solution, L is the length of the tube or channel, d_h is the hydraulic diameter of channel calculated by $(2wh/(w + h))$ in which w and h are the width and the height of the channel. Re and Sc in Eq. (19) are Reynolds and Schmidt numbers, respectively.

$$\mathrm{Re} = \frac{\mathrm{d}_h \cdot v}{\nu} \tag{20}$$

$$\mathrm{Sc} = \frac{\nu}{D} = \frac{\mu}{\rho D} \tag{21}$$

In these equations, ν is the kinematic viscosity, μ is the dynamic viscosity, v is the flow velocity, and ρ is the flow density [20]. The mass transfer coefficient correlations are classified based on the channel flow geometry and various flow regimes in **Table 2**.

Eqs. (22)(a) and (b) are widely used to calculate the mass transfer coefficient in both feed and draw side of FO membrane. However, the implementation of these empirical equations in forward osmosis process has brought some controversial debates. These equations were derived based on ultrafiltration (UF) process which suffers more severely from concentration polarization phenomenon as compared to FO process. Hence, they are not necessarily valid for the evaluation of dilutive and concentrative ECP in FO [20]. Moreover, UF membranes differ from FO ones structurally as the former is porous while the latter is mainly dense composite membranes. Besides, the Sh number is correlated to the frictional factor which might be different for FO and UF processes [14, 19].

It is worth mentioning that Eq. (22)a is only valid where the length of the entry region is equal to $0.029d_h Re$ while in most lab-scale FO cells the length of the channel is shorter than the entry

Flow geometry	Laminar regime (Re < 2000)-(a)	Turbulent regime (Re > 2000)-(b)	Equation
Rectangular channels w/o spacers	$Sh = 1.85\left(\mathrm{ReSc}\frac{d_h}{L}\right)^{0.33}$	$Sh = 0.04\mathrm{Re}^{0.75}\mathrm{Sc}^{0.33}$	(22)
Rectangular channels w/ spacers	$Sh = 0.46(\mathrm{ReSc})^{0.36}$	$Sh = 0.0096\mathrm{Re}^{0.5}\mathrm{Sc}^{0.6}$	(23)
Tube	$Sh = 1.62\left(\mathrm{ReSc}\frac{d_h}{L}\right)^{0.33}$	$Sh = 0.023\mathrm{Re}^{0.83}\mathrm{Sc}^{0.33}$	(24)
Radial cross flow system	$Sh = 1.05\left(\mathrm{ReSc}\frac{h}{R_c}\right)^{0.38}$	$Sh = 0.275\left(\mathrm{Re}^{1.75}\,\mathrm{Sc}\frac{2h}{L}\right)^{0.33}$	(25)*
Stirred cell	$Sh = 0.23\mathrm{Re}^{0.567}\mathrm{Sc}^{0.33}$	$Sh = 0.03\,\mathrm{Re}^{0.66}\,\mathrm{Sc}^{0.33}\,\mathrm{Pe}_{test}^{0.16}$	(26)**

*R_c is the radius of the flow channel and h is the half channel height.

**Pe_{test} is the test Peclet number which is equal to $(J_w h/D)$.

Table 2. Mass transfer coefficient for different flow regimes and geometry [22].

length. Therefore, this equation does not seem to be suitable for investigating the transport phenomena in the lab-scale. Regarding Eq. (22)b, it was derived based on the pressure drop during turbulent flow in RO and UF experiments, whereas in FO process, the pressure drop is insignificant due to the absence of hydraulic pressure [13, 19]. Hence, a significant amount of research is underway to modify the common Sh number equations in the literature. For example, Tan and Ng [19] proposed an exact solution to evaluate the local mass transfer coefficient for the hydrodynamic boundary layer in FO process. The local Sh number was derived from the Navier-Stokes equations for the fluid flow parallel to a flat and non-porous surface as follows:

Laminar boundary layer

$$Sh = 0.332\,\mathrm{Re}_y^{1/2}\,\mathrm{Sc}^{1/3} \quad \mathrm{Re} \leq 2\times 10^5 \tag{27}$$

Turbulent boundary layer

$$Sh = 0.0292\,\mathrm{Re}_y^{0.8}\,\mathrm{Sc}^{0.33} \quad \mathrm{Re} > 2\times 10^5 \tag{28}$$

Hence, mean mass transfer coefficient, k_c, can be obtained as follows:

$$k_c = \frac{\int_0^L k\mathrm{d}y}{\int_0^L \mathrm{d}y} = \frac{0.664D\,\mathrm{Re}_t^{1/2}\,\mathrm{Sc}^{1/3} + 0.0365D\,\mathrm{Sc}^{1/3}\left[\mathrm{Re}_L^{4/5} - \mathrm{Re}_t^{4/5}\right]}{L} \tag{29}$$

where, Re_t and Re_L are the transition Reynolds number and Reynolds number at L, respectively, and L is the length of the channel. Experimental investigation showed that the mass transfer coefficient developed from boundary layer concept (k_c) provided more accurate results as compared to that obtained from UF experiments in Eq. (22).

4.2. Evaluating mass transfer coefficient by RO experiment

The film theory is generally applied to capture the effect of the ECP on a membrane surface. Using this theory, the concentration profile near the membrane surface is obtained as a function of permeation flux and mass transfer coefficient:

$$J_w = k\ln\left(\frac{c_m}{c_b}\right) \tag{30}$$

where c_m and c_b are the concentration at the membrane surface and in bulk, respectively. By estimating the concentration at the membrane surface the value of mass transfer coefficient is calculated using Eq. (30). The concentration at the membrane surface can be calculated from the osmotic pressure difference across the membrane. By measuring the water flux and salt rejection in an RO experiment and coupling these with the pure water flux, an estimate of the osmotic pressure difference across the membrane can be made, and consequently, the mass transfer coefficient is calculated by the following equation:

$$k = \frac{1}{J_w}\ln\left(\frac{\Delta\pi}{2R_g T c_b R_j}\right) \tag{31}$$

where R_j is the salt rejection by the membrane, R_g is the universal gas constant, and T is absolute temperature. The detailed procedure to derive Eq. (31) is described elsewhere [23].

4.3. Evaluating mass transfer coefficient in the PRO mode

Using DI water as the feed solution in the PRO mode, the water flux through the membrane can be calculated by a reduced form of Eq. (18) as follows:

$$J_w = A\left[\pi_{D,b} \exp\left(-\frac{J_w}{k}\right)\right] \tag{32}$$

The mass transfer coefficient can be calculated by rearranging this equation.

5. Flux prediction

The current models developed are mainly focused on finding an accurate value of solute resistivity (K), and very less attention has been paid to find a proper value of mass transfer coefficient for FO [7, 11, 12]. There are no direct techniques to determine the value of the structural parameters of a membrane, primarily porosity and tortuosity, so its value is typically evaluated by fitting the experimental data to the transport model [24]. In this technique, the value of K directly depends upon the mass transfer coefficient. Hence, there is a crucial need for finding an accurate value of mass transfer coefficient.

Investigating the current models developed for FO, it was also observed that they are insensitive to a change in the feed flow rate, while our earlier investigations demonstrated that the flux changes moderately with the flow rate [25]. In the previous sections, it was shown that the mass transfer coefficients could be obtained by three methods. Hence, it is suggested that the researchers critically compare the results obtained from the three sets of mass transfer coefficient and utilize the one that increases the sensitivity of the flux results to the feed flow rate.

To start with the modeling of the FO, the hydraulic permeability (A) and salt resistivity of the support layer (K) needs to be determined. The hydraulic permeability of the membrane is determined by the RO setup as discussed earlier. Salt resistivity coefficient depends upon the structural parameters of the membrane, such as porosity, tortuosity, and thickness and on the diffusion coefficient of salt (D). Since the structural parameter is an intrinsic property of the membrane, it is assumed to be constant for a particular membrane [7, 11]. The salt diffusion coefficient is also constant at a specific temperature and is not changing significantly in a narrow range of molarities [26, 27]. Hence, the value of K at a particular temperature is constant and can be evaluated by rearranging Eq. (15). As an example, **Table 3** presents the experimental FO data that is used to determine the value of K for a thin film composite FO membrane. In this experimental research, DI water and NaCl are used as feed and draw solutions, respectively. The detailed properties of the membrane are presented elsewhere [28]. All experiments were conducted at 23°C and the values of mass transfer coefficients obtained

Draw concentration (M)	Osmotic pressure (draw side) (bar)	Feed concentration (M)	Osmotic pressure (feed side) (bar)	Flux (LMH)	*K* (s/m) × 10^5
1.5	75.4	0.05	2.05	10.2	6.99
1.5	75.4	0.1	4.13	9.0	6.70
1.5	75.4	0.25	10.57	6.4	6.8
1.5	75.4	0.5	21.7	4.2	6.73
1.5	75.4	1.0	47.9	1.5	7.14
Average	—	—	—	—	6.9

Table 3. FO experiments for calculation of *K*. Tests were conducted at 23°C.

from RO test were used for the calculation of *K*. As expected, the *K* values were almost constant for different feed concentrations. Hence, the average *K* value of 6.9 can be reasonably used for prediction of water flux.

Obtaining hydraulic permeability (*A*), salt resistivity of the support layer (*K*), and mass transfer coefficients of both feed and permeate side (k_D and k_F) the model water flux is calculated by Eq. (15). A typical representation of matching between theoretical and experimental results is to plot the model predicted values of water flux along with the experimental values as a function of the driving force (osmotic pressure of the draw solution), as shown in **Figure 4**.

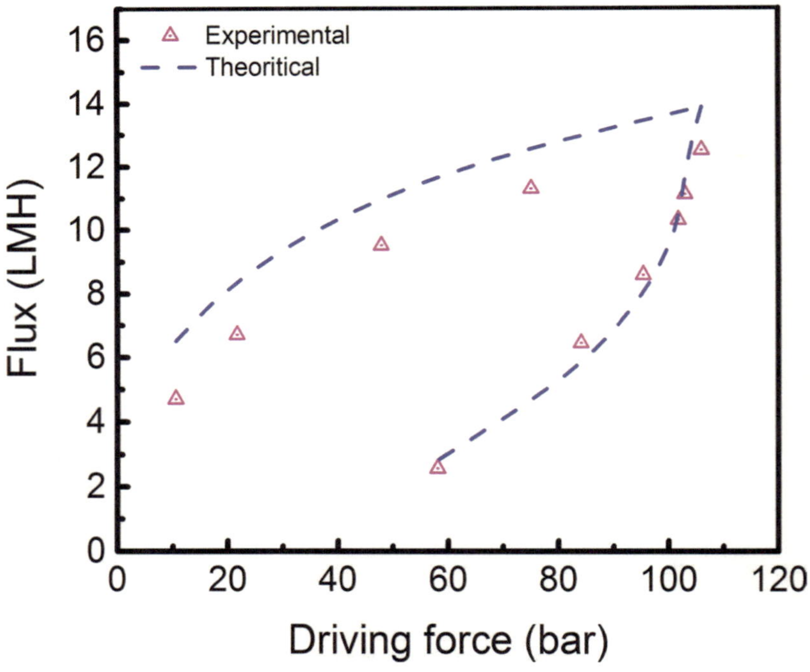

Figure 4. Typical comparison of experimental FO data and predicted fluxes by the model as a function of osmotic driving forces.

Draw concentration	Feed concentration	Experimental flux (LMH)		Theoretical flux (LMH) (*k* from RO experiment)		Theoretical flux (LMH)* (*k* from Sh number in Eq. (22))	
		Feed flow 1 LPM	Feed Flow 3 LPM	Feed flow 1 LPM	Feed flow 3 LPM	Feed flow 1 LPM	Feed flow 3 LPM
0.25 M	0.05 M	3.9	4.5	3.8	4.2	4.1	4.1
0.5 M	0.05 M	5.6	7.7	5.8	6.4	6.1	6.1
1 M	0.05 M	8.7	10.1	8.2	9.0	8.5	8.5
1.5 M	0.05 M	9.9	11.6	9.6	10.6	9.9	9.9

*The value of *K* for this case was found using Eq. (15).

Table 4. The sensitivity of the model to predict change in flux with the change in the values of ***k*** with the experimental results.

It is well known that increasing the feed flow rate increases the water flux through the membrane by enhancing the mixing near the membrane surface, thereby reducing the effect of ECP (concentrative ECP in the case of FO). The change in the flow rate is reflected through the change in the mass transfer coefficient. Hence it is recommended to test the sensitivity of the FO developed model to the variation of feed flow rate. As a case study, the experimental results and the model predictions obtained using two mass transfer coefficients, one from Eq. (22) and the other one from RO tests, are presented in **Table 4**. As can be observed, mass transfer coefficients yield results that match well with experimental data. However, using the values of k obtained from Eq. (22), the fluxes were found to be insensitive to flow rates, whereas k values evaluated by RO experiment resulted in more reasonable predictions at higher feed flow rate as well.

6. Conclusion

In this chapter, the governing equations of transport through an FO membrane were presented based on the mass balance in the concentration boundary layers on both sides of the membrane (ECP) and inside the support layer (ICP). Although ICP is reported in the literature to play a significant role in the reduction of the effective osmotic driving force, the impact of ECP is usually underestimated. The ECP primarily depends upon the value of mass transfer coefficient (k), and the exponential nature of concentration profile in the boundary layer makes ECP very sensitive to the value of k. Hence, another theme of this chapter was to provide appropriate methods for the estimation of mass transfer coefficient. Previous studies were all based on using the Sh number correlation developed from the UF experiments to predict the flux in the FO process. The main shortcoming of these studies was the insensitivity of the model predictions to a change in the feed flow rate. Hence other experimental methods based on RO and PRO were proposed to provide a better estimation of mass transfer coefficient. In summary, the results obtained in this study suggest that both

ECP and ICP play a key role in the performance of FO membrane and thus the mass transfer coefficient (*k*) which mainly affects the ECP is as important as solute resistivity (*K*) which is reflected in the ICP effect.

Acknowledgements

The authors gratefully acknowledge the financial support provided by the Natural Sciences and Engineering Research Council of Canada (NSERC) and Canada's Oil Sands Innovation Alliance (COSIA).

Nomenclature

A	pure water permeability ($Lm^{-2}\,h^{-1}\,bar^{-1}$)
B	solute permeability ($Lm^{-2}\,h^{-1}$)
c	concentration of solute ($mol\,L^{-1}$)
d_h	hydraulic diameter (m)
D	diffusion coefficient ($m^2\,s^{-1}$)
n	van't Hoff factor
J	flux ($Lm^{-2}\,h^{-1}$)
k	mass transfer coefficient (ms^{-1})
K	solute resistivity (m)
L	length of channel (m)
p	hydraulic pressure (bar)
R_g	universal gas constant ($Jmol^{-1}\,K^{-1}$)
R_j	salt rejection
Re	Reynolds number
Re_L	Reynolds number at the end of membrane channel
Re_t	transition Reynolds number
Sc	Schmidt number
Sh	Sherwood number
T	absolute temperature (K)

Greek symbols

δ	thickness of ECP boundary layer (m)
ε	porosity of membrane support
μ	dynamic viscosity (Pa.s)
ν	kinematic viscosity ($m^2 s^{-1}$)
ρ	density of water (kg m^{-3})
τ	tortuosity of membrane support
π	osmotic pressure (bar)
σ	reflection coefficient
v	velocity (ms^{-1})

Subscripts

b	bulk solution
D	draw solution
F	feed solution
i	interface between support layer and active layer of membrane
m	membrane surface
s	solute
v	pure water
w	water

Author details

Amrit Bhinder, Simin Shabani and Mohtada Sadrzadeh*

*Address all correspondence to: sadrzade@ualberta.ca

Department of Mechanical Engineering, Donadeo Innovation Center for Engineering, Advanced Water Research Lab (AWRL), University of Alberta, Edmonton, AB, Canada

References

[1] Klaysom C, Cath TY, Depuydt T, Vankelecom IFJ. Forward and pressure retarded osmosis: Potential solutions for global challenges in energy and water supply. Chemical Society Reviews. 2013;**42**:6959-6989. DOI: 10.1039/c3cs60051c

[2] Chung T-S, Li X, Ong RC, Ge Q, Wang H, Han G. Emerging forward osmosis (FO) technologies and challenges ahead for clean water and clean energy applications. Current Opinion in Chemical Engineering. 2012;**1**:246-257. DOI: 10.1016/j.coche.2012.07.004

[3] Cath T, Childress A, Elimelech M. Forward osmosis: Principles, applications, and recent developments. Journal of Membrane Science. 2006;**281**:70-87. DOI: 10.1016/j.memsci.2006.05.048

[4] Hickenbottom KL, Hancock NT, Hutchings NR, Appleton EW, Beaudry EG, Xu P, et al. Forward osmosis treatment of drilling mud and fracturing wastewater from oil and gas operations. Desalination. 2013;**312**:60-66. DOI: 10.1016/j.desal.2012.05.037

[5] Lutchmiah K, Verliefde a RD, Roest K, Rietveld LC, Cornelissen ER. Forward osmosis for application in wastewater treatment: A review. Water Research. 2014;**58**:179-197. DOI: 10.1016/j.watres.2014.03.045

[6] McCutcheon JR, McGinnis RL, Elimelech M. A novel ammonia—Carbon dioxide forward (direct) osmosis desalination process. Desalination. 2005;**174**:1-11. DOI: 10.1016/j.desal.2004.11.002

[7] Mccutcheon JR, Elimelech M. Modeling water flux in forward osmosis: Implications for improved membrane design. AICHE Journal. 2007;**53**:1736-1744. DOI: 10.1002/aic

[8] Mehta GD, Loeb S. Internal polarization in the porous substructure of a semipermeable membrane under pressure retarded osmosis. Journal of Membrane Science. 1978;**4**:261-265. DOI: 10.1016/S0376-7388(00)83301-3

[9] Lee KL, Baker RW, Lonsdale HK. Membranes for power generation by pressure-retarded osmosis. Journal of Membrane Science. 1981;**8**:141-171. DOI: 10.1016/S0376-7388(00)82088-8

[10] Loeb S, Titelman L, Korngold E, Freiman J. Effect of porous support fabric on osmosis through a Loeb-Sourirajan type asymmetric membrane. Journal of Membrane Science. 1997;**129**:243-249. DOI: 10.1016/S0376-7388(96)00354-7

[11] McCutcheon JR, Elimelech M. Influence of concentrative and dilutive internal concentration polarization on flux behavior in forward osmosis. Journal of Membrane Science. 2006;**284**:237-247. DOI: 10.1016/j.memsci.2006.07.049

[12] Suh C, Lee S. Modeling reverse draw solute flux in forward osmosis with external concentration polarization in both sides of the draw and feed solution. Journal of Membrane Science. 2013;**427**:365-374. DOI: 10.1016/j.memsci.2012.08.033

[13] van den Berg GB, Racz IG, Smolders CA. Mass transfer coefficients in cross-flow ultrafiltration. Journal of Membrane Science 1989;**47**:25-51. doi:10.1016/S0376-7388(00)80858-3

[14] Gekas V, Hallstrom B. Mass transfer in the membrane concentration polarization layer under turbulent cross flow. Journal of Membrane Science. 1987;**30**:153-170. DOI: 10.1016/S0376-7388(00)81349-6

[15] Belfort G, Nagata N. Fluid mechanics and cross-flow filtration: Some thoughts. Desalination. 1985;**53**:57-79. DOI: 10.1016/0011-9164(85)85052-9

[16] Koutsou CP, Yiantsios SG, Karabelas AJ. A numerical and experimental study of mass transfer in spacer-filled channels: Effects of spacer geometrical characteristics and Schmidt number. Journal of Membrane Science. 2009;**326**:234-251. DOI: 10.1016/j.memsci.2008.10.007

[17] Rodrigues C, Rodrigues M, Semiao V, Geraldes V. Enhancement of mass transfer in spacer-filled channels under laminar regime by pulsatile flow. Chemical Engineering Science. 2015;**123**:536-541. DOI: 10.1016/j.ces.2014.11.047

[18] Phillip WA, Yong JS, Elimelech M. Reverse draw solute permeation in forward osmosis: Modeling and experiments. Environmental Science and Technology. 2010;**44**:5170-5176. DOI: 10.1021/es100901n

[19] Tan CH, Ng HY. Modified models to predict flux behavior in forward osmosis in consideration of external and internal concentration polarizations. Journal of Membrane Science. 2008;**324**:209-219. DOI: 10.1016/j.memsci.2008.07.020

[20] Mulder M. Basic Principles of Membrane Technology. 2nd ed. Dordrecht: Kluwer Academic; 1996. DOI: 10.1007/978-94-009-1766-8

[21] Heldman DR, Moraru CI. Encyclopedia of Agricultural, Food, and Biological Engineering. 2nd ed. New York: Taylor & Francis; 2010. DOI: 10.1081/E-EAFE2

[22] Wang J, Dlamini DS, Mishra AK, Pendergast MTM, Wong MCY, Mamba BB, et al. A critical review of transport through osmotic membranes. Journal of Membrane Science. 2014;**454**:516-537. DOI: 10.1016/j.memsci.2013.12.034

[23] Al Mamun MA, Sadrzadeh M, Chatterjee R, Bhattacharjee S, De S. Colloidal fouling of nanofiltration membranes: A novel transient electrokinetic model and experimental study. Chemical Engineering Science. 2015;**138**:153-163. DOI: 10.1016/j.ces.2015.08.022

[24] Manickam SS, Mccutcheon JR. Model thin film composite membranes for forward osmosis: Demonstrating the inaccuracy of existing structural parameter models. Journal of Membrane Science. 2015;**483**:70-74. DOI: 10.1016/j.memsci.2015.01.017

[25] Bhinder A, Fleck BA, Pernitsky D, Sadrzadeh M. Forward osmosis for treatment of oil sands produced water: Systematic study of influential parameters. Desalination and Water Treatment. 2016;**57**:22980-22993. DOI: 10.1080/19443994.2015.1108427

[26] Lobo V. Mutual diffusion coefficients in aqueous electrolyte solutions. Pure and Applied Chemistry. 1993;**65**:2613-2640. DOI: 10.1351/pac199365122613

[27] Vitagliano V, Lyons PA. Diffusion coefficients for aqueous solutions of sodium chloride and barium chloride. Journal of the American Chemical Society. 1956;**76**:1549-1552. DOI: 10.1021/ja01589a011

[28] Khorshidi B, Bhinder A, Thundat T, Pernitsky DJ, Sadrzadeh M. Developing high throughput thin film composite polyamide membranes for forward osmosis treatment of SAGD produced water. Journal of the American Chemical Society. 2016;**511**:29-39. DOI: 10.1016/j.memsci.2016.03.052

Chapter 5

Temperature Effect on Forward Osmosis

Sangwoo Shin and Albert S. Kim

Additional information is available at the end of the chapter

http://dx.doi.org/10.5772/intechopen.72044

Abstract

Forward osmosis, or simply, osmosis, refers to a process by which a solvent moves across a semipermeable membrane due to the difference in the solute concentration established across the membrane. Because of its spontaneous nature, forward osmosis has received immense attention during the last few decades, particularly for its diverse applications, which include municipal wastewater treatment, seawater desalination, membrane bioreactor, potable water purification, food processing, drug delivery, energy generation, and so forth. Of many parameters that determine the performance of the forward osmosis process, the most fundamental factor that impacts performance is temperature. Considering the importance of the temperature on the forward osmosis process, there have been only a limited number of studies about the effect of temperature on the osmosis-driven process. In this chapter, we discuss the temperature effect on the forward osmosis process from two main aspects. First, we provide an extensive and in-depth survey on the currently available studies related to the anisothermal osmosis phenomena. Second, we then discuss a state-of-the-art theoretical framework that describes the anisothermal forward osmosis process that may shed light on achieving an enhanced performance via temperature control.

Keywords: forward osmosis, temperature, thermal effect, concentration polarization, water flux, solute flux, membrane scaling

1. Introduction

Osmosis, one of the most fundamental transport processes responsible for homeostasis in living organisms, has a rich history of applications—ranging from food preservation to water treatment and drug delivery. Osmosis occurs when a solute concentration difference is established across a semipermeable membrane. Due to the chemical potential imbalance, the water molecules will spontaneously migrate across the membrane toward the higher solute concentration side. Such a process has been regarded as one of the most central mechanisms that dictates the membrane-based water treatment technologies. The most widely utilized process, in our opinion, is reverse osmosis (RO) for solute removal, which requires an external hydraulic pressure to overcome the osmotic pressure difference across the membrane. In contrast to RO, the process that exploits the spontaneous transport of solvent molecules driven by the osmotic phenomenon is referred to as forward osmosis (FO) or direct osmosis (DO), which is, in principle, the same as the original osmosis.

FO was first conceptualized by Batchelder as a means for water treatment since the 1960s [1]. Since then, there has been a growing interest in applying FO to wastewater treatment technologies either as a stand-alone or in combination with other technologies such as membrane distillation, thermal distillation, or reverses osmosis [2]. Particularly, FO has been utilized in space stations for wastewater reclamation due to its excellent long-term stability and low energy consumption [3, 4]. Not limited to wastewater treatment, FO has also been explored extensively for many useful applications such as seawater desalination [5–7], portable hydration bags [8], food processing [9–11], pharmaceutical systems [12–14], and energy conversion [15, 16].

Unlike RO, FO is purely an osmosis-driven process, which is thermodynamically spontaneous. The osmotic pressure difference $\Delta\pi$, which is a driving force for the FO process, may be expressed using van't Hoff's law as

$$\Delta\pi = RT\Delta C \tag{1}$$

for weakly interacting molecules, where ΔC is the solute concentration difference, R is the gas constant, and T is the temperature. From the equation, it can be noted that the temperature is one of the most critical factors determining the rate of osmosis. In addition, temperature further changes viscosity, diffusivity, and density, which are important parameters in momentum and energy transfer phenomena. Despite the importance of temperature on FO process and despite the fact that there exist a number of papers that address the temperature effect, the reported data are widely scattered and does not show an agreeable consensus. In this chapter, we aim to provide a holistic understanding of the temperature effect on an osmotic phenomenon. Our intention is not to give an exhaustive review of the FO process in detail but to focus on the temperature effect and hopefully to provide insight for better control over the osmotic phenomenon. Readers who wish to learn about the FO process more in detail may refer to the following review papers [2, 8, 17].

2. Operating principle

2.1. Mechanism

In the FO process, the solvent (water) transport is driven solely by osmotic pressure difference without the need of any external hydrostatic pressure, allowing for lower energy consumption compared to RO. To extract water from the feed solution, the osmotic pressure at the opposite side of the membrane must be higher, which requires a highly concentrated solution; this concentrated solution is typically referred to as the draw solution. Draw solutes need to be inert and easily removable. A semipermeable membrane separates the feed solution and the concentrated draw solution where the chemical potential difference allows the water to flow through the membrane while leaving behind the solutes in the feed stream. Regions of high and low solute concentrations refer to those of low and high solvent chemical potentials, respectively. As the semipermeable membrane restricts the solute transport and maintains chemical potential differences of both solute and solvent, water migrates from its high solvent chemical-potential region (i.e., of low solute concentration) to low solvent chemical-potential region (i.e., of high solute concentration). Such a water transport leads to dilution of the draw solution where the diluted draw solution can be further recycled such that the initial solute concentration is recovered. Particularly for desalination applications, the solutes in the draw solution (osmotic agent or draw solutes) are chosen to be inert, nontoxic, and easily removed to obtain the desalinated water with ease. One example includes NH_4CO_2, which can be easily removed by decomposing at a moderately elevated temperature (~60°C) followed by low-temperature distillation [18, 19]. Extra energy is, however, necessary to re-dissolve NH_4CO_2 into the draw solution for a continuous FO operation.

2.2. Concentration polarization

The water flux across the membrane results in concentration of the feed solution and dilution of the draw solution since the membrane mainly allows passage of water molecules. This phenomenon, referred to as concentration polarization (CP), has an adverse impact on the efficacy of the FO process since such an effect reduces the effective osmotic pressure difference across the membrane, thus hindering water transport.

CP is highly influenced by the morphology of the membranes. The membranes used in the FO process consist of a thin, dense layer that rejects the solutes (active layer) followed by a coarse, thick porous layer (support layer or porous substrate) to reinforce the mechanical stability against fluid pressure and shear. This configuration makes the membrane asymmetric in which the orientation of the membrane with respect to the direction of the water flux (i.e., from low to high osmotic pressure) leads to significantly different transport dynamics [20].

Typically in the FO process, the active layer is placed against the feed stream in order to minimize fouling since the support layer is more susceptible to colloidal fouling due to the large pores. This configuration is called FO mode, as shown in **Figure 1(a)**. However, the downside of placing it in this way is that there is a significant dilutive internal concentration polarization (ICP) in the thick porous substrate. This is because the support layer is in contact with the concentrated draw solution hindering the solute diffusion, which significantly reduces the water flux (**Figure 1(a)**).

In contrast, when the active layer is placed against the draw stream, one can expect a higher water flux since this configuration can avoid the dilutive ICP at the expense of accelerated membrane fouling. This configuration is called the pressure-retarded osmosis (PRO) mode, as shown in **Figure 1(b)**, typically realized in standard PRO systems. To avoid any confusion, we will refer to the membrane configuration in which the active layer is placed against the feed solution as the *FO mode*, whereas the opposite case is the *PRO mode* during FO processes.

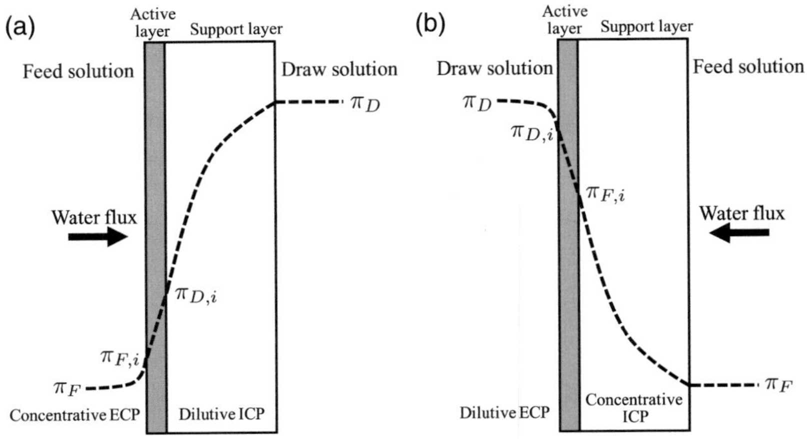

Figure 1. Influence of CP on the osmotic pressure distribution in the FO process. The membrane is configured in (a) FO mode and (b) PRO mode.

3. System temperature effect on FO

The first quantitative experiments on temperature-dependent osmosis go back almost a century ago [21]. Traxler demonstrated the osmosis of pyridine by using a thin rubber sheet as a semipermeable membrane within a uniform system temperature, ranging from 5 to 85°C (**Figure 2(a)**). He showed that as the temperature is increased, the transport of pyridine across the membrane is also increased (**Figure 2(b)**). In this chapter, such a uniform temperature will be refered to as 'system temperature' indicating the absence of local or transmembrane temperature gradient.

From the van't Hoff equation, the osmotic pressure is directly proportional to the system temperature, which is an indispensable factor for the FO process. However, temperature not only influences the osmotic pressure but also impacts many other key properties that are important to the transport process such as viscosity, diffusivity, solubility, density, and so forth. Such a change in the properties not only influences the water flux but also alters the solute rejection/diffusion and membrane fouling. In this section, we provide a summary of how the system temperature influences the water transport, solute rejection, and membrane fouling. We note that the experimental studies that will be covered in the following sections employ a circulating crossflow type setup (in contrast to a dead-end type as seen in Traxler's experiments in **Figure 2**).

3.1. Water flux

The most direct consequence of raising the system temperature is the increased water flux across the membrane due to lowered water viscosity and increased water diffusivity, which effectively increases the water permeability across the membrane. Since the transport of water through the active layer of the membrane follows the solution-diffusion mechanism [22], it is

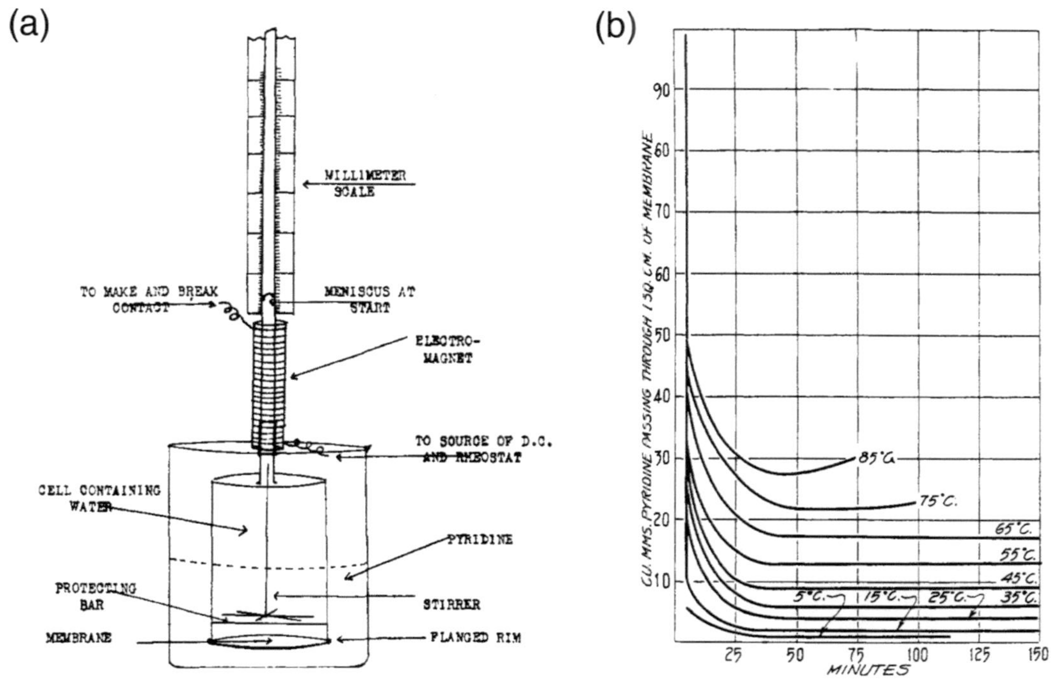

Figure 2. The first quantitative experiments reported on the effect of temperature on the osmosis phenomenon. (a) A schematic of the experimental setup that allows temperature control via a thermostat. (b) Transport of pyridine across a rubber membrane under various temperature conditions. Reprinted with permission from Ref. [21]. © 1928 American Chemical Society.

commonly believed (and also observed) that the diffusivity D exhibits an Arrhenius relation, that is, $D \sim \exp(-s/T)$, where s is an empirical constant related to the activation energy [23, 24]. However, we also note a counterexample where Petrotos et al. failed to show such a behavior [25].

On the basis of our survey, the available literature related to the temperature-dependent FO reported increased water flux with temperature. **Table 1** provides a summary of experimental conditions and resulting water flux from the available literature [23–31]. Here, we define a new quantity to indicate how much solvent flux increases with respect to the system temperature, as indicated in the last column of **Table 1**:

$$j_M = \frac{J_{w,M} - J_{w,0}}{T_M - T_0}, \tag{2}$$

where $J_{w,M}$ and $J_{w,0}$ are the water fluxes at a given maximum system temperature T_M and at base temperature T_0, respectively. The survey shows that raising the temperature does increase the water flux, but the extent of such an increase varies across the literature, especially depending on the membrane orientation. This observation implies that the CP phenomena are uniquely influenced by the temperature, leading to variations in the water flux.

Reference	Feed solution (concentration)	Draw solution (concentration)	Membrane[1]	Mode[2]	Temperature (°C)	$J_{w,0}$ (LMH)	j_M[3] (LMH/°C)
[25]	Tomato juice (0.13 M)	NaCl (3.9 M)	PA		26–58	1.5	0.030
[26]	NaCl (0–86 mM)	KCl (0.5–3 M)	CT	FO	25–45	19	0.43
[27]	Deionized water	NaCl (0.5 M)	CT		20–40	5.5	0.14
			PA			17	0.49
[23]	Sucrose (0–1.65 M)	NaCl (2–4 M)	CT		20–30	24	0.91
	Sucrose (0–0.7 M)	NaCl (4 M)	PA			2.5	0.15
[28]	NaCl (0.1 M)	NaCl (1 M)	CT	PRO	20–40	11	0.89
				FO		9.4	0.59
[29]	NaCl (0–1 M)	NaCl (1.5 M)	CT	PRO	20–40	43	1.4
				FO		18	0.63
[24]	NaCl (60 mM)	Na_2SO_4 (1.5 M)	CT		25–45	15	0.35
[30]	NaCl (0.2–0.5 M)	NH_4HCO_3 (3 M)	CE	PRO	30–50	5.4	0.10
[31]	Deionized water	NaCl (1.2 M)	CT	FO	20–30	14	0.61

[1]PA: polyamide; CT: cellulose triacetate; CE: cellulose ester
[2]FO mode: active layer placed against feed solution; PRO mode: active layer placed against draw solution
[3]$j_M = J_{w,M} - J_{w,0} / T_M - T_0$; $J_{w,M}$: water flux at maximum temperature T_M; $J_{w,0}$: water flux at base temperature T_0

Table 1. A summary of influence of temperature on the water flux.

McCutcheon and Elimelech were the first to study the influence of temperature on the CP phenomena [29]. Raising the temperature increases the water flux because of the decreased water viscosity in solutions (and/or solubility) and increased water solubility and diffusivity within the membrane. At the same time, however, the higher flux also increases both the ICP and ECP, which essentially limit the water flux as a feedback hindrance. Therefore, such a self-limiting behavior driven by two counteracting effects leads to the fact that the temperature has a "modest" effect on the water flux at high water flux conditions [29]. This self-hindering effect of the solvent flux is unavoidable in most membrane separation processes. It is similar to the fact that, in RO, applying high pressure initiates increasing permeate flux, which will eventually bring more solutes from the bulk phase to the membrane surface, enhancing the CP. Therefore, additional gain of the RO permeate flux is not as much as anticipated when the pressure is increased.

The change in the temperature influences the CP phenomena in different ways depending on the orientation of the membrane. This is because the formation of the ICP, which is the most critical factor that limits the driving force, is dependent on the membrane configurations. In the PRO mode, the concentrative ICP is developed in the feed side (see **Figure 1(b)**). By reducing the ICP using deionized water as the feed, the water flux was shown to be highly dependent on the temperature, confirming the impact of ICP on the FO process [29].

In the presence of solutes in the feed side so that the ICP is present, however, the water flux was shown to be almost insensitive to the temperature, at least in the operating temperature range (20–40°C). This behavior is attributed to the coupled interaction between ICP and ECP.

Although the increased solute diffusion at higher temperature mitigates the concentrative ICP in the support layer so that the water flux can be increased, such an increased water flux carries more solutes from the feed bulk phase to the vicinity of the support layer surface and enhances the dilutive ECP, thereby reducing the osmotic driving force. Therefore, the two opposing effects on the water transport effectively limit the enhancement of the water flux such that the temperature has a marginal effect on the overall water flux. If both water and solute diffusivities increase in a similar behavior, the net diffusive transport must be more or less the same.

In the FO mode, however, the water flux was shown to be significantly influenced by the temperature. Overall, the water flux was observed to be lower than the PRO mode due to the presence of the dilutive ICP. This was proven mathematically using the method of proof by contradiction [32]. Such a low water flux effectively suppresses the extent of concentrative ECP in the feed side. Also, the influence of concentrative ECP on the water flux is less important than the dilutive ECP in the draw solution side because the initial solute concentration in the bulk phase is much lower at the feed solution than the draw solution. This implies that the ECP has a minor effect on the driving force in the FO mode. Therefore, when the membrane is placed in the FO mode, the water flux is significantly influenced by the temperature since the ICP is the only major factor that determines the driving force.

One assumption McCutcheon and Elimelech had made while analyzing their data were the insignificant solute diffusion across the membrane [29], which otherwise leads to further ICP. Obviously, commercially available membranes are known to permit diffusion of the solutes, which can impact the formation of the CP effect. Since the solute diffusion is also sensitive to the temperature, the transmembrane solute flux should also lead to a change in the water flux. We discuss the effect of temperature on the solute diffusion and rejection in the following section.

3.2. Diffusion and rejection of solutes

It is of general consensus that the transmembrane solute diffusion increases with temperature. A number of groups have recently investigated experimentally the temperature effect on the transmembrane solute diffusion and the solute rejection [26–28].

Xie et al. recognized that the effective size of the solute molecules was the most important parameter for the transmembrane solute diffusion [27], which was predicted theoretically using the integral equation theory [33]. Hydration of charged organic solutes results in an increase in the effective solute size, which directly influences the solute diffusion and rejection rate, as it was well understood that the rejection of the charged organic solutes would be much higher than the neutral organic solutes. In this regard, neutral solutes were more likely to diffuse across the pores than the charged solutes in both the cellulose triacetate membranes and polyamide membranes. This implies that increasing the temperature leads to higher solute diffusion due to the increased solute diffusivity. Moreover, increasing the temperature leads to faster dissolution of the solutes into the membrane such that even hydrophobic neutral solutes absorb into the membrane at an order of magnitude higher rate at elevated temperatures.

Notably, the ratio between the water flux J_w and the solute flux J_s was shown to be more or less constant regardless of the system temperature [27]. Such a constant ratio implies that the structural properties may not change, at least in the operating temperature range (20–40°C). In fact, although it is documented in the literature that the RO membrane properties such as pore sizes may change when the temperature is above 40°C [34], it was reported in various FO studies that the membrane structural properties do not change significantly below 45°C [26, 27]. However, it is more reasonable to say that the structural properties of FO membranes change with temperature in a way that the ratio between solvent and solute fluxes remain almost constant. In a solution-diffusion model, permeabilities of solvent and solutes, A and B, respectively, are believed to increase with membrane temperature. The permeate concentration is controlled by only their ratio, A/B. If A and B increase with T while A/B remains less sensitive to T, then the solute diffusion can be seen phenomenologically insensitive to temperature. This is because although higher T increases both the solute and solvent fluxes, it is only the ratio that influences the concentration of solutes passing through the membrane. This topic is discussed theoretically in detail in Section 5.

Meanwhile, You et al. showed that the transmembrane solute diffusion was also shown to be dependent on the membrane orientation regardless of the temperature in which the PRO mode was shown to exhibit higher solute flux across the membrane than the FO mode, which is similar to the behavior of the water flux [28].

3.3. Membrane scaling

Membrane scaling occurs when the solute concentration is high enough to initiate precipitation. This is directly related to solute rejection and the CP phenomena, implying that membrane scaling should also be temperature-dependent.

Zhao and Zou studied how the temperature influences the membrane scaling over time, which is important in long-term operations [24]. Due to the fast water flux at elevated temperature,

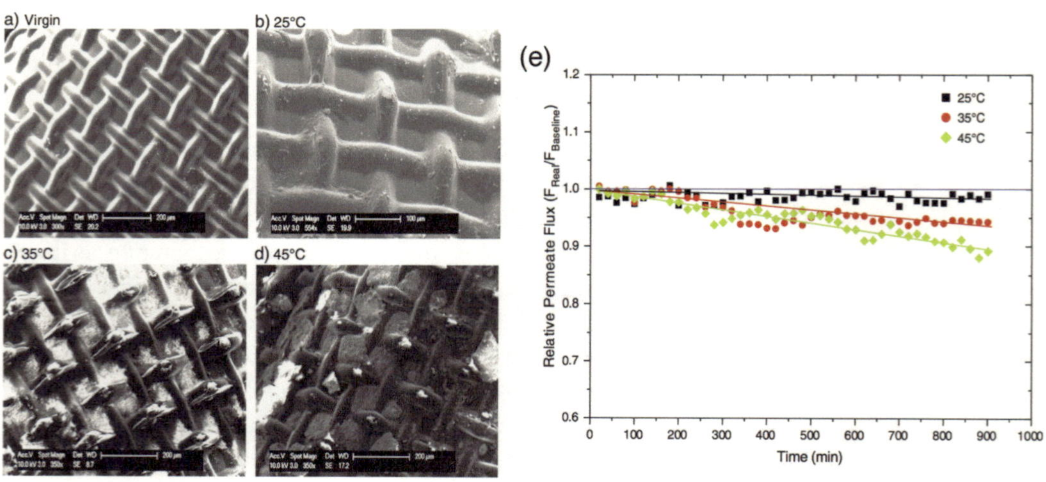

Figure 3. Temperature-dependent membrane fouling and associated water flux decline. (a–d) Scanning electron microscope images of the (a) virgin and (b–d) fouled membranes at various temperatures; (b) 25°C; (c) 35°C; (d) 45°C; and (e) water flux ratio over time at each temperatures. Reprinted with permission from Ref. [24]. © 2011 Elsevier.

increase in the final concentration of the feed solution (i.e., concentration after 28 hours of running) was accelerated by more than 100% when the temperature was raised from 25 to 45°C, which led to faster membrane scaling. Concentrative polarization is also enhanced when the water flux is increased, which results in an accelerated membrane scaling. This was confirmed by directly visualizing the fouled membrane by using a scanning electron microscope (**Figure 3(a)–(d)**) and also by measuring the decrease in flow rate over time (**Figure 3(e)**), showing faster decline of water flux over time at elevated temperatures due to the scaling. In addition to higher solute concentration near the membrane surface driven by the temperature-enhanced solvent flux, the changes in solubility limits for inorganic species may contribute to the accelerated fouling behavior.

4. Transmembrane temperature gradient in FO

One step further, we can also consider a case where the temperature is unevenly distributed across the membrane. In such a case, the temperature gradient may allow independent control of transport on either side of the membrane. In practice, temperature gradients can occur frequently; temperature of the feed solution can increase due to the heat released from the hydraulic pumping or when the solution is pretreated. Likewise, the temperature of the draw solution may change due to the post-treatment process for recovery and recycling of draw solutes such as thermal and membrane distillation. Since heating only on one side of the solution requires lower energy than heating up the entire system, imposing a temperature gradient across the membrane may offer an energy-efficient control over the osmotic phenomena.

In the presence of a temperature gradient, van't Hoff's law (of Eq. (1)) cannot be used directly to calculate the osmotic pressure difference since it relies on the assumption of the constant system temperature. A full theory accounting for the temperature gradient in osmosis may result in highly nonlinear effects on the FO performance. Furthermore, the temperature gradient may provide an additional complexity to the coupled mass and heat transfer phenomena within the membrane. In this section, we provide a summary of how the temperature difference between the feed and the draw solution influences the FO performance, including the water transport and solute diffusion/rejection.

4.1. Water flux

Although the temperature dependence on the water flux shows an agreeable consensus as shown in **Table 1**, the anisotropic temperature effect is shown to differ largely across various studies. When the temperature on either side of the solutions is increased, the water flux becomes higher than that at the base temperature, but lower than when the temperatures of both sides of the solutions are increased. It is, however, left unclear which side of the solution has more influence on the FO process when heated as this does not have an agreeable consensus. **Table 2** provides a summary of the effect of temperature difference on the FO process under various experimental parameters. For simplicity, we define

$$j_F = \frac{J_{w,MF} - J_{w,0}}{T_{MF} - T_0} \tag{3}$$

and

$$j_D = \frac{J_{w,MD} - J_{w,0}}{T_{MD} - T_0} \tag{4}$$

as included on the right-hand side of **Table 2**. Eqs. (3) and (4) refer to the water flux increase per temperature change when the *feed side* or the *draw side* is heated only, respectively. Phuntsho et al. calculated using a commercial software (OLI Stream Analyzer) where the osmotic pressure difference across the membrane can be higher when the draw side is heated in contrast to heating the feed side [26]. However, the temperature difference not only changes the osmotic pressure difference but also gives spatial nonlinearity to other important transport properties such as the solution viscosity as well as solvent/solute diffusivity in bulk phases and their solubilities in the membrane phase, which may impact the CP phenomena in various ways depending on the membrane orientation.

In general, regardless of either the feed or draw, raising the temperature on either side leads to increase in both the water flux and the solute flux. Xie et al. stated that raising the feed solution temperature leads to enhanced diffusivity of the water molecules, whereas raising the draw solution temperature leads to decreased draw solution viscosity and increased draw solute diffusivity, both of which lead to increased water flux and reverse solute flux [27]. However, the degree to which the water flux and solute flux are increased varies across the literature [10, 26–28, 31, 35, 36] (see **Table 2**).

Reference	Feed solution	Draw solution	Membrane	Mode	Temperature (°C)	$J_{w,0}$ (LMH)	j_F[1] (LMH/°C)	j_D[2] (LMH/°C)
[10]	Pineapple juice (0.37 M)	Sucrose (40 wt%)+NaCl (12 wt%)	CT		25–45	1.2	0.045	
[26]	NaCl (0–86 mM)	KCl (0.5–3 M)	CT	FO	25–45	19	0.048	0.12
[27]	Deionized water	NaCl (0.5 M)	CT		20–40	5.5	0.045	0.065
			PA			17	0.125	0.175
[35]	NaCl (0–0.5 M)	NH_4HCO_3 (1–4 M)	CT	PRO	25–45	2.5		0.028
				FO		1.9		0.018
[28]	NaCl (0.1 M)	NaCl (1 M)	CT	PRO	20–40	11	0.54	0.19
				FO		9.4	0.41	0.18
[36]	Anthocyanin (24 μM)	NaCl (6 M)	CT	PRO	25–40	4.9	0.013	
				FO		13	0.53	
[31]	Deionized water	NaCl (1.2 M)	CT	FO	20–30	14	0.22	0.54

[1] $j_F = J_{w,MF} - J_{w,0} / T_{MF} - T_0$; Feed side heated. $J_{w,MF}$: water flux at the maximum feed temperature T_{MF}.
[2] $j_D = J_{w,MD} - J_{w,0} / T_{MD} - T_0$; Draw side heated. $J_{w,MD}$: water flux at the maximum draw temperature T_{MD}.

Table 2. A summary of influence of temperature difference on the water flux.

Phuntsho et al. showed that increasing the *draw solution* temperature resulted in more water flux compared to increasing the temperature of the feed solution [26]. Their membrane was oriented in the PRO mode where the active layer was facing the draw solution. They argued that increasing the draw temperature led to reduced solution viscosity and increased draw solute diffusivity. This change resulted in the reduction of dilutive ICP on the draw side, thereby increasing the water flux. Again, such a behavior is attributed to the fact that the dilutive ICP plays a more significant role than the concentrative ECP in determining the water flux [26]. Such a preferential water flux increase due to the increased draw temperature was also observed by Xie et al. [27] and Cath et al. [31].

You et al. showed, however, that regardless of the membrane orientation, the water flux increased more when the *feed solution* temperature is increased rather than the draw solution [28], which is in a disagreement with the observations made by Phuntsho et al. [26], Xie et al. [27], and Cath et al. [31]. You et al. argued that the water diffusion kinetics is more important than the thermodynamic driving force (i.e., osmotic pressure difference) of the solution in determining the water flux, thus the feed temperature governs the water flux rather than the draw solution temperature [28].

Interestingly, in Nayak and Rastogi's study [36], the water flux in the FO mode was shown to be higher than the water flux in the PRO mode particularly when the molecular size of the feed solute is large enough such that the external concentration polarization cannot be ignored. They also showed that this is indeed true for concentrating anthocyanin, which is a large sugar molecule. In their work, the water flux in the FO mode was measured to be 260% higher than that in the PRO mode.

4.2. Solute diffusion/rejection

As mentioned in the preceding section, Xie et al. showed that the neutral solutes are more likely to diffuse through the membrane than the charged ones due to their smaller hydrodynamic size [27]. In this sense, transmembrane temperature differences barely influenced the solute rejection rate for the *charged* solutes, whereas the *neutral* solutes were significantly influenced by the temperature difference. It was shown that raising the draw temperature (from 20 to 40°C) led to more neutral solute rejection, even more compared to the isothermal condition at base temperature (20°C) [27]. The reason being is that raising the draw temperature leads to increased water flux, which contributes to the increased solute rejection. At the same time, keeping the feed temperature low reduces the deposition of the solutes on to the membrane, thus preventing the neutral feed solutes from dissolving into the membrane and diffusing across the membrane [27].

5. Theoretical perspectives

To the best of our knowledge, effects of temperature and its gradient on the osmosis phenomena and FO processes have been investigated only phenomenologically without fundamental understanding. The theoretical research is currently in a burgeoning state in explaining the transmembrane temperature gradient effect on the FO performance. In this section, we first briefly review the conventional FO theories [37, 38] based on the solution-diffusion model and van't Hoff's law. Then, we revisit statistical mechanics to identify the baseline of the osmosis-diffusion theories, where the isothermal condition was first applied. We then develop a new, general theoretical framework on which FO processes can be better understood under the influence of the system temperature, temperature gradient, and chemical potentials.

5.1. Revisit to the solution-diffusion model

The solution-diffusion model is widely used to describe the FO process, which was originally developed by Lonsdale et al. to explain the RO phenomena using isothermal-isobaric ensemble [39]. In the model, the chemical potential of water is represented as a function of temperature, pressure, and solute concentration, i.e. $\mu_w = \mu_w(T, P, C)$, and its transmembrane gradient is

$$\Delta\mu_w = \int \left(\frac{\partial \mu_w}{\partial C}\right)_{T,P} dC + \int \left(\frac{\partial \mu_w}{\partial P}\right)_{T,C} dP, \tag{5}$$

where the integration is over the membrane region. From the basic thermodynamic relationship,

$$\int \left(\frac{\partial \mu_w}{\partial P}\right)_{T,C} = \overline{V}_w \tag{6}$$

is used where $\overline{V}_w$ is the molar volume of water. In the isothermal-isobaric equilibrium $(\Delta\mu_w = 0)$, the applied pressure ΔP is balanced with the transmembrane difference of the osmotic pressure, i.e. $\Delta P = \Delta\pi$. This condition gives

$$0 = \int \left(\frac{\partial \mu_w}{\partial C}\right)_{T,P} dC + \overline{V}_w \Delta\pi \tag{7}$$

and hence we derive $\Delta\mu_w = \overline{V}_w(\Delta p - \Delta\pi)$. It is assumed that the water transport within the membrane is phenomenologically Fickian, having the transmembrane chemical potential difference of water as a net driving force. The water flux is given as

$$J_w = \frac{D_w C_w}{RT} \frac{d\mu_w}{dx} \simeq \frac{D_w C_w}{RT} \frac{\Delta\mu_w}{\delta_m}, \tag{8}$$

which becomes

$$J_w = A(\Delta p - \Delta\pi), \tag{9}$$

where $A(= D_w C_w / RT\delta_m)$ is the solvent permeability that can be obtained experimentally. The solute flux is similarly given as

$$J_s = -D_s \frac{dC'}{dx} \simeq D_s \frac{\Delta C'}{\delta_m} = D_s \left(\frac{\Delta C'}{\Delta C}\right) \frac{\Delta C}{\delta_m} = \frac{D_s K_m}{\delta_m} \Delta C = B\Delta C, \tag{10}$$

where $\Delta C'$ and ΔC are the concentration differences across the interior and exterior of the membrane, respectively, and $K_m = \Delta C'/\Delta C$ is the partition coefficient, which is assumed to be constant, and $B(= D_s K/\delta_m)$ is the solute permeability.

Figure 4(a) shows a schematic representing the PRO and FO modes altogether. Concentrations in the PRO and FO modes are denoted as C and n, respectively. In the PRO mode, C_1 and C_5 are the draw and feed concentrations, and C_2, C_3, and C_4 are concentrations at interfaces between the draw solution and the active layer, the active layer and the porous substrate, and the porous substrate and the feed solution, respectively. In the FO mode, n_1 and n_5 are the draw and feed concentrations, respectively, and similarly, n_2, n_3, and n_5 have the meanings corresponding to those in the PRO mode. To systematically compare the performances of the PRO and FO modes, we set $n_1 = C_1$ and $n_5 = C_5$, which are the draw (C_d) and feed (C_f) concentrations, respectively. Solvent and solute fluxes in the PRO mode are denoted as J_w^{PRO} and J_s^{PRO}, and those of the FO mode are J_w^{FO} and J_s^{FO}, respectively. In each mode, solvent and solute fluxes are oriented in opposite directions, influencing each other's driving forces. The active layer and porous substrate have thicknesses of δ_m and δ_s, respectively, as located in regions of $-\delta_m < x < 0$ and $0 < x < \delta_s$, respectively. Solute molecules migrate with molecular diffusivity D_0 in the porous substrate that is characterized using its thickness δ_s, porosity ε, and tortuosity τ.

In the PRO mode, the solvent flux (in magnitude) is

$$J_w = A(\pi_2 - \pi_3) \tag{11}$$

where π_2 and π_3 are osmotic pressures at concentration C_2 and C_3, respectively. In a steady state, the water flux J_w is constant in both the active and porous regions. The solute flux in the active layer is:

$$J_s = B(C_2 - C_3) \quad \text{for} \quad -\delta_m < x < 0 \tag{12}$$

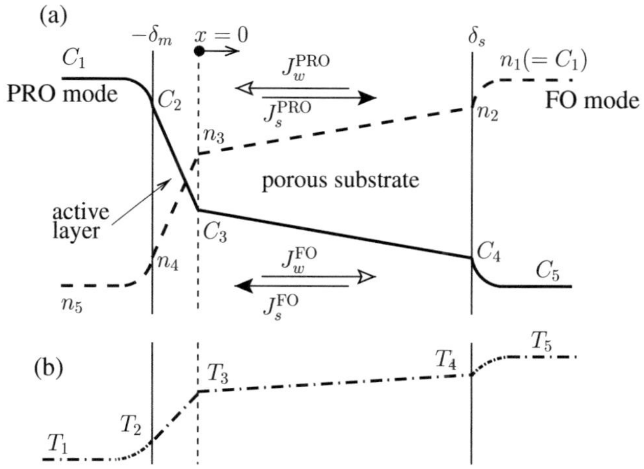

Figure 4. A schematic representation of (a) concentration polarization across a skinned membrane during FO process in the PRO and FO modes, represented using the solid and dashed lines, respectively and (b) arbitrary temperature profile increasing from the active layer to the porous substrate.

and that in the porous substrate:

$$J_s = -\frac{\epsilon}{\tau} D \frac{dC}{dx} - J_w C \quad \text{for} \quad 0 < x < \delta_s. \tag{13}$$

In a steady state, J_s of Eqs. (12) and (13) are equal to each other. Flux equations for the FO mode can be easily obtained by replacing subscript 2 by 4 in Eqs. (11) and (12) and replacing C by n in Eqs. (12), (13). Fluxes of the PRO and FO modes are calculated as

$$J_w^{PRO} \simeq \frac{1}{K} \ln \left[\frac{B + A\pi_d - J_w^{PRO}}{B + A\pi_f} \right] \tag{14}$$

and

$$J_w^{FO} \simeq \frac{1}{K} \ln \left[\frac{B + A\pi_d}{B + A\pi_f + J_w^{FO}} \right], \tag{15}$$

respectively, where π_d and π_f are the osmotic pressure of the draw and feed concentrations, respectively, and

$$K = \frac{\delta_s \tau}{D_0 \epsilon} = \frac{S}{D_0} \tag{16}$$

is interpreted as the characteristic mass transfer resistance, proposed by Lee et al. [37]. Following the convention of standard mass transfer theory, K^{-1} can be interpreted as the mass transfer coefficient of FO processes. In Eq. (16), $S (= \delta_s \tau / \epsilon)$, defined as the structural parameter having units in length, represents the actual path length of molecules passing through the tortuous porous substrate, which is by definition longer than the thickness δ_s. For mathematical simplicity, one can write the flux equation for both modes:

$$J_w = \frac{1}{K} \ln \left[\frac{B + A\pi_d - \varphi J_w}{B + A\pi_f + (1 - \varphi) J_w} \right] \tag{17}$$

where

$$\varphi = \begin{pmatrix} 1 & \text{for PRO mode} \\ 0 & \text{for FO mode} \end{pmatrix} \tag{18}$$

is an integer to toggle between the two modes. Any theoretical development can be initiated from Eq. (17) to consider universally both the FO and PRO modes, and then a proper value of φ can be chosen.

5.1.1. Underlying assumptions and approximations

In the theory, there are several key assumptions during derivations of Eqs. (14) and (15). These assumptions are summarized in the following for the PRO mode for simplicity, but conceptually are identical to those in the FO mode.

1. Mass transfer phenomena are described using the solution-diffusion model in which the solvent and solute transport are proportional to the transmembrane differences in the osmotic pressures and solute concentrations, respectively [39]. If one sees these combined phenomena as diffusion, the solvent transport can be treated as semibarometric diffusion. In other words, under the influence of pressure, the solute transport can be treated as Fickian diffusion, driven by the concentration gradient. In a universal view, the net driving forces of the solvent and solutes are their chemical potential differences.

2. In the flux equations, π_d and π_f are, respectively, overestimated and underestimated because their ture values are those at the draw-membrane and feed-membrane interfaces, i.e. π_2 and π_4, which are difficult to obtain. This approximation does not cause obvious errors if the flow velocities of the draw and feed solutions are fast enough to suppress formation of any significant external concentration polarizations. A necessary condition, which is less discussed in theories, is the high diffusivity or low molecular weight of solutes.

3. The osmotic pressure is presumed to be linear with the solute concentration C. In the PRO mode, one can indicate

$$\pi_2 - \pi_3 = \left(\frac{\pi_2 - \pi_3}{\pi_2 - \pi_4}\right)(\pi_2 - \pi_4) = \left(\frac{1 - C_3/C_2}{1 - C_4/C_2}\right)(\pi_2 - \pi_4) \tag{19}$$

using $\pi_2 - \pi_k = \pi_2(1 - C_k/C_2)$ for $k = 3, 4$. Eq. (19) can be erroneous if the draw concentration is extraordinarily high or pair-wise interactions between solutes are very strong so that the weak solution approach fails. A study on nonlinearity of π with respect to C can be found elsewhere [37, 38].

4. Rigorously saying, mass transport phenomena are assumed to be in a steady state and equilibrium thermodynamics are used to explain the filtration phenomena. Although the FO phenomenon occurs in an open system, transient behavior is barely described in the literature.

5. In the porous substrate, the bulk porosity is assumed to be uniform, which implies isotropic pore spaces. Moreover, the interfacial porosity between the active and porous layers is assumed to be equal to the bulk porosity. An in-depth discussion on the interfacial porosity can be found elsewhere [40]. In the same vein, the tortuosity is a characteristic geometric constant of the substrate, which is hard to measure independently. More importantly, tortuosity is included in the definition of the structural parameter S, which is used to fit the experimental data to the flux equations.

6. The solute diffusivity D_0 is assumed to be constant, that is, independent of the solute concentration such that the concentration profile is further implied to be linear within the porous substrate.

7. Finally, temperatures of the draw and the feed streams are assumed equal although hydraulic and thermal conditions of these two streams can be independently controlled. As a consequence, heat transfer across the membrane is barely discussed in the literature.

In practice, solvent and solute permeability A and B are measured experimentally in the RO mode using feed solution of zero and finite concentrations, respectively. The applied pressure is selected as a normal pressure to operate the RO, and the solute concentrations are usually in the range of that of a typical brackish water. Variations in A and B with C_d and C_f are presumed to be negligible, similar to those of RO cases. In Eq. (17), J_w is directly related to the interfacial concentration, i.e. C_3 and n_3 in the PRO and FO modes, respectively, and therefore it can be predicted only if K is known. Mathematically, one FO flux equation has two unknowns, which are J_w and K. In most cases, the permeate flux J_w is measured experimentally and then used to back-calculate K. This experiment-based prediction often results in an imbalance of mass transfer [41, 42]. A recent study assumes that the interfacial porosity between the active and porous layers is different from the bulk porosity of the porous substrate, which successfully resolves the origin of the imbalance between theoretical and measured K values [40].

This chapter aims to explain how the temperature across the FO membrane, which consists of the active and porous layers, may affect the performance of the mass transfer at the level of statistical physics. The transmembrane temperature gradient prevents from using the abovementioned assumptions and approximations, which are widely used in the FO analysis. First, the SD model is purely based on isothermal-isobaric equilibrium in a closed system. Second, the external concentration polarizations in the draw and feed sides cannot be neglected at the same level because the temperature gradient causes a viscosity difference across the membrane. Third, the weighting factor connecting $\pi_2 - \pi_3$ and $\pi_2 - \pi_4$ cannot be represented only by concentrations but instead should include temperatures at the interfaces. Fourth, even if one can achieve a perfect solute rejection, i.e. $B = 0$, steady heat transfer across the membrane should be included since porous membrane is not a perfect thermal insulator. Fifth, the temperature gradient may change the (effective) properties of the active and porous layers such as A, B, ϵ, and τ in principle and the molecular diffusivity $D_0 \to D(T)$. Sixth, Fick's law should include additional thermal diffusion or temperature effects for determining the collective diffusion. Seventh, of great necessity is a novel, quantitative equation to calculate the osmotic pressure under the gradients of concentration as well as temperature, which generalizes van't Hoff's equation (1).

5.2. Heat transfer

Figure 4(b) shows an arbitrary temperature profile across the FO membrane, increasing from the active layer side to the porous layer side. In bulk phases of the active and porous sides, temperatures are maintained at T_1 and T_4, respectively. For simplicity, we set $T_1 < T_4$. Stream temperature on the active side increases to T_2, and within the membrane, temperature elevates from T_2 to T_3. Since the active layer is often made thin, a linear variation of temperature can be readily assumed. From the active-porous interface to the porous layer surface to the solution, the temperature increases from T_3 to T_4. A similar external temperature polarization occurs in the PL-side bulk phase, generating the temperature change from T_4 to T_5. The overall temperature profile is conceptually akin to the concentration profile in the FO mode. Having the same bulk temperatures, i.e. T_1 and T_5, the flow direction can noticeably change values from T_2 to T_4. For logical consistency, a steady state is assumed while investigating the heat transfer across the FO membrane in this chapter. Thus, heat fluxes of the four regions are

$$q_{BA} = h_{BA}(T_2 - T_1) \tag{20}$$

$$q_{AL} = h_{AL}(T_3 - T_2) \tag{21}$$

$$q_{PL} = h_{PL}(T_4 - T_3) \tag{22}$$

$$q_{BP} = h_{BP}(T_5 - T_4), \tag{23}$$

where subscripts BA and BP indicate bulk phases in the active and porous layer sides, respectively, and AL and PL mean the active layer and porous layer, respectively. The net temperature difference across the membrane is $T_4 - T_2$, which is to be approximated as $T_5 - T_1$. In the steady state, the heat flux q should be equal in each region, that is, $q = q_{BA} = q_{AL} = q_{PL} = q_{BP}$. Dividing each equation of (20)–(23) by the heat transfer coefficient $h's$, one derives

$$q = h_{\text{eq}}(T_5 - T_1) \tag{24}$$

$$\frac{1}{h_{\text{eq}}} = \frac{1}{h_{BA}} + \frac{1}{h_{AL}} + \frac{1}{h_{PL}} + \frac{1}{h_{BP}}. \tag{25}$$

Note that Eq. (24) assumes that the heat transfer is solely based on thermal conduction without thermal convection, that is, transfer rate of heat by solvent flux. In the FO process with the transmembrane thermal gradient, Eqs. (21) and (22) should be revised as

$$q_{AL} = h_{AL}(T_3 - T_2) \pm H_w J_w \tag{26}$$

$$q_{PL} = h_{PL}(T_4 - T_3) \pm H_w J_w, \tag{27}$$

where H_w and J_w are the enthalpy and flux of the solvent, respectively, and the sign is plus when the concentration and temperature profiles both increase and decrease together, otherwise it is negative. For example, for the temperature profile shown in **Figure 4**, the FO concentration profile has the same trend to that of the temperature, and therefore signs in Eqs. (26) and (27) are positive. In this case, Eq. (25) needs to be modified to

$$\frac{1}{h_{\text{eq}}} = \frac{1}{h_{BA}} + \frac{1}{h'_{AL}} + \frac{1}{h'_{PL}} + \frac{1}{h_{BP}}, \tag{28}$$

where

$$h'_{AL} = h_{\text{AL}} \pm \frac{H_w J_w}{T_3 - T_2} \tag{29}$$

$$h'_{LL} = h_{PL} \pm \frac{H_w J_w}{T_4 - T_3} \tag{30}$$

This heat balance analysis is very similar to that of membrane distillation [43, 44], but the FO process does not have any solvent phase transition so that the latent heat is not considered.

5.3. Mass transfer mechanisms

5.3.1. Anisothermal osmotic pressure

In statistical mechanics, Gibbs energy is the master function of the isothermal-isobaric ensemble. Consider a box in which two regions are separated by a semipermeable membrane. In equilibrium, the maximum entropy condition requires that the chemical potential divided by the temperature should be constant, i.e.

$$\Delta\left(\frac{\mu(T,P,N)}{T}\right) = 0, \tag{31}$$

which converts to the constant chemical potential for the isothermal environment, i.e. $\Delta\mu = 0$ for constant T. Note that in the conventional solution-diffusion model, the chemical potential of water μ_w in the external phase is assumed as a function of solute concentration C and pressure P. From Eq. (31), van't Hoff's osmotic pressure difference is derived as

$$\Delta\pi = RT\Delta C, \tag{32}$$

which can perhaps be extended intuitively to $\Delta\pi = R\Delta(CT)$ in the temperature gradient. Here we assume that the membrane properties do not change significantly with solute concentration C and local temperature T. In the presence of a concentration gradient only, van't Hoff's equation indicates that water (solvent) molecules tend to move from a lower solute concentration region to a higher solute concentration region. This is due to the water chemical potential being higher in the lower C region. Now we replace the concentration gradient by the temperature gradient. Diffusion of water molecules is purely based on their kinetic energy as proportional to T and the temperature gradient across the membrane, as shown in **Figure 4(a)**. For simplicity, we consider only the active layer of which A and B values are assumed to be insensitive to temperature. Therefore, similar to the direct contact membrane distillation, two solutions of high and low temperatures are in contact with the membrane surfaces. Since solutes are absent, the water motion is purely diffusive under the chemical potential gradient induced by the temperature gradient. Water molecules in the high temperature region move faster than those in the low temperature region. Therefore, water transfer must follow the direction of the temperature gradient. If one side of the membrane has a solution of both high temperature and concentration, then the net osmotic pressure must be less than that of the concentration gradient only, that is,

$$\Delta\pi = a\Delta C - b\Delta T \tag{33}$$

where a must be equal to RT and b is a positive constant. To the best of our knowledge, $a(T) = RT$ has not been rigorously proven, and $b(c)$ is so far unknown. The theoretical development of the anisothermal osmotic pressure, $\pi = \pi(C,T)$, as a natural extension from van't Hoff's equation is of urgent importance to the current literature in water transport theories, which are to be utilized not only in desalination and fresh water production but also in a broad applications of separation and filtration.

5.3.2. Anisothermal diffusion

Fick's law is a phenomenological equation based on experimental observations. The equation states that the diffusive flux J is proportional to the concentration gradient

$$\vec{J} = -D\,\vec{\nabla}\,C. \tag{34}$$

In the dilute limit, the diffusivity is independent of concentration C, i.e. $D \neq D(C)$, and if the solute molecules are Brownian, D is proportional to temperature T: $D \propto T$. If and only if the molecular motion is dragged by the viscous force, which is directly related to their relative velocity to the solvent (often stationary), then the drag force can be written as

$$\widetilde{F}_{\text{drag}} = -\beta \vec{v}, \tag{35}$$

where $\vec{v}$ is the molecular velocity relative to that of the solvent medium, and β is the drag coefficient independent of $\vec{v}$. The Brownian diffusivity is proven to be $D = k_B T/\beta$, where k_B is Boltzmann's constant. Stokes proved that $\beta = 3\pi\eta_w d_p$ where η_w is the solvent viscosity and d_p is the particle (molecule) diameter.

In the presence of the spatial variation of T, Eq. (34) is generalized as [45]

$$\vec{J} = \frac{D}{T} \vec{\nabla} (CT). \tag{36}$$

Thus, substitution of the Stokes-Einstein diffusivity into Eq. (36) gives

$$\vec{J} = \frac{k_B}{\beta} \vec{\nabla} (CT) = \frac{1}{\beta} \vec{\nabla} (\pi), \tag{37}$$

which is valid if the solvent viscosity η_w is a weak function of T such as water. For a homogeneous system, the diffusive flux may in general be

$$\vec{J} = -\alpha \vec{\nabla} \mu - \beta \vec{\nabla} T, \tag{38}$$

where one can write the chemical potential gradient as

$$\vec{\nabla} \mu = \left(\frac{\partial \mu}{\partial C}\right)_{P,T} \vec{\nabla} C + \left(\frac{\partial \mu}{\partial T}\right)_{C,P} \vec{\nabla} T + \left(\frac{\partial \mu}{\partial P}\right)_{C,T} \vec{\nabla} P. \tag{39}$$

Substitution of Eq. (39) into (38) gives

$$\vec{J} = -D\left(\vec{\nabla} C + k_T \vec{\nabla} \ln T + k_P \vec{\nabla} \ln P\right), \tag{40}$$

which defines the thermal diffusion coefficient $k_T D$, where k_T is the thermal diffusion ratio, which is a dimensionless quantity. The coefficient $k_P D$ is the barodiffusion coefficient. In the dilute limit, k_T vanishes as it is proportional to C. The barodiffusion is often negligible as the diffusion is characterized in a stationary fluid that will have finite velocity if the hydraulic pressure is applied.

5.3.3. *Solute diffusivity matters*

In the conventional isothermal theory of FO, one can write a conceptual relationship between the water flux and the transmembrane osmotic pressure difference as [33]

$$J_w \propto D \ln \Delta\pi, \tag{41}$$

which clearly indicates that J_w increases with both D and $\Delta\pi$, but $\Delta\pi$ increases much slower than D due to the logarithmic dependence. To double the flux J_w, there are two mathematical choices: $D \rightarrow 2D$ (linear) and $\Delta\pi \rightarrow (\Delta\pi)^2$ (geometric in a specific unit, or $\Delta C \rightarrow (\Delta C)^2$). The first way of increasing the solute diffusivity is related to finding or developing novel draw solutes, while the second option is practically challenging as it makes the draw recovery more energy consuming. Especially when selecting the draw solutes, their diffusivity is the most critical parameter in FO processes, as solutes of high diffusivity significantly decrease the ECP and ICP.

If we write intuitively the anisothermal osmotic pressure as

$$\Delta\pi = RT_m\Delta C - b\Delta T \tag{42}$$

across the membrane with $\Delta T = T_1 - T_2$ and $T_m = \frac{1}{2}(T_1 + T_2)$, it would be interesting to know the particular transmembrane temperature difference that can nullify the net osmotic pressure gradient:

$$\Delta T = b^{-1}R\overline{T}\Delta C. \tag{43}$$

As both T_1 and T_2 increase while keeping ΔT constant, $\Delta\pi$ increases. Moreover, increased T_m may noticeably enhance the solvent as well as solute diffusion. This thought process strongly supports the experimental literature in FO research, equivocally showing that the solvent flux is proportional to the system temperature. Note that Eq. (41) includes the permeability coefficients of solvent (A) and solute (B). As we discussed in the previous section, we know

$$\frac{\partial A}{\partial T} \text{ and } \frac{\partial B}{\partial T} \gtrsim 0 \tag{44}$$

so that both the solvent and solute fluxes increase with the mean temperature T_m of the membrane where ΔT is maintained constant.

On the basis of our investigation, temperature effects on the osmotic phenomena are not as simple as expected from the linear van't Hoff equation, but highly correlated through the temperature-dependent material constants of solvent (η, A), solutes (D, B), and their strong linkage to the osmotic pressure: $\pi \rightarrow \pi(C, T)$.

6. Concluding remarks

This chapter provides a comprehensive review on the effect of temperature on the FO process. Although the motivation for studying the temperature effect comes from the fact that osmosis is a thermodynamically spontaneous process, changing the system temperature either locally or globally can offer more effective ways of engineering the FO process with lower energy consumption. However, as evidenced by the scattered data across the literature and a lack of theoretical descriptions, more robust and systematic studies are warranted for deeper understanding of the phenomena. For example, most of the temperature-dependent FO studies relate the changes in the water and the solute flux to the change in the physical properties of the bulk solution only, neglecting any changes in the membrane properties such as water permeability *A*, solute permeability *B*, and mass transfer resistance *K*. Furthermore, a holistic theory accounting for the effect of transmembrane temperature gradient on the FO process is still missing, hence to be constructed in the near future.

Author details

Sangwoo Shin[1] and Albert S. Kim[2]*

*Address all correspondence to: albertsk@hawaii.edu

1 Department of Mechanical Engineering, University of Hawaii at Manoa, Honolulu, Hawaii, USA

2 Department of Civil and Environmental Engineering, University of Hawaii at Manoa, Honolulu, Hawaii, USA

References

[1] Batchelder GW. Process for the Demineralization of Water. Google Patents; 1965. US Patent 3,171,799

[2] Chekli L, Phuntsho S, Kim JE, Kim J, Choi JY, Choi JS, et al. A comprehensive review of hybrid forward osmosis systems: Performance, applications and future prospects. Journal of Membrane Science. 2016;**497**:430-449 https://doi.org/10.1016/j.memsci.2015.09.041

[3] Cath TY, Gormly S, Beaudry EG, Flynn MT, Adams VD, Childress AE. Membrane contactor processes for wastewater reclamation in space: Part I. Direct osmotic concentration as pretreatment for reverse osmosis. Journal of Membrane Science. 2005;**257**(1):85-98 https://doi.org/10.1016/j.memsci.2004.08.039

[4] Cath TY, Adams D, Childress AE. Membrane contactor processes for wastewater reclamation in space: II. Combined direct osmosis, osmotic distillation, and membrane distillation for treatment of metabolic wastewater. Journal of Membrane Science. 2005;**257**(1): 111-119 https://doi.org/10.1016/j.memsci.2004.07.039

[5] Kravath RE, Davis JA. Desalination of sea water by direct osmosis. Desalination. 1975;**16**(2): 151-155 https://doi.org/10.1016/S0011-9164(00)82089-5

[6] Kessler JO, Moody CD. Drinking water from sea water by forward osmosis. Desalination. 1976;**18**(3):297-306 https://doi.org/10.1016/S0011-9164(00)84119-3

[7] Moody CD, Kessler JO. Forward osmosis extractors. Desalination. 1976;**18**(3):283-295 https://doi.org/10.1016/S0011-9164(00)84118-1

[8] Cath TY, Childress AE, Elimelech M. Forward osmosis: Principles, applications, and recent developments. Journal of Membrane Science. 2006;**281**(1):70-87 https://doi.org/10.1016/j.memsci.2006.05.048

[9] Warczok J, Gierszewska M, Kujawski W, Güell C. Application of osmotic membrane distillation for reconcentration of sugar solutions from osmotic dehydration. Separation and Purification Technology. 2007;**57**(3):425-429 https://doi.org/10.1016/j.seppur.2006.04.012

[10] Ravindra Babu B, Rastogi NK, Raghavarao KSMS. Effect of process parameters on transmembrane flux during direct osmosis. Journal of Membrane Science. 2006;**280**(1):185-194 https://doi.org/10.1016/j.memsci.2006.01.018

[11] Wrolstad RE, McDaniel MR, Durst RW, Micheals N, Lampi KA, Beaudry EG. Composition and sensory characterization of red raspberry juice concentrated by direct-osmosis or evaporation. Journal of Food Science. 1993;**58**(3):633-637 https://doi.org/10.1111/j.1365-2621.1993.tb04344.x

[12] Theeuwes F, Yum SI. Principles of the design and operation of generic osmotic pumps for the delivery of semisolid or liquid drug formulations. Annals of Biomedical Engineering. 1976;**4**(4):343-353 https://doi.org/10.1007/BF02584524

[13] Wright JC, Johnson RM, Yum SI. DUROS® osmotic pharmaceutical systems for parenteral & site-directed therapy. Drug Delivery Technology. 2003;**3**(1):64-73

[14] Su YC, Lin L. A water-powered micro drug delivery system. Journal of Microelectromechanical Systems. 2004;**13**(1):75-82. DOI: 10.1109/JMEMS.2003.823215

[15] McGinnis RL, Elimelech M. Global challenges in energy and water supply: The promise of engineered osmosis. Environmental Science & Technology. 2008;**42**(23):8625-8629. DOI: 10.1021/es800812m

[16] Achilli A, Cath TY, Childress AE. Power generation with pressure retarded osmosis: An experimental and theoretical investigation. Journal of Membrane Science. 2009;**343**(1):42-52 https://doi.org/10.1016/j.memsci.2009.07.006

[17] Lutchmiah K, Verliefde ARD, Roest K, Rietveld LC, Cornelissen ER. Forward osmosis for application in wastewater treatment: A review. Water Research. 2014;**58**:179-197 https://doi.org/10.1016/j.watres.2014.03.045

[18] McCutcheon JR, McGinnis RL, Elimelech M. A novel ammonia–carbon dioxide forward (direct) osmosis desalination process. Desalination. 2005;**174**(1):1-11 https://doi.org/10.1016/j.desal.2004.11.002

[19] McCutcheon JR, McGinnis RL, Elimelech M. Desalination by ammonia–carbon dioxide forward osmosis: Influence of draw and feed solution concentrations on process performance. Journal of Membrane Science. 2006;**278**(1):114-123 https://doi.org/10.1016/j.memsci.2005.10.048

[20] Gray GT, McCutcheon JR, Elimelech M. Internal concentration polarization in forward osmosis: Role of membrane orientation. Desalination. 2006;**197**(1-3):1-8 https://doi.org/10.1016/j.desal.2006.02.003

[21] Traxler RN. The effect of temperature on rate of osmosis. Journal of Physical Chemistry A. 1928;**32**(1):127-141. DOI: 10.1021/j150283a010

[22] Wijmans JG, Baker RW. The solution-diffusion model: A review. Journal of Membrane Science. 1995;**107**(1-2):1-21 https://doi.org/10.1016/0376-7388(95)00102-I

[23] Garcia-Castello EM, McCutcheon JR, Elimelech M. Performance evaluation of sucrose concentration using forward osmosis. Journal of Membrane Science. 2009;**338**(1):61-66 https://doi.org/10.1016/j.memsci.2009.04.011

[24] Zhao S, Zou L. Effects of working temperature on separation performance, membrane scaling and cleaning in forward osmosis desalination. Desalination. 2011;**278**(1):157-164 https://doi.org/10.1016/j.desal.2011.05.018

[25] Petrotos KB, Quantick P, Petropakis H. A study of the direct osmotic concentration of tomato juice in tubular membrane–module configuration. I. The effect of certain basic process parameters on the process performance. Journal of Membrane Science. 1998;**150**(1): 99-110 https://doi.org/10.1016/S0376-7388(98)00216-6

[26] Phuntsho S, Vigneswaran S, Kandasamy J, Hong S, Lee S, Shon HK. Influence of temperature and temperature difference in the performance of forward osmosis desalination process. Journal of Membrane Science. 2012;**415**:734-744 https://doi.org/10.1016/j.memsci.2012.05.065

[27] Xie M, Price WE, Nghiem LD, Elimelech M. Effects of feed and draw solution temperature and transmembrane temperature difference on the rejection of trace organic contaminants by forward osmosis. Journal of Membrane Science. 2013;**438**:57-64 https://doi.org/10.1016/j.memsci.2013.03.031

[28] You SJ, Wang XH, Zhong M, Zhong YJ, Yu C, Ren NQ. Temperature as a factor affecting transmembrane water flux in forward osmosis: Steady-state modeling and experimental validation. Chemical Engineering Journal. 2012;**198**:52-60 https://doi.org/10.1016/j.cej.2012.05.087

[29] McCutcheon JR, Elimelech M. Influence of concentrative and dilutive internal concentration polarization on flux behavior in forward osmosis. Journal of Membrane Science. 2006;**284**(1):237-247 https://doi.org/10.1016/j.memsci.2006.07.049

[30] Ng HY, Tang W, Wong WS. Performance of forward (direct) osmosis process: Membrane structure and transport phenomenon. Environmental Science & Technology. 2006;**40**(7): 2408-2413 https://doi.org/10.1021/es0519177

[31] Cath TY, Elimelech M, McCutcheon JR, McGinnis RL, Achilli A, Anastasio D, et al. Standard methodology for evaluating membrane performance in osmotically driven membrane processes. Desalination. 2013;**312**:31-38 https://doi.org/10.1016/j.desal.2012.07.005

[32] Kim AS, Lee SW. Intrinsic flux inequality in forward osmosis (FO) and pressure-retarded osmosis (PRO) processes. Membrane Journal. 2015;**25**(4):367-372. http://db.koreascholar.com/article.aspx?code=306432

[33] Kim AS, Kim SW. Performance analysis of forward osmosis processes from the integral equation theory. Desalination and Water Treatment. 2013;**51**(25-27):5289-5297 http://dx.doi.org/10.1080/19443994.2013.768757

[34] Dale MC, Okos MR. Reverse osmosis membrane performance as affected by temperature and pressure. Industrial & Engineering Chemistry Product Research and Development. 1983;**22**(3):452-456 https://doi.org/10.1021/i300011a013

[35] Chanukya BS, Patil S, Rastogi NK. Influence of concentration polarization on flux behavior in forward osmosis during desalination using ammonium bicarbonate. Desalination. 2013;**312**:39-44 https://doi.org/10.1016/j.desal.2012.05.018

[36] Nayak CA, Rastogi NK. Forward osmosis for the concentration of anthocyanin from Garcinia indica Choisy. Separation and Purification Technology. 2010;**71**(2):144-151 https://doi.org/10.1016/j.seppur.2009.11.013

[37] Lee KL, Baker RW, Lonsdale HK. Membranes for power generation by pressure-retarded osmosis. Journal of Membrane Science. 1981;**8**(2):141-171 https://doi.org/10.1016/S0376-7388(00)82088-8

[38] Loeb S. Production of energy from concentrated brines by pressure-retarded osmosis. Journal of Membrane Science. 1976;**1**:49-63 https://doi.org/10.1016/S0376-7388(00)82257-7

[39] Lonsdale HK, Merten U, Riley RL. Transport properties of cellulose acetate osmotic membranes. Journal of Applied Polymer Science. 1965 Apr;**9**(4):1341-1362 https://doi.org/10.1002/app.1965.070090413

[40] Kang PK, Lee W, Lee S, Kim AS. Origin of structural parameter inconsistency in forward osmosis models: A pore-scale CFD study. Desalination. 2017;**421**:47-60 https://doi.org/10.1016/j.desal.2017.05.018

[41] Manickam SS, Gelb J, McCutcheon JR. Pore structure characterization of asymmetric membranes: Non-destructive characterization of porosity and tortuosity. Journal of Membrane Science. 2014;**454**:549-554 https://doi.org/10.1016/j.memsci.2013.11.044

[42] Manickam SS, McCutcheon JR. Model thin film composite membranes for forward osmosis: Demonstrating the inaccuracy of existing structural parameter models. Journal of Membrane Science. 2015;**483**:70-74 https://doi.org/10.1016/j.memsci.2015.01.017

[43] Khayet M, Matsuura T. Membrane Distillation: Principles and Applications. New York: Elsevier; 2011

[44] Kim AS. Cylindrical cell model for direct contact membrane distillation (DCMD) of densely packed hollow fibers. Journal of Membrane Science. 2014;**455**:168-186 https://doi.org/10.1016/j.memsci.2013.12.067

[45] Efros A. Theory of thermal diffusion of Brownian particles. Soviet Physics. Journal of Experimental and Theoretical Physics. 1966;**23**(3):536-541

Chapter 6

Pressure Dependency of the Membrane Structure Parameter and Implications in Pressure Retarded Osmosis (PRO)

Torleif Holt, Edvard Sivertsen, Willy R. Thelin and Geir Brekke

Additional information is available at the end of the chapter

http://dx.doi.org/10.5772/intechopen.72444

Abstract

Pressure retarded osmosis (PRO) can be used to exploit the mixing energy *e.g.* between river water and sea water. A PRO membrane must be highly permeable for water, whereas salt ions should be retained. Furthermore, the structure parameter of the membrane support and backing structure must be low. This paper summarises an assessment of the pressure dependency of the structure parameter for flat sheet membranes, and a transport model for PRO and procedures for determination of the pressure dependency of the structure parameter are presented. The results from laboratory experiments show that that the structure parameter increases significantly with increasing trans-membrane pressure. The increase in the structure parameter was observed to depend on both characteristics of the membrane and the fresh water spacer. Using a finely textured tricot spacer reduced the pressure dependency on the structure parameter, compared to a coarser spacer. Applying a non-woven backing material between the membrane and the fresh water spacer also reduced the impact of pressure. The results show that membranes suitable for river water/sea water PRO must have a sufficiently low structure parameter and additionally resist severe deformation at relevant operating pressures.

Keywords: osmotic power, pressure retarded osmosis, structure parameter, pressure dependence

1. Introduction

Pressure retarded osmosis (PRO) is one feasible technology that can be used to exploit the mixing energy from salt gradients which is commonly referred to as salinity gradient power or osmotic power [1, 2]. In PRO the transport of water through the membrane is caused by the difference in osmotic pressure across the membrane skin, and the net volume increase on the high saline side due to mass transport against a pressure gradient can be utilised to run a turbine. It should be mentioned that indirect alternatives to exploit the osmotic power, such as osmotic energy recovery in desalination of sea water, have gained increasing attention recently [3–5].

The mass transport of salt and water in PRO can be characterised by three parameters, the water permeability, A, the salt permeability, B, and the structure parameter, S [1, 6]. The parameters must be optimised in order to maximise produced power, implying that the water permeability should be high, and both the salt permeability and the structure parameter should be low. Membrane development has been a prioritised research area for more than a decade, and significant improvements in PRO membrane performance have been achieved over the last years [6–11]. Membrane and element configuration has also been a focus area, and both flat sheet and hollow fibre configurations should be further investigated [12–14].

Resent research has showed that various transport models [14–17] fail to accurately model PRO performance as a function of pressure increase. Kim and Elimelech [18] have related the deviation between observed and modelled performance to adverse effects between the membrane support and the feed channel spacer. Both membrane deformation and obstruction of water permeation were proposed mechanisms to explain the reduced membrane performance at increasing pressures. In case of membrane obstruction, *i.e.* the spacer blocks part of the active membrane area This effect was referred to as the *spacer shadow effect*. Kim and Elimelech showed that the water permeability remained almost independent of the trans-membrane pressure when a diamond shaped feed spacer was applied. On the other hand, the salt permeability increased significantly when the trans-membrane pressure exceeded a certain value (in the range 9–12 bar).

She *et al.* [19] have also studied the impact of spacer characteristics on PRO performance. They showed that mechanical deformation of the PRO membrane did occur during PRO operation. Subsequently, they determined water and salt permeabilities obtained after deformation as a function of trans-membrane pressure in RO experiments, using the same types of feed spacers. Finally, the structure parameter was determined from calculations using the observed water fluxes from the PRO experiments. The variations in the estimated membrane parameters, as well as the mechanical deformation, were found to depend on spacer characteristics.

The interaction between the membrane and the feed spacer is found to reduce the PRO performance of flat sheet membranes. Hollow fibres are self-supporting structures, meaning that the use of spacers is avoided. Any pressure dependency of the PRO performance of hollow fibre membranes must therefore be related to other mechanisms than interactions between membrane and spacer. Chou *et al.* [7] observed a discrepancy between modelled and measured performances for fibres with the skin applied on the bore side. They determined the structure parameter at several pressure steps, and observed that this parameter decreased with increasing pressure. It was suggested that this was due to expansion of the polymer network resulting in reduced tortuosity of the membrane support when the inside of the fibres was pressurised.

The objective of this paper is to present a hypothesis for the interaction between the membrane and spacer which partly builds on the hypothesis of Kim and Elimelech [18]. Based on characterisation experiments we have demonstrated good correlation between measured and modelled membrane performances by applying a pressure dependent structure parameter.

http://dx.doi.org/10.5772/intechopen.72444

Further, the implications of membrane and spacer interactions in PRO will be discussed and related to the need for optimisation of the characteristic parameters of PRO membranes.

2. Theory

2.1. PRO modelling

A simplified flow diagram indicating main components in a PRO process such as pre-treatment stage, membrane modules, and pressure exchanger, is given in **Figure 1**. In PRO, water will be transported against a pressure gradient due to the difference in osmotic pressure between the draw solution and the feed solution. The net volume increase on the high saline side, which are operated at elevated pressure, can *e.g.* be converted to power in a turbine. The produced power, P, equals the volume flux, J_V, through the membrane, multiplied with the hydraulic pressure difference over the membrane, Δp,

$$P = J_v \Delta p \tag{1}$$

Since the volume of salt transported through the membrane is negligible compared to the volume of water, the volume flux can be replaced by the water flux, J_{w}.

Different model frameworks describing the transport of salt and water through osmotic membranes have been developed by several authors [15, 20–24]. This paper is based on the stagnant boundary layer model presented by Thorsen and Holt [15], and the basic equations are given below.

Figure 2 shows the cross section of an osmotic membrane in a cross-flow cell, indicating the concentration profile of salt at a given position in the cell, from the fresh water side, through the membrane and to the sea water side.

The transport of water and salt (J_s) through the membrane skin is described by two flux equations, where the positive flux directions are indicated by the arrows in **Figure 2**

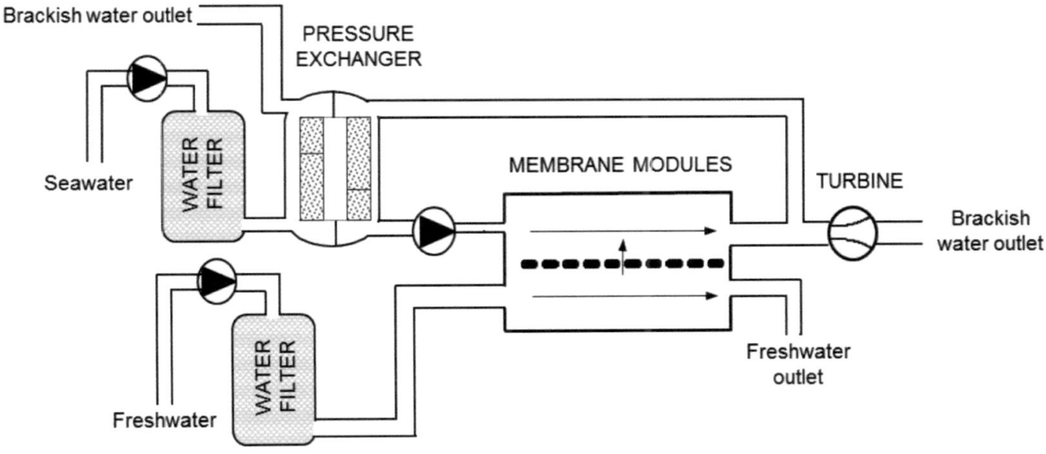

Figure 1. Simplified flow diagram of a PRO power plant.

$$J_w = A(\Delta\pi_{skin} - \Delta p) \tag{2}$$

and

$$J_s = B\Delta s_{skin} = B(c_{sm} - c_p) \tag{3}$$

A and B are the water and salt permeability of the skin, respectively. The osmotic pressure across the skin, $\Delta\pi_{skin}$, is related to the concentration difference $(c_{sm} - c_p)$ of salt over the skin by

$$\Delta\pi_{skin} = iRT(c_{sm} - c_p) = iRT\Delta c_{skin} \tag{4}$$

where i is the van't Hoff coefficient that equals 2.0 for ideal solutions of NaCl. A value of 1.9, which are based on published data for osmotic pressures in NaCl solutions, have been used in the present calculations [25]. R is the universal gas constant and T is the absolute temperature.

The coupled transport of salt in the support membrane and the boundary layers can be expressed by the mass balance

$$-J_s = \frac{\varphi}{\tau} D \frac{dc}{dx} - J_w c \tag{5}$$

where the porosity, φ, and the tortuosity, τ, in the boundary layers on the membrane surfaces equals unity. D is the diffusion coefficient of salt (NaCl). Inserting the water flux in Eq. (2) and the salt flux in Eq. (3) into the mass balance in Eq.(5) and evaluating the transport of water

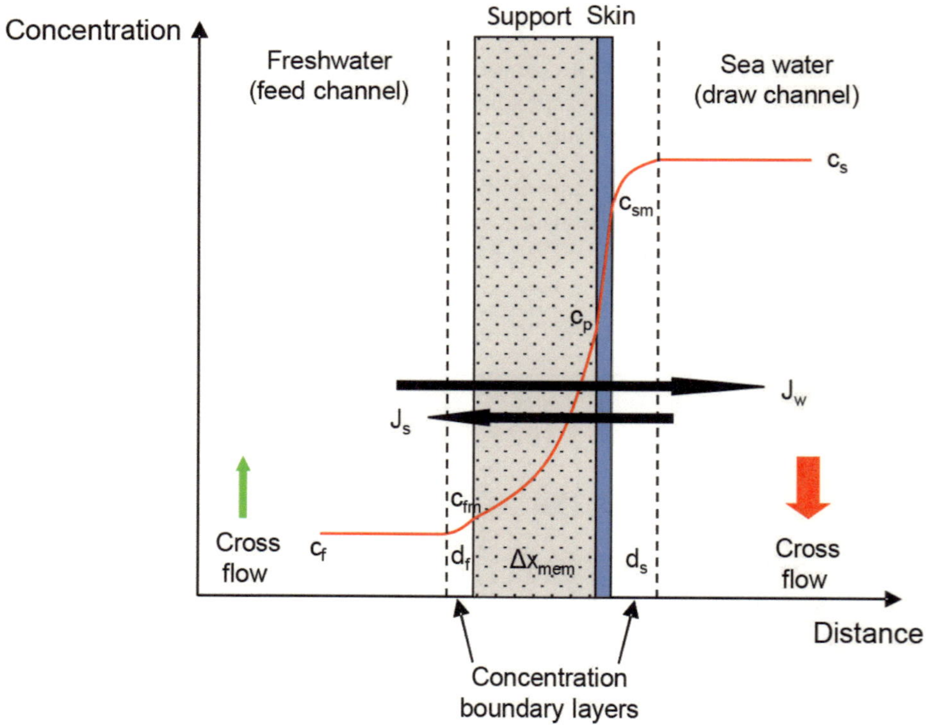

Figure 2. Concentration profile over the membrane and boundary layers.

and salt in the different transport zones results in five equations containing five unknown parameters, J_s, J_w, c_{fm}, c_p and c_{sm}. After some rearrangement, the following expression for the concentration difference across the skin, Δc_s can be found:

$$\Delta c_{skin} = \frac{c_s - c_f e^{\left\{\frac{(S+d_s+d_f)J_w}{D}\right\}}}{e^{\left(\frac{dJ_w}{D}\right)} + \frac{B}{J_w}\left(e^{\left\{\frac{(S+d_s+d_f)}{D}\right\}} - 1\right)} \tag{6}$$

The equation relates the concentration difference of salt over the membrane skin to the bulk concentrations of salt, and furthermore to the characteristic membrane parameters, as well as the boundary layer thickness on each side, d_s and d_f, respectively. The structure parameter, S, of the membrane support is defined as

$$S = \frac{\tau}{\phi}\Delta x_{mem} \tag{7}$$

where Δx_{mem} is the thickness of the support membrane that for practical purpose will equal the measured membrane thickness. The salt flux can be found by multiplying Eq. (6) by B.

The water flux can be found by combining Eqs. (2), (4) and (6) giving

$$J_w = A\left(iRT\frac{c_s - c_f e^{\left\{\frac{(S+d_s+d_f)J_w}{D}\right\}}}{e^{\left(\frac{dJ_w}{D}\right)} + \frac{B}{J_w}\left(e^{\left\{\frac{(S+d_s+d_f)}{D}\right\}} - 1\right)} - \Delta p\right) \tag{8}$$

which is valid when the salt water faces the skin side of the membrane, *i.e.* PRO mode.

2.2. Pressure dependency of the structure parameter

The left sketch in **Figure 3** illustrates the cross section of a PRO membrane at zero trans-membrane pressure. The support membrane rests on the top of the filaments of the feed spacer. The contact area between the membrane and the spacer will in such case be low, and the presence of the spacer material has little or no impact on the mass transfer. An eventual impact will be included in the structure parameter determined by modelling of isobaric experiments.

When pressure is applied on the skin side in a PRO experiment the pressure will exert a force on the membrane, such that feed spacer will be squeezed into the support membrane. This situation is illustrated in **Figure 3** (right sketch). As a result, the membrane may be deformed, and the contact area between the membrane and the feed spacer might increase. Furthermore, the properties of the support structure, such as porosity and interconnections between pores, might be affected. The net effect of these phenomena can be modelled as an increased structure parameter.

A simple equation has been developed in order to illustrate the pressure dependency on the structure parameter and the implicit effect on the water flux:

$$S = S_0\frac{1}{1-\left(F\Delta p/\Delta p_{ref}\right)} \tag{9}$$

where S_0 is the structure parameter at zero trans-membrane pressure, Δp. Arbitrarily values for the constant F were selected, and constant Δp_{ref} was set to 10.6 bar. As shown in **Figure 4**, the increase in the structure parameter is modest at low trans-membrane pressures, but increases rapidly at higher pressures. The water flux will be reduced when the structure parameter increases. The effect is more pronounced for higher F values.

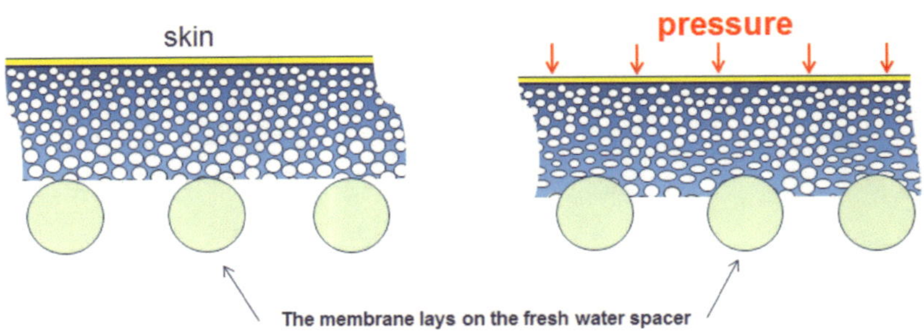

Figure 3. PRO membrane and spacer material at isobaric conditions (left) and pressurised conditions (right). The pores (illustrated by white circles) in the support membrane and possible reinforcement are interconnected giving continuous transport paths.

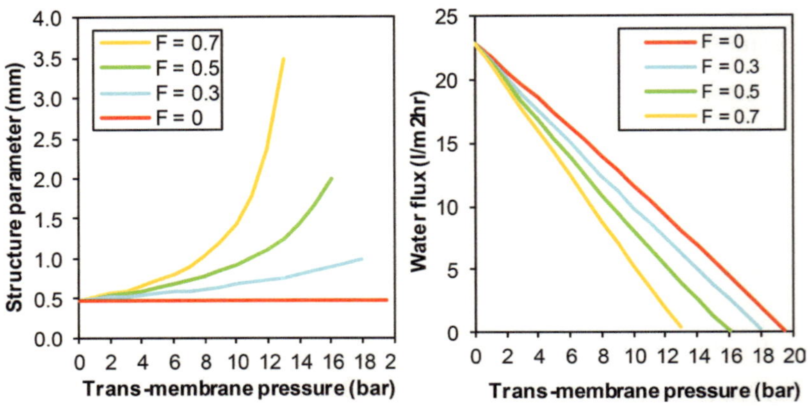

Figure 4. Structure parameters modelled as function of trans membrane pressure according to Eq. (9).

3. Experimental

3.1. Apparatus

All results presented in this work were obtained from measurements performed with two small cross-flow units as illustrated in **Figure 5**. Two membrane cells with different effective membrane area of 6.1 and 9.5 cm^2, respectively, were applied. The channel width was 1.1 cm and the depth of the draw channel was 0.07 cm for both cells. The depths of the feed channels for the two different cells were 0.1 and 0.05 cm, respectively. The draw channels were filled with a 0.07 cm thick diamond spacer, whereas different types of spacers were used in the feed channels.

Both feed and draw solution were pumped through the cross-flow cell using dual-piston pumps with displacement volumes of approximately 10 ml/stroke. The fluids were fed into the pumps from reservoirs placed on balances, and subsequently recycled back to the reservoirs. The cross-flow cells and up-stream tubing were immersed in temperature controlled water baths to maintain the temperature at 20°C during the experiments. The pressures, p, the

http://dx.doi.org/10.5772/intechopen.72444

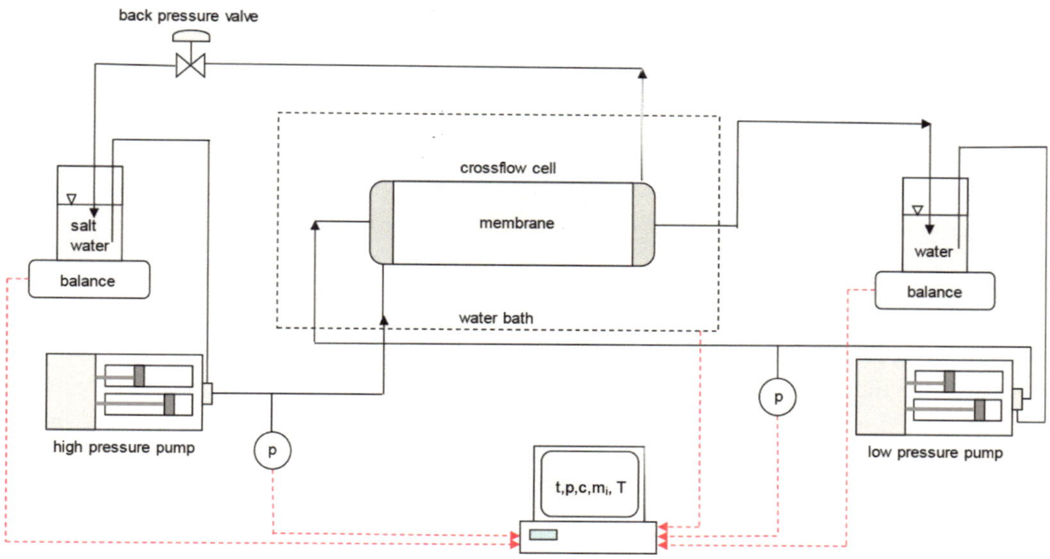

Figure 5. Simplified flow diagram for the two cross flow apparatuses used in the study.

temperature in the water bath, T, and the readings of the balances, m_i, were monitored and logged at regular intervals, t. An inline conductivity cell enabling the determination of the salt concentration in the fresh water, c, was not used in the present experiments.

3.2. Membranes and feed water spacers

The membranes used in this study include one CTA membrane and two TFC membranes (TFC1 and TFC2) from Hydration Technology Inc. and one TFC membrane (TFC3) from Nitto Denko. It should be noted that TFC1 and TFC2 are the first and second generation of the same membrane.

A relatively coarse tricot spacer with 0.5 mm thickness has been used as feed water spacer in our standard test protocols. In addition, some experiments were performed with a finer tricot spacer with 0.25 mm thickness. Photos of both types of spacers are shown in **Figure 4**.

3.3. Test protocol

The salt water solutions were made by dissolving NaCl (p.a.) in degassed (vacuum) and purified water. Degassed and purified water was also used as feed solution in the PRO experiments, and for all pre-treatment and rinsing steps.

If prescribed by the manufacturer, the membranes were pre-treated by immersion in a fluid of composition specified by the membrane manufacturer (often 50 vol. % methanol) for a prescribed time (typically 30 to 300 seconds). Subsequently, the samples were immersed in purified water for minimum 60 minutes prior to assembly in one of the membrane cells. The membranes that were not pre-conditioned were immersed in purified water prior to assembly in one of the membrane cells, in some cases combined with vacuum degassing of the sample.

After assembly, a hydraulic water permeability test was performed. The water flux was measured for minimum four pressure steps, ranging from 1 to 10 bar. Each pressure step lasted for minimum 1 hour. Subsequently, two independent osmotic flow experiments were performed at isobaric conditions. The first experiment was performed in FO mode, *i.e.* draw solution against the membrane support, followed by a second experiment in PRO mode, *i.e.* draw solution against the membrane skin. The cross-flow cell and tubing was flushed with purified water between each experiment.

3.4. Experimental conditions

For the osmotic experiments, the flow velocities (based on open channel) were 1.08 and 0.76 cm/s for the draw channel and the feed channel, respectively, unless stated otherwise. These flow velocities are in the same order as expected flow velocities in a full-scale membrane module for sea water/fresh water PRO. For the hydraulic water permeability experiments, purified water was supplied to both sides of the membrane.

During the osmotic experiments both sides of the membrane were conditioned at ambient pressure by bypassing the back-pressure valve shown in **Figure 5**. During the hydraulic water permeability experiments, and some of the PRO experiments, the back-pressure valve was used to regulate the applied pressure on the draw side. However, most PRO experiments were performed using a closed draw solution loop instead of the back-pressure valve. The closed draw solution loop was continuously pressurised by the volume increase in the draw solution loop.

4. Data analyses and modelling

4.1. Flux and permeability calculations

The water flux was determined based on mass changes in the feed reservoirs. The reported water fluxes were estimated based on the initial phase in each experiment, *i.e.* during the first 1 to 2 hours, before dilution of the draw solution and salt accumulation in the feed solution influenced the mass transport. Hydraulic water permeabilities were calculated from the hydraulic permeability experiments. The salt fluxes were determined by potentiometric analyses of Cl^- ions in a sample collected in the feed reservoir at the end of each experiment. The measured average salt fluxes were corrected to initial conditions using the ratio between initial and average salt concentration differences across the membrane.

4.2. Determination of A, B and S from isobaric osmotic flow experiments

A, *B* and *S* were determined for each membrane by modelling of two isobaric osmotic flow experiments ($\Delta p = 0$). The two experiments, one performed in FO mode and one in PRO mode, produced one water flux and one salt flux each that were used as input to the transport model described in Section 2. Further, *A*, *B* and *S* was determined as the combination of parameters resulting in the minimum sum of squared relative deviations between measured and modelled fluxes. Of the four fluxes that were obtained from the two osmotic flow experiments, three of them are independent, which corresponds to the minimum degrees of freedom required for the parameter estimation. All experiments were modelled by using a boundary layer thickness of 40 μm [26].

4.3. Modelling of PRO experiments

In order to assess the specific power production as function of applied pressure each experiment was divided into pressure steps. The water and salt fluxes, and the salt concentrations on both sides of the membrane, were calculated by mass balances for each pressure step, using the membrane parameters determined for the applied membrane, according to Section 4.2.

4.4. Determination of pressure dependency of the structure parameter

In order to assess the pressure dependency of the structure parameter, S was allowed to increase with pressure. Thus, the modelling procedure described in Section 4.2 was repeated for each pressure step. However, with the distinction that A and B were kept constant and equal to the values determined at isobaric conditions, whereas only the structure parameter was fitted to minimise the sum of squared relative deviations between measured and modelled fluxes.

5. Results and discussion

5.1. Modelling of PRO experiments with constant S

Figure 7 shows the water flux and specific power as function of trans-membrane pressure for two CTA membranes with imbedded reinforcement. The membranes originated from two different production batches. Symbols correspond to experimental data, whereas lines correspond to modelled values which are based on the characteristic membrane parameters determined from the osmotic flow experiments.

It can be observed from **Figure 7** that the measured water fluxes, and thus the specific power, were not very high, which is typical for asymmetric membranes. Furthermore, a significant deviation between measured and modelled performance was observed at increasing trans-membrane pressure.

5.2. Modelling of PRO experiments with pressure dependent S

Figure 8 shows the same experiments as presented in **Figure 7** with the distinction that the modelled values were obtained by using a pressure dependent structure parameter. The pressure dependent structure parameter obtained for the two CTA membranes is plotted as function of trans-membrane pressure in **Figure 9**.

It can be observed that the structure parameter increases significantly with increasing pressure for both membrane samples. Further, the observed variation in the structure parameter with trans-membrane pressure resembles the proposed behaviour given by Eq. (10).

Figure 10 shows the water flux and specific power as function of trans-membrane pressure for two parallel runs with a TFC membrane with imbedded reinforcement, denoted as TFC1. The modelled values were obtained by using a pressure dependent structure parameter. Note that the difference in salt concentration across the membrane skin at maximum specific power was 26.4 and 28.2 g/l for Experiment 1 and Experiment 2, respectively, which explains the observed difference in performance for the two experiments.

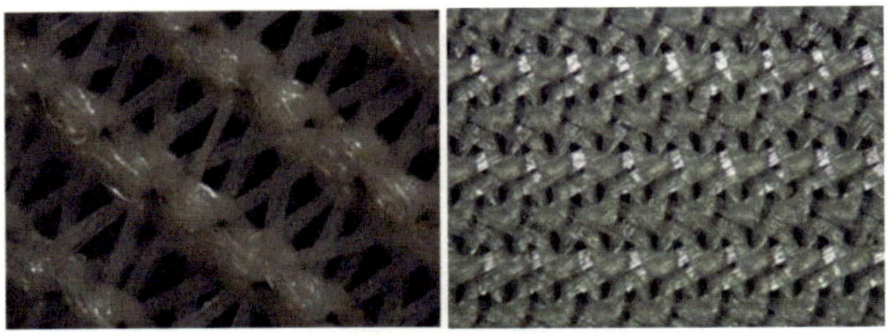

Figure 6. Top view of the tricot spacers used in the feed channel. Left: coarse spacer of 0.5 mm thickness. Right: Fine spacer of 0.25 mm thickness. Photos are shown at the same scale.

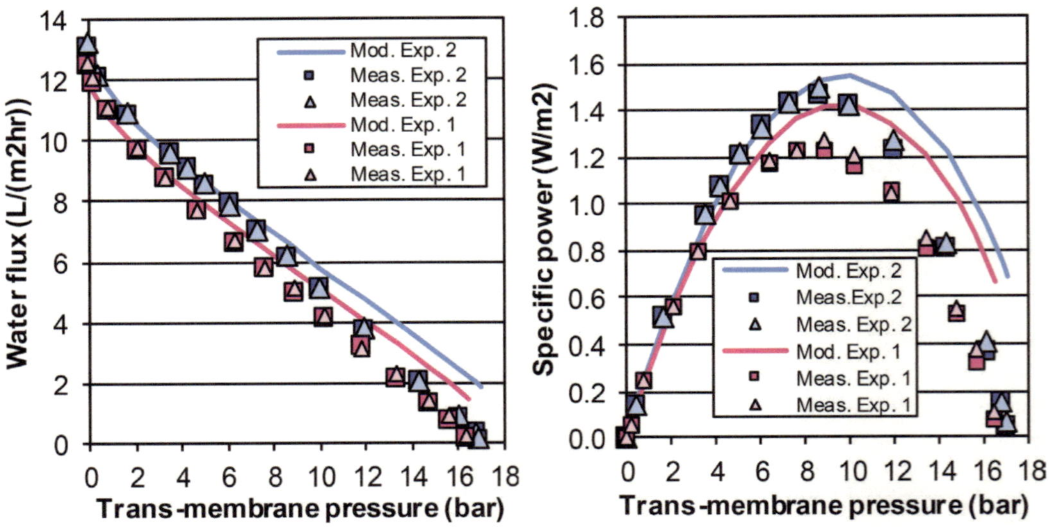

Figure 7. Water flux and specific power as function of trans-membrane pressure for CTA membranes. Modelled values were obtained by applying constant structure parameter determined at isobaric conditions.

Figure 11 shows the pressure dependency of the structure parameter for the two experiments performed with TFC1. Generally, it was observed that the structure parameter of the TFC1 membrane was less affected by pressure than the CTA membrane. *E.g.* at 10 bar the *S* value of the TFC1 membrane was doubled compared to isobaric conditions, whereas the increase in *S* value at 10 bar for the CTA membrane was in the range of 400%.

5.3. Impact of flow velocity on the pressure dependency of S

Table 1 summarises a series of PRO experiments, each performed with different cross-flow velocities and using the membrane denoted TFC2. Experiments 5 and 9 were both performed at standard conditions. The maximum specific power, P_{max}, and the difference in salt concentration across the membrane at maximum specific power, Δc at P_{max}, are given in the table, as well as the pressure found by extrapolation of the water flux vs. the trans-membrane pressure curve to zero water flux, p_{osm}. The latter is commonly referred to as the practical osmotic pressure.

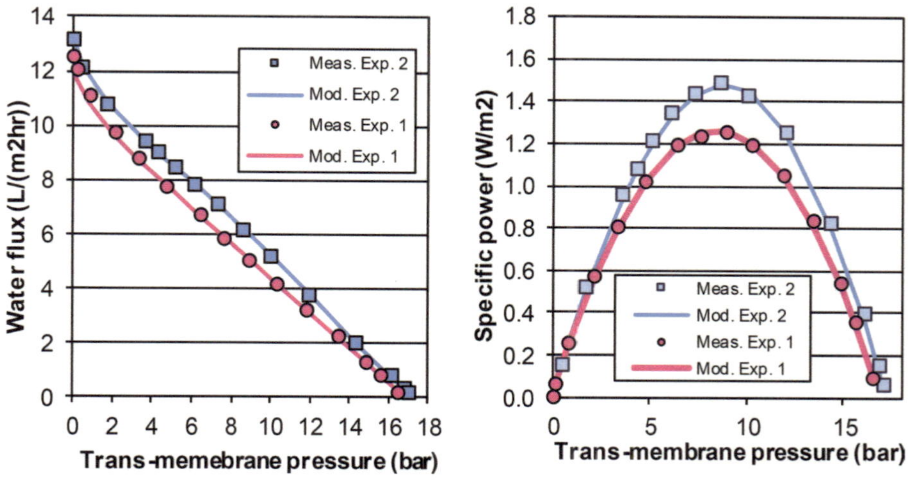

Figure 8. Water flux and specific power as functions of trans-membrane pressure for CTA membranes. Modelled values were obtained by applying a pressure dependent structure parameter.

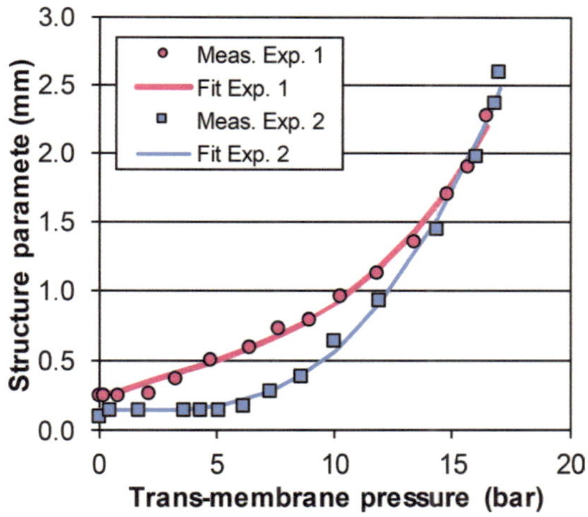

Figure 9. Modelled structure parameter as function of trans-membrane pressure for CTA membranes.

All experiments were modelled according to the procedure described in Section 4.4, and the respective pressure dependent structure parameters are shown in **Figure 12**.

From the reported specific power data in **Table 1** it was observed that the TFC2 membrane (second generation) performed significantly better than the TFC1 membrane (first generation). Further, the results from Experiment 5 and Experiment 9 performed at identical conditions are very similar and indicate good reproducibility.

Comparing the pressure dependency of the structure parameter for the two different TFC membranes in **Figure 12** and **Figure 11**, it was observed that the structure parameter of the TFC2 membrane was less influenced by increasing trans-membrane pressure. Further, the

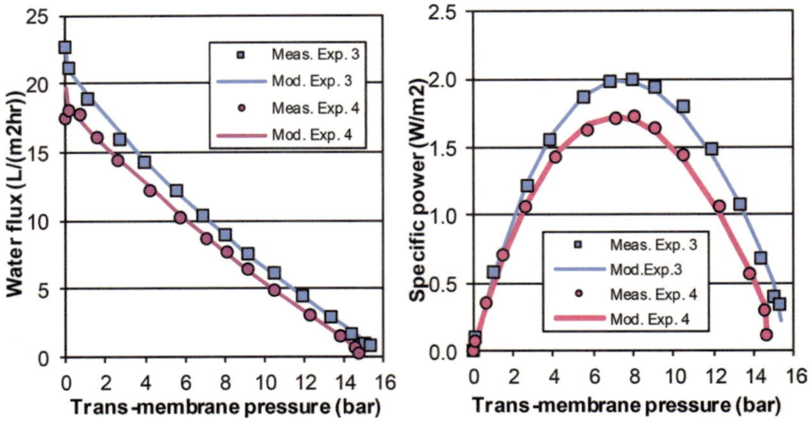

Figure 10. Water flux and specific power as functions of trans-membrane pressure for the TFC1 membrane. Modelled values were obtained by applying a pressure dependent structure parameter.

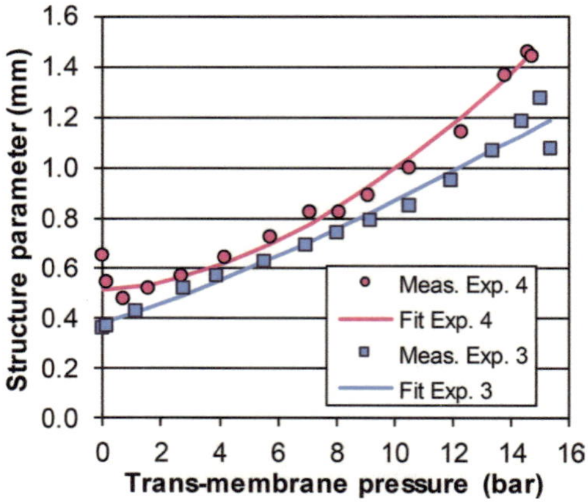

Figure 11. Structure parameters as functions of trans-membrane pressure for the TFC1 membrane.

Exp.	u_{draw} (cm/s)	u_{feed} (cm/s)	P_{max} (W/m²)	Δc at P_{max} (g/L)	p_{osm} (bar)
5	1.08	0.76	3.4	27.2	18.5
6	1.62	1.14	3.8	27.4	18.8
7	2.16	1.52	4.0	27.2	19.5
8	3.25	2.27	4.2	27.4	19.5
9	1.08	0.76	3.4	27.4	18.5

Table 1. Summary of PRO experiments performed with the TFC2 membrane and variable cross-flow velocities.

increase in structure parameter with increasing trans-membrane pressure was observed to have relatively identical slopes for all experiments performed with the TFC2 membrane. Additionally, the structure parameter was observed to decrease at increasing flow velocities.

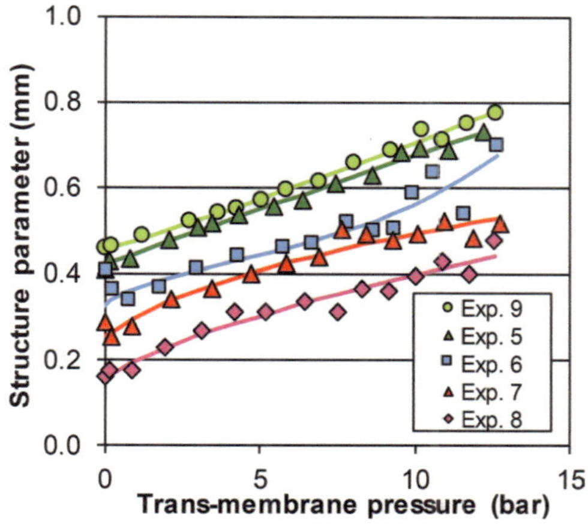

Figure 12. Pressure dependent structure parameters as functions of trans-membrane pressure for the TFC2 membrane.

Figure 13 illustrates the membrane and the feed channel in a cross-flow cell. Since the support membrane is a porous structure some water might be anticipated to flow in the longitudinal direction inside the support membrane as illustrated by the red arrows. The flow velocity inside the support membrane and the penetration depth for the longitudinal flow inside the support will depend on the cross-flow rate, as well as both the flow resistance in the spacer material and in the support membrane, respectively.

For low cross-flow rates and for feed spacers with low pressure drop, the pressure gradient in the feed channel will be small, and little or no water will flow in the longitudinal direction inside the support membrane. At higher cross-flow rates, the pressure gradient in the membrane support will increase, and a significant flow of water inside the support membrane may occur. This will reduce the magnitude of the structure parameter since the effective diffusion length will be reduced when the support structure become more saturated.

Even if high cross-flow velocities may improve mass transfer through the membrane by the effects discussed above, such measure will require increased pumping energy and additionally result in lower utilisation of the feed solution. It should be noted that large pressure losses are unacceptable in sea water/fresh water PRO, and sufficiently low pressure losses are important factors to be considered during development and design of membrane modules for application in PRO plants.

5.4. Impact of spacer selection on the pressure dependency of S

Table 2 summarises results from PRO experiments performed with the TFC3 membrane that was produced without fabric reinforcement.

Two different feed spacers having different thickness and structure were tested. Both spacers were of the tricot type. The feed spacer of 0.25 mm thickness had a much finer structure with smaller distance between the filaments (*cf.* **Figure 6**). Note that experiments 12–15 were performed with the same membrane sample, and between each experiment the membrane cell was opened in order to enable replacement of the feed spacer. Further, a Hirose Histar 15-TH48 (HH 15-TH48) non-woven fabric was placed between the membrane and the feed spacer in

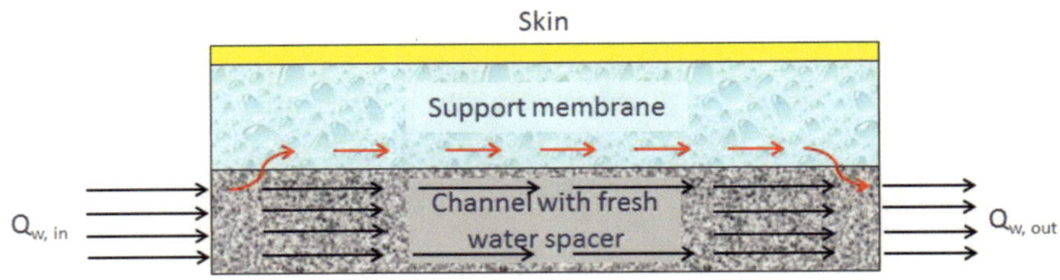

Figure 13. Flow conditions in the fresh water channel and support membrane.

Exp.	P_{max} (W/m²)	Δc at P_{max} (g/L)	p_{osm} (bar)	Feed water spacer	Extra fabric	Comment
10	3.5	28.1	14.7	0.50 mm	None	
11	4.5	27.4	18.0	2·0.25 mm	None	
12	4.6	27.1	19.0	2·0.25 mm	HH 15-TH48	
13	4.9	28.1	21.5	2·0.25 mm	HH 15-TH48	
14	4.3	28.3	20.0	0.50 mm	HH 15-TH48	
15	4.6	28.1	21.7	2·0.25 mm	HH 15-TH48	Spacer inv.

Table 2. Summary of PRO experiments with the TFC3 membrane.

order to assess if improved support to the membrane did influence membrane compaction, and the resulting increase in the structure parameter. Experiments 10 and 11 were performed with different membrane samples. The pressure dependent structure parameter was calculated for each experiment according to the procedure described in Section 4.4 and is shown in **Figure 14**.

The modelled structure parameter in the experiments performed with the 0.5 mm thick spacer was observed to increase more rapidly with increasing pressure compared to the experiments performed with the less coarse spacer of 0.25 mm thickness. Comparing the two experiments performed with the 0.5 mm spacer it was observed that the introduction of the extra non-woven fabric reduced the observed pressure dependency of the structure parameter. This indicates that improving the support for the membrane does influence the compaction of the membrane structure and the resulting increase in the structure parameter at elevated pressures.

The positive impact on the pressure dependency of the structure parameter by including the extra non-woven fabric was also observed for the experiments performed with the 0.25 mm feed spacer. The structure parameter of the fabric was estimated to 0.13 mm by performing independent salt diffusion experiments. The additional transport resistance exerted by the non-woven fabric can be recognised in the modelled structure parameter as the reinforcement layer ideally should add 0.13 mm to the isobaric structure parameter. This increment was not observed in Experiment 12; however, the deviation is within the expected uncertainty found in the pressure dependent structure parameters.

In Experiment 15, the feed spacer was inverted such that the "flat" side was facing the membrane, resulting in a slightly higher structure parameter compared to the experiments performed with normal orientation of the spacer.

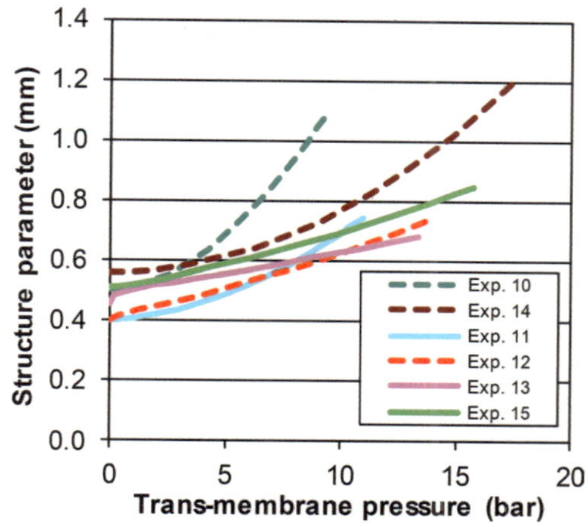

Figure 14. Structure parameter as function of trans-membrane pressure for the TFC3 membrane.

These results show that the extent of support for the membrane is crucial for the PRO performance. It was observed that the coarser spacer resulted in both a faster and a larger increase in the structure parameter at increasing pressure, compared to the more fine-structured spacer providing more support to the membrane. Similar behaviour has been observed in multiple experiments performed with different types of PRO membranes, and agrees well with recent literature [27]. The effect of introducing the extra reinforcement was observed to be larger for the coarser spacer.

5.5. Proposed measures to improve PRO performance at elevated pressures

In order to promote a high specific power in sea water/fresh water PRO, the structure parameter must be low, preferably less than 0.5 mm. The isobaric structure parameters measured for many existing membranes are well below this value. However, when pressurised, an excessive increase in the structure parameter have been observed for many potentially good PRO membranes. An improved strength of the support membrane which is more resistant to compression will therefore be required.

The results in **Figure 14** suggest that one approach to reduce the pressure dependency of the structure parameter might be to apply fine textured feed spacers. However, this will result in increased pressure drop in the feed channel, which might drastically reduce the net produced power in a PRO plant. Even the relatively coarse 0.5 mm feed spacer used in the present work will result in an unacceptable pressure loss. Thus, it should be investigated if it is possible to cast the support membrane directly on a feed spacer, possibly a fine textured tricot type. Supposing that this is viable, two membrane sheets may be separated by *e.g.* a simple diamond type spacer ensuring reasonable low frictional losses.

5.6. Uncertainty in experiments and modelling

The calculation of pressure dependent structure parameters in this paper were based on the assumption that the water and salt permeability were independent of the applied pressure, which may appear to be somewhat contradictory to part of the literature [18, 19]. Nevertheless, our assumption is based on several arguments. (1) In the presented work, the

Exp.	A_{leak}/A (%)	Exp.	A_{leak}/A (%)	Exp.	A_{leak}/A (%)
1	4.9	6	0.1	11	−3.1
2	4.2	7	−0.5	12	0.1
3	5.7	8	−0.6	13	−0.3
4	2.8	9	−1.0	14	0.1
5	0.2	10	−2.5	15	−0.1

Table 3. Hydraulic leakage relative to water permeability.

PRO experiments were performed with a tricot type feed spacer that has been found to result in the lowest variation in the modelled membrane parameters due to variations in trans-membrane pressures [19], and (2) initial water permeability tests using pressures up to 10 bar were performed prior to all PRO experiments. Thus, any membrane deformation that could be expected to influence the skin properties of the membrane as a result of pressurisation should have occurred during the water permeability tests. And (3) the obtained water permeability that were calculated at several pressure steps for each membrane were found to be independent of trans-membrane pressure.

At the end of each PRO experiment the amount of salt on the feed side was determined and compared with the amount of salt calculated by using the transport model. If an excess of salt was determined this was attributed to a hydraulic leakage, and subsequently a leakage volume was calculated by assuming zero salt rejection for the leakage. A leakage permeability, A_{leak}, was calculated based on the leakage volume, duration of the experiment, and average pressure during the experiment. The leakage permeabilities determined for the various experiments are given in **Table 3**. The total leakage volume was distributed for each pressure step based on the duration and average pressure of the step. The salt concentrations on each side of the membrane were subsequently recalculated resulting in new water and salt fluxes, and finally an updated value of excess salt was determined. The calculations converged quickly, and the excess salt was normally low, indicating no (negative leakage volume and thus negative ratio) or only minor leakages.

6. Conclusions

The pressure dependency of the structure parameter in PRO has been investigated for flat sheet membranes, and a transport model including procedures for determination of the pressure dependency of the structure parameter have been presented.

The results from laboratory experiments show that the structure parameter increases significantly with increasing trans-membrane pressure. This was the case both for the CTA membrane and the three TFC membranes that were tested, however, the impact of pressure on the structure parameter was found to be larger for the CTA membrane. Furthermore, the increase in the structure parameter was observed to depend on the type of feed spacer. Using a finely textured spacer of the tricot type reduced the impact of pressure on the structure parameter in comparison to a coarser spacer material. Applying a non-woven backing material between the membrane and the fresh water spacer was also observed to reduce the impact of pressure on the structure parameter. These results show that developing membranes with sufficiently low structure parameter for pressures relevant for PRO will rely on the membrane's ability to resist deformation during compression. The type of feed spacer is another factor which is crucial to avoid deformation and the resulting increase in the structure parameter at elevated pressures.

The results also showed that increased flow velocities in the feed channel and the draw channel, respectively, will improve the mass transfer of water through the membrane. This might be partly ascribed to reduced concentration polarisation on the membrane surfaces. It is also suggested that high pressure gradients in the feed channel may result in convective flow in parts of the support membrane, improving the mass transfer conditions further. However, large frictional losses in the flow channels, will drastically reduce the net produced power in a sea water/ fresh water PRO plant, and must be avoided. This will limit the choice of feed spacers that can be used for PRO.

Nomenclatures

A	water permeability (m/s/Pa)
A_{leak}	hydraulic leakage permeability (m/s/Pa)
B	salt permeability (m/s)
c_f	bulk concentration at the fresh water side (g/l)
c_{fm}	surface concentration at the fresh water side (g/l)
c_p	concentration at the interface between the skin and the porous support (g/l)
c_s	bulk concentration at the salt water side (g/l)
c_{sm}	surface concentration at the salt water side (g/l)
Δc_{skin}	concentration difference across the membrane skin ($= c_{sm} - c_p$) (g/l)
D	diffusion coefficient (m^2/s)
d_s	salt water film thickness (m)
d_f	freshwater film thickness (m)
F	constant in Eq. (10)
i	corrected van't Hoff coefficient (–)

J_V	volume flux (m/s)
J_w	water flux (m/s)
J_s	salt flux (mol/m²/s)
P	specific power (W/m²)
P_{max}	maximum specific power (W/m²)
p_{osm}	practical osmotic pressure (bar)
Δp	trans-membrane pressure (bar)
Δp_{max}	maximum trans-membrane pressure (bar)
Δp_{ref}	reference pressure in Eq. (10) (bar)
$Q_{water,in}$	volumetric flow of water entering a module (–)
$Q_{water,out}$	volumetric flow of water exiting a module (–)
R	universal gas constant (J/K/mol)
S	structure parameter (m)
S_0	isobaric structure parameter (m)
T	absolute temperature (K)
u_{draw}	empty channel velocity at draw side (cm/s)
u_{feed}	empty channel velocity at feed side (cm/s)
x	direction perpendicular to the membrane surface (m)
Δx_{mem}	membrane thickness (m)

Greek letters

τ	tortuosity (–)
φ	porosity (–)
$\Delta \pi_{skin}$	osmotic pressure difference across the membrane skin (bar)

Author details

Torleif Holt[1*], Edvard Sivertsen[2], Willy R. Thelin[2] and Geir Brekke[3]

*Address all correspondence to: Torleif.Holt@sintef.no

1 SINTEF Petroleum, Trondheim, Norway

2 SINTEF Building and Infrastructure, Trondheim, Norway

3 Statkraft AS, Oslo, Norway

References

[1] Skilhagen SE, Dugstad JE, Aaberg RJ. Osmotic power—Power production based on the osmotic pressure difference between waters with varying salt gradients. Desalination. 2008;**220**:476-482. DOI: 10.1016/j.desal.2007.02.045

[2] Logan BE, Elimelech M. Membrane-based processes for sustainable power generation using water. Nature. 2012;**488**:313-319. DOI: 10.1038/nature11477

[3] Touati K, Tadeo F, Elfil H. Osmotic energy recovery from reverse osmosis using two-stage pressure retarded osmosis. Energy. 2017;**132**:213-224. DOI: 10.1016/j.energy.2017.05.050

[4] Bajraktari N, Helix-Nielsen C, Madsen HT. Pressure retarded osmosis from hypersaline sources—A review. Desalination. 2017;**413**:65-85. DOI: 10.1016/j.desal.2017.02.017

[5] Attarde D, Jain M, Singh PK, Gupta SK. Energy-efficient seawater desalination and wastewater treatment using osmotically driven membrane processes. Desalination. 2017; **413**:86-100. DOI: 10.1016/j.desal.2017.03.010

[6] Sivertsen E, Holt T, Thelin WR, Brekke GM. Iso-watt diagrams for evaluation of membrane performance in pressure retarded osmosis. Journal of Membrane Science. 2015;**489**:299-307. DOI: 10.1016/j.memsci.2015.04.042

[7] Chou S, Wang R, Fane AG. Robust and high performance hollow fiber membranes for energy harvesting from salinity gradients by pressure retarded osmosis. Journal of Membrane Science. 2013;**448**:44-54. DOI: 10.1016/j.memsci.2013.07.063

[8] Yip YN, Tiraferri A, Phillip WA, Schiffman JD, Hoover LA, Kim YC, Elimelech M. Thin-film composite pressure retarded osmosis membranes for sustainable power generation from salinity gradients. Environmental Science & Technology. 2011;**45**:4360-4369. DOI: 10.1021/es104325z

[9] Han G, Zhang S, Li X, Chung TS. High performance thin film composite pressure retarded osmosis (PRO) membranes for renewable salinity-gradient energy generation. Journal of Membrane Science. 2013;**440**:108-121. DOI: 10.1016/j.memsci.2013.04.001

[10] Han G, Wang P, Chung TS. Highly robust thin-film composite pressure retarded osmosis (PRO) hollow fiber membranes with high power densities for renewable salinity-gradient energy generation. Environmental Science and Technology. 2013;**47**:8070-8077. DOI: 10.1021/es4013917

[11] Zhang S, Sukitopaneenit P, Chung TS. Design of robust hollow fibre membranes with high power density for osmotic energy production. Chemical Engineering Journal. 2014; **241**:457-465. DOI: 10.1016/j.cej.2013.10.063

[12] Kim YC, Elimelech M. Potential of osmotic power generation by pressure retarded osmosis using seawater as feed solution. Analysis and experiments. Journal of Membrane Science. 2013;**429**:330-337. DOI: 10.1016/j.memsci.2012.11.039

[13] Sivertsen E, Holt T, Thelin WR, Brekke GM. Pressure retarded osmosis efficiency for different hollow fibre membrane module flow configurations. Desalination. 2013;**312**:107-123. DOI: 10.1016/j.desal.2012.11.019

[14] She QH, Jin X, Tang CY. Osmotic power production from salinity gradient resource by pressure retarded osmosis: Effects of operating conditions and reverse solute diffusion. Journal of Membrane Science. 2012;**401**:262-273. DOI: 10.1016/j.memsci.2012.02.014

[15] Thorsen T, Holt T. The potential for power production from salinity gradients by pressure retarded osmosis. Journal of Membrane Science. 2009;**335**:103-110. DOI: 10.1016/j.memsci.2009.03.003

[16] Yip NY, Elimelech M. Thermodynamic and energy efficiency analysis of power generation from natural salinity gradients by pressure retarded osmosis. Environmental Science & Technology. 2012;**46**:5230-5239. DOI: 10.1021/es300060m

[17] Tiraferri A, Yip NY, Phillip WA, Schiffman JD, Elimelech M. Relating performance of thin-film composite forward osmosis membranes to support layer formation and structure. Journal of Membrane Science. 2011;**367**:340-352. DOI: 10.1016/j.memsci.2010.11.014

[18] Kim YC, Elimelech M. Adverse impact of Feed Channel spacers on the performance of pressure retarded osmosis. Environmental Science & Technology. 2012;**46**:4673-4681. DOI: 10.1021/es3002597

[19] She Q, Hou D, Liu J, Tan KH, Tang CY. Effect of feed spacer induced membrane deformation on the performance of pressure retarded osmosis (PRO): Implications for PRO process operation. Journal of Membrane Science. 2013;**445**:170-182. DOI: 10.1016/j.memsci.2013.05.061

[20] McCutcheon JR, Elimelech M. Modeling water flux in forward osmosis: Implications for improved membrane design. AICHE Journal. 2007;**53**:1736-1744. DOI: 10.1002/aic.11197

[21] McCutcheon JR, Elimelech M. Influence of concentrative and dilutive internal concentration polarization on flux behavior in forward osmosis. Journal of Membrane Science. 2006;**284**:237-247. DOI: 10.1016/j.memsci.2006.07.049

[22] McCutcheon JR, Elimelech M. Forward (direct) osmosis desalination using polymeric membranes. Abstracts of Papers of the American Chemical Society. 2004;**228**:261

[23] Tang CY, She QH, Lay WCL, Wang R, Fane AG. Coupled effects of internal concentration polarization and fouling on flux behavior of forward osmosis membranes during humic acid filtration. Journal of Membrane Science. 2010;**354**:123-133. DOI: 10.1016/j.memsci.2010.02.059

[24] Tang WL, Ng HY. Concentration of brine by forward osmosis: Performance and influence of membrane structure. Desalination. 2008;**224**:143-153. DOI: 10.1016/j.desal.2007.04.085

[25] Dytnerskij JI. Membranprozesse—Theorie und Berechnung. Moskau: Verlag Chimija; 1986. p. 254

[26] Sivertsen E, Holt T, Thelin WR, Brekke GM. Modelling mass transport in hollow fibre membranes used for pressure retarded osmosis. Journal of Membrane Science. 2012;**417-418**: 69-79. DOI: 10.1016/j.memsci.2012.06.014

[27] Sim J, Koo J, Nam S, Kim E, Hwang TM. Effect of combined positions of feed spacer-type tricot on the performance in pressure retarded osmosis (PRO). Desalination and Water Treatment. 2017;**77**:63-68. DOI: 10.5004/dwt.2017.20667

Chapter 7

Nonideal Solution Behavior in Forward Osmosis Processes Using Magnetic Nanoparticles

Jimmy D. Roach, Mandy M. Bondaruk and
Zain Burney

Additional information is available at the end of the chapter

http://dx.doi.org/10.5772/intechopen.72474

Abstract

Despite the tremendous progress made toward the realization of wider application for forward osmosis (FO) technologies, lack of suitable draw solutes that provide high water flux, low reverse solute flux, and facile recovery has hindered commercial development. An extensive variety of osmotic agents have been investigated during the past decade, and while simple inorganic salts remain the most widely used, organic-coated magnetic nanoparticles (MNPs) offer exploitable properties that hold great promise. In addition to size-mitigated reverse flux and low-cost recovery via magnetic separation, devitalized MNPs provide enhanced osmotic performance when compared to that of the ungrafted coating material at similar concentration levels, a consequence of greater nonideal solution behavior. This nonideality has been assessed using a simple, semiempirical model and is largely attributable to the increased solvent-accessible surface area and enhanced hydration. When attached to MNPs, polymers appear to behave osmotically as much smaller molecules, providing higher osmotic pressures and improved FO performance.

Keywords: forward osmosis, nonideality, draw solute, magnetic nanoparticles, counterion binding

1. Introduction

Forward osmosis (FO) exploits the natural osmotic pressure gradient between two fluids separated by a semi-permeable membrane to induce the net transport of solvent from a solution of lower osmotic pressure to that of higher osmotic pressure. The FO process appears to provide a low-energy, low-cost alternative to more conventional membrane-based separation methods and offers a myriad of potential applications in industries as diverse as desalination, oil and gas, and food processing [1, 2]. Despite advances made in FO during the past decade, several challenges must still be overcome before more widespread relevance of the technology can be realized [3]. Recently, Shaffer et al. [4] provided a thermodynamic argument showing that FO-reverse osmosis (RO) desalination schemes cannot provide energy savings when compared to standalone RO. Although FO technology has been applied to a variety of water treatment strategies, draw solute inadequacies restrict its wider application [5, 6]. Mitigation of these inadequacies requires identification of draw solutions that achieve high osmotic pressure while minimizing reverse solute flux and also providing ease of recovery; the need for osmotic agents that allow for facile, inexpensive recovery remains paramount [7].

During the past decade, researchers have primarily focused their efforts in two areas, FO membrane production and draw solute identification. While considerable progress has been made toward the development of inexpensive and more robust membranes [8, 9], few commercially viable osmotic agents have been identified [10]. Desirable properties of the ideal osmotic agent are that it be nontoxic, inexpensive, stable, and highly water-soluble. In addition, the agent should have limited reverse draw solute flux, reduce internal concentration polarization (ICP), and be easily recoverable. Some osmotic agents and recovery schemes investigated to date include using inorganic salts with recovery by RO [11]; using poly(sodium acrylate) with recovery by ultrafiltration (UF) [12]; using thermoresponsive chitosan derivatives with recovery by aggregation at elevated temperature [13]; using ammonia-carbon dioxide with recovery by thermal separation [14]; using poly(N-isopropylacrylamide-co-acrylic acid) with recovery by heating and centrifugation [15]; using surfactants with recovery by UF [16]; and, using polyelectrolyte-based hydrogels with recovery by elevated temperature and pressure [17]. A critical review of what the authors term non-responsive and responsive draw solutes was recently provided by Cai and Hu [7].

Because they meet several of the aforementioned criteria, low reverse draw flux and easy recovery in particular, functionalized magnetic nanoparticles (MNPs) have garnered much attention as potential osmotic agents [18]. These MNPs typically incorporate a superparamagnetic core of Fe_3O_4, with a magnetization value of 75.0 emu g^{-1} [19], onto which organic content is coated. Among the grafting agents that have been affixed to MNPs and investigated in FO processes are 2-pyrrolidine, triethylene glycol, and poly(acrylic acid) [20]; dextran [21]; poly(ethylene glycol) diacid [22]; poly(sodium acrylate) [23–25]; poly(sodium styrene-4-sulfonate) and poly(N–isopropylacrylamide) [26]; citrate [27]; hyperbranched polyglycerol [28]; and, citric acid and oxalic acid [19]. A primary advantage of using MNPs is their ease of recyclability through magnetic separation, although particle aggregation has been shown to diminish FO water flux values after multiple regeneration cycles [10]. Another benefit of derivatized MNPs is that they have been shown to provide higher osmotic pressures when compared to solutions of the organic grafting agents alone [20], an enhancement attributable to increased solution nonideality.

A solution behaves ideally when: (1) solute/solute, solvent/solvent, and solvent/solute interactions are identical and (2) all solute and solvent molecules occupy the same volume. Real solutions deviate from ideality due to an energetic nonequivalence in one or more of these interactions and/or volume occupancies are not identical. In aqueous solution, water molecules exhibit particularly strong hydrogen bonding with various organic functional groups, carboxylate moieties in particular [29]. Factors such as hydration, ion-pairing, and dimerization can be significant contributors to thermodynamic nonideality [30] and can dramatically impact the osmotic performance of FO draw solutions.

A variety of models have been developed to explain the interesting osmotic behavior of concentrated solutions of proteins and other biological molecules [31–34]. The nonideal solution behavior of large biological molecules can lead to extreme changes in osmotic pressure. As an example, at a fixed protein concentration, the osmotic pressures of bovine serum albumin (BSA) solutions display greater than fivefold changes in the range $3 < pH < 8$ [32]. Such nonideality is generally attributable to variations in solvent-accessible surface area and polymeric segmental motion [35]. Models that adequately describe nonideal behavior in BSA and other polymer solutions provide a basis for explaining the unique osmotic properties of MNPs used in FO.

2. Osmotic theory

In order to function effectively as a draw agent in FO, the osmotic pressure of the draw solution must far exceed that of the feed solution. In terms of desalination, the draw must have an osmotic pressure significantly in excess of 7.7 atm in the case of a brackish feed, and in excess of 27 atm in the case of a seawater feed [4]. Because of their abilities to achieve high osmotic pressures while maintaining low solution viscosities, simple inorganic salts remain the most widely used draw agents. In addition, small ions tend to have greater diffusivity values thus moderating the effect of concentrative ICP. The strong affinity of small inorganic ions for water is revealed in their highly exothermic enthalpies of hydration [36]. This strong affiliation serves to significantly lower the chemical potential of water in draw solutions. Strong solvent/solute interactions provide high solution osmotic pressures while paradoxically making the regeneration of draw solute more difficult. Resolving this paradox has spurn interest in the development of easily removable draw agents that allow for regeneration through exploitation of solute size, thermal sensitivity, or magnetic properties. Of course, to be effective in FO processes these solutes must still provide appreciable osmotic pressure. Interestingly, structural features of various macromolecular species and molecular aggregates that allow for easy removal from aqueous solution can also serve to enhance osmotic pressure through nonideal solvent/solute interactions.

2.1. Osmotic pressure and FO water flux

The effects of osmotic pressure, solution viscosity, and molecular/ionic diffusivity on water flux (J_w) are shown in Eq. (1),

$$J_w = \frac{D\varepsilon}{t\tau} \ln \frac{B + A\pi_{D,m} - J_w}{B + A\pi_{F,b}} \tag{1}$$

where D is the diffusion coefficient of the solute (which decreases with solution viscosity); ε, t, and τ are the porosity, thickness, and tortuosity of the membrane support layer, respectively; B is the salt permeability coefficient of the membrane active layer; A is the pure water permeability coefficient; $\pi_{D,m}$ is the osmotic pressure of the draw solution at the membrane surface; and, $\pi_{F,b}$ is the osmotic pressure of the feed solution in the bulk [37]. Water flux increases with increasing osmotic pressure difference ($\pi_{D,m} - \pi_{F,b}$), however the relationship is nonlinear because of ICP. As Eq. (1) demonstrates, draw solution osmotic pressure is the principal driving force in FO processes.

2.2. Thermodynamic basis of osmotic pressure

Consider an FO process using a polymer solution as the osmotic agent. If a polymer solution is separated from pure water by a semipermeable membrane the movement of water through the barrier is explained in terms of the chemical potential of the water, μ_w, under isothermal conditions, as given in Eq. (2),

$$\mu_w(P, X) = \mu_w^o(P, X^o) + RTln(\alpha_w) \tag{2}$$

where P is pressure, X is solution composition, R is the gas constant, T is temperature, α_w is the activity of water in the solution, and the superscript "o" denotes standard conditions. For the derivation that follows α_w will be replaced with the mole fraction of water in solution, X_w. In **Figure 1**, water spontaneously moves from the left side to the right side because $\mu_{w,left} > \mu_{w,right}$. Alternatively, it is possible to prevent net water flow by increasing the external pressure on the polymer solution such that $\mu_{w,left} = \mu_{w,right}$. The amount by which the external pressure is increased to prevent net flow is termed the osmotic pressure, π, of the draw solution.

As Eq. (2) implies, it is reasonable to differentiate μ_w in terms of P and X_s (the mole fraction of solute) to obtain Eq. (3).

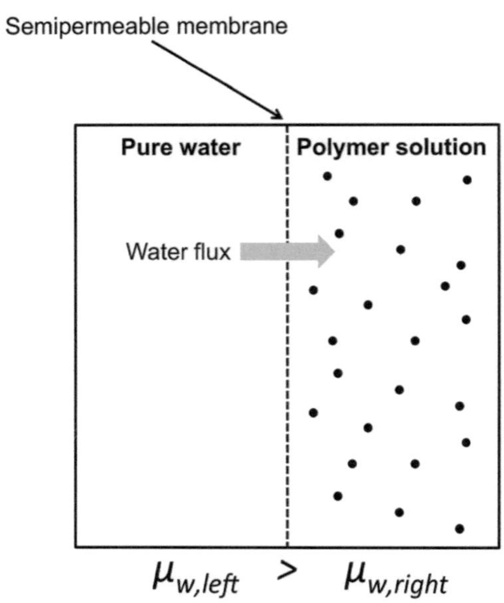

Figure 1. Osmotic behavior of an aqueous polymer solution.

$$d\mu_w = \left(\frac{\partial \mu_w}{\partial P}\right)_{T,X_s} dP + \left(\frac{\partial \mu_w}{\partial X_s}\right)_{T,P} dX_s \tag{3}$$

The definitions of Gibbs free energy and chemical potential are given by Eqs. (4) and (5), respectively,

$$G = H - TS \tag{4}$$

$$\mu_w = \left(\frac{\partial G}{\partial n_w}\right)_{T,P,n_s} \tag{5}$$

where H is enthalpy, S is entropy, n_w is moles of water, and n_s is moles of solute. Application of fundamental thermodynamics to a two-component solution of water and polymer solute, s, provides Eq. (6), in which V is the volume of solution.

$$dG = -SdT + VdP + \left(\frac{\partial G}{\partial n_w}\right)_{T,P,n_s} dn_w + \left(\frac{\partial G}{\partial n_s}\right)_{T,P,n_w} dn_s \tag{6}$$

Eq. (6) reveals that under conditions of constant temperature and solution composition, the derivative of Gibbs free energy with respect to pressure is given by Eq. (7).

$$\left(\frac{\partial G}{\partial P}\right)_{T,X} = V \tag{7}$$

By differentiating Eq. (5) with respect to pressure, while holding other variables constant, Eq. (8) is obtained.

$$\left(\frac{\partial \mu_w}{\partial P}\right)_{T,X} = \frac{\partial^2 G}{\partial P \partial n_w} \tag{8}$$

Similarly, by differentiating Eq. (7) with respect to amount of water Eq. (9) is obtained, in which V_{m_w} is the partial molar volume of water.

$$\frac{\partial^2 G}{\partial n_w \partial P} = \left(\frac{\partial V}{\partial n_w}\right) = V_{m_w} \tag{9}$$

Because of the symmetry of second derivatives, meaning the order of differentiation is inconsequential, the partial molar volume of water is also given by Eq. (10).

$$V_{m_w} = \left(\frac{\partial \mu_w}{\partial P}\right)_{T,X} \tag{10}$$

Next, differentiation of an analogous form of Eq. (2) with respect to X_w provides Eq. (11).

$$\left(\frac{\partial \mu_w}{\partial X_w}\right)_{T,P} = \frac{RT}{X_w} \tag{11}$$

Because $X_w = 1 - X_s$ and therefore $\frac{dX_w}{dX_s} = -1$, Eq. (12) can be obtained.

$$\left(\frac{\partial \mu_w}{\partial X_s}\right)_{T,P} = \left(\frac{\partial \mu_w}{\partial X_w}\right)_{T,P} \frac{dX_w}{dX_s} = -\frac{RT}{1 - X_s} \tag{12}$$

If there is no net flow of water in an apparatus like that depicted in **Figure 1**, $d\mu_w = 0$ providing Eq. (13).

$$\left(\frac{\partial \mu_w}{\partial P}\right)_{T,X_s} dP = -\left(\frac{\partial \mu_w}{\partial X_s}\right)_{T,P} dX_s \tag{13}$$

Substituting Eqs. (10) and (12) into Eq. (13) and then integrating provides Eq. (14).

$$\int_{P_o}^{P_o+\pi} V_{m_w} dP = RT \int_0^{X_s} \frac{dX_s}{1 - X_s} \tag{14}$$

Assuming the solution is incompressible (meaning that partial molar volume is independent of pressure) allows for simple integration providing Eq. (15).

$$\pi = -\frac{RT}{V_{m_w}} \ln(1 - X_s) = -\frac{RT}{V_{m_w}} \ln(X_w) \tag{15}$$

For dilute solutions ($X_s \ll 1$ and $n_s \ll n_w$) the approximations in Eqs. (16) and (17) are justified,

$$\ln(1 - X_s) \approx -X_s \tag{16}$$

$$X_s = \frac{n_s}{n_s + n_w} \approx \frac{n_s}{n_w} \tag{17}$$

which upon substitution into Eq. (15) provides the familiar van't Hoff equation, Eq. (18).

$$\pi V = n_s RT \tag{18}$$

Deviations of solution osmotic pressure data from Eq. (18) are generally attributable to nonideal solvent-solute and solute-solute interactions. One way of expressing the extent to which a solution deviates from ideality is through the osmotic coefficient, ϕ, which is defined on an amount fraction basis in Eq. (19).

$$\phi = \frac{\mu_w^o - \mu_w}{RT \ln X_w} \tag{19}$$

The osmotic coefficient is analogous to the activity coefficient and can be defined in terms of other concentration units. It is often used in conjunction with i, which accounts for dissociation/ion-pairing, to provide Eq. (20), where C_s is the molar concentration of associated solute.

$$\pi = i\varphi C_s RT \tag{20}$$

Alternatively, and in particularly for polymer solutions, solution osmotic pressure is often expressed as a power series expansion in C_s as in Eq. (21),

$$\pi = RT\left(\frac{C_s}{M_r} + A_2 C_s^2 + A_3 C_s^3 + \dots\right) \tag{21}$$

where M_r is molar mass and A_2 and A_3 are the second and third virial coefficients, respectively. These coefficients are temperature dependent, empirically determined constants for a given solvent system. In terms of the activity of water, α_w, osmotic pressure is perhaps best expressed as shown in Eq. (22).

$$\pi = -\frac{RT}{V_{m_w}}\ln(\alpha_w) \tag{22}$$

An empirical, semi-empirical, or theoretical methodology can then be used to relate α_w in Eq. (22) to X_w in Eq. (15). Given the significance of Eqs. (15) and (22), it is important to discuss the factors that effectively reduce the mole fraction of *free* water through hydration of solute species. The hydration number of a solute, h, influences X_w as shown in Eq. (23).

$$X_w = \frac{n_w - hn_s}{n_w - hn_s + in_s} \tag{23}$$

In terms of solute molality (C_{sm}), a concentration unit often reported in FO studies, the hydration number of a solute, h, can be incorporated as shown in Eq. (24),

$$C_{sm} = \frac{n_s}{M_w - (hn_s \times 0.018015)} \tag{24}$$

where M_w is the total mass of water in the solution in kg. Solutes with greater h values produce solutions with higher osmotic pressures at a given concentration and are potentially better draw agents in FO processes, though viscosity considerations are also very important.

2.3. Osmotic pressure of aqueous solutions of inorganic salts

Wilson and Stewart [38] have provided a good discussion of how solution osmotic pressure is affected by the hydration of simple ionic compounds. The short range interactions between electron pairs in water molecules and cations lead to h values that can range from, for example, 1.8 for NH_4^+ to 13 for Mg^{2+} [39]. To illustrate the influence of hydration, consider the comparison of aqueous solutions of NaCl and KCl as osmotic agents. Achilli et al. [11] determined the concentrations of NaCl and KCl required to achieve a solution osmotic pressure of 44 atm and also the corresponding J_w values for these solutions. **Table 1** provides the results of using Eqs. (15) and (23), with literature values [40] for h and i, to calculate osmotic pressures. The sodium ion's smaller size and corresponding higher charge density impart a larger h value,

Compound	Molarity	h	i	X_w	π (atm)	J_w (m/s)
NaCl	0.869	3.9	1.84	0.968	44	3.38×10^{-6}
KCl	0.943	1.7	1.85	0.968	44	3.74×10^{-6}

Table 1. Osmotic properties of aqueous solutions of NaCl and KCl [11, 40].

allowing NaCl solutions to achieve a given osmotic pressure at a lower concentration than KCl solutions.

In terms of osmotic pressure and corresponding FO performance there are diminishing returns on using ever-higher concentrations of ionic compounds, especially when increased solution viscosity is also considered. While hydration numbers tend to increase with increasing cation charge density, they decrease with increasing concentration, owing in part to increased ion-pairing, effectively reducing i. The hydration of molecular aggregates or macromolecular species and its corresponding effect on solution osmotic pressure has also been extensively studied, especially for systems consisting of poly(ethylene glycol) (PEG), DNA, chondroitin sulfate, and BSA [31–35, 41, 42]. These studies provide valuable insights into FO processes using molecular aggregates or macromolecular species as draw agents, especially those incorporating MNPs.

2.4. Osmotic properties of aqueous solutions of large organic molecules

In their studies of BSA, Kanal et al. [32] observed that osmotic pressure decreases as solution pH increases from 3 to approximately 4.6 and then increases with pH. Increases in osmotic pressure on either side of the minimum are attributed to increased electrostatic repulsive interactions. At pH values below the isoelectric point (pI_{BSA} = 5.4), the protein adopts a net positive charge along its surface. At pH values above pI_{BSA}, it is net negative. Electrostatic repulsion leads to a less compact protein conformation, greater segmental motion, more effective hydration, and higher osmotic pressures. Near the isoelectric point, the net-neutral protein strands adopt a more compact configuration, are less hydrated, and even tend to aggregate due to reduced intermolecular repulsion. The osmotic nonideality of BSA solutions is generally attributable to two sources: (1) large solvent/solute interactions that effectively increase polymer hydration (h) and (2) segmental motion of small portions of the polymer chains that effectively increase the number of particles in solution (i). Similar sources of nonideal behavior were also used to describe the osmotic properties of aqueous solutions of PEG [31, 43, 44].

The hydration of PEG of molecular weight 2000 Da (PEG^{2000}), both unattached and attached to distearoyl phosphoethanolamine liposomes ((DSEP)-PEG^{2000}), was investigated by Tirosh et al. [43]. Using differential scanning calorimetry, PEG^{2000} was found to bind 136 $\pm$ 4 water molecules, while (DSEP)-PEG^{2000} binds 210 $\pm$ 6 water molecules. In terms of hydration number per monomeric unit (approximately 46 units in 2000 Da PEG), these binding values correspond to hydration numbers of 3.0 and 4.6 for PEG^{2000} and (DSEP)-PEG^{2000}, respectively. The increase in water molecule binding is attributed to conformational changes, a coil configuration in PEG^{2000} and a brush configuration in (DSEP)-PEG^{2000}. When grafted to the liposome surface, the close proximity of the polymeric strands causes them to repel each other and to adopt a more extended, easily hydrated, form. Such behavior has been exploited in the development of draw agents that incorporate superparamagnetic magnetite (Fe_3O_4) onto which polymers were grafted [19–28].

3. MNPs as FO draw agents

A summary of some recent applications of derivatized MNPs as draw agents in FO processes is provided in **Table 2**, which includes approximate concentrations of the repeating (monomeric) units used as capping agents on the MNPs. Other researchers have demonstrated that the osmotic properties of aqueous polymer solutions are perhaps best interpreted in terms of monomer concentration [31, 45].

Coating agent	Size (nm)	[Monomer] (M)	J_w (LMH)	π (atm)	Ref.
2-Pyrrolidine	28	0.15	4.6	17	[20]
TREG	24	0.20	5.8	23	
PAA^{1800}	21	1.0	7.6	36	
Dextran	10	11	8.9	N/A	[21]
PEG^{250}-$(COOH)_2$	11.7	0.37	N/A	73	[22]
PEG^{600}-$(COOH)_2$	13.5	0.88	9.1	66	
PEG^{4000}-$(COOH)_2$	17.5	5.9	N/A	55	
PAA^{1800}	5	1.5	11.2	70	[46]
PAA^{1800}	20	N/A	N/A	18	[23]
$PNaAA^{1800}$	20	N/A	2.1	32	
$PCaAA^{1800}$	20	N/A	1.8	27	
PNaSS-PNIPAM	5	2.3	14.9	55.0	[26]
	9	2.5	9.9	40.8	
Citrate	3–8	0.015	16	N/A	[27]
HPG	20.9	2.1	6.7	15	[28]
$PNaAA^{2100}$	9	0.0083	5.3	11.4	[24]
Citric acid	40	0.52	12.7	64	[19]
Oxalic acid	35	0.84	10.3	47	
PNaAA	160	12.4	N/A	19.5	[25]
Si-COOH	12.7	0.046	1.7	6.3	[47]
Si-PEG^{530}	13.6	0.43	2.0	7.6	

Abbreviations: TREG: triethylene glycol; PAA: poly(acrylic acid); PEG-$(COOH)_2$: poly(ethylene glycol) diacid; PNaAA: poly(sodium acrylate); PCaAA: poly(calcium acrylate); PNaSS-PNIPAM: poly(sodium styrene-4-sulfonate) and poly(N-isopropylacrylamide) [15% PNaSS, 85% PNIPAM]; HPG: hyperbranched polyglycerol; Si-COOH: N-(trinethoxysilylpropyl)ethylenediamine triacetic acid; Si-PEG: 2-[methoxy- (polyethyleneoxy)propyl]trimethoxysilane. Superscripts represent the average molecular weights of polymeric stands.

Table 2. Summary of MNP-based draw agents used in FO processes.

3.1. Osmotic behavior of draw agents alone vs. grafted onto MNPs

Some investigators have studied the FO properties of osmotic agents that are both alone in aqueous solution and grafted onto MNPs [20, 24]. Ling et al. [20] compared 2-pyrrolidine, TREG, and PAA as draw solutes. When grafted onto MNPs, 2-pyrrolidine exhibited a near sixfold increase in osmolality when compared to the ungrafted solute. TREG and PAA exhibited approximately threefold and thirtyfold increases in osmolalities, respectively, at similar concentrations when grafted onto MNPs. Dey and Izake [24] found that 3.5 wt.% PNaAA provided a FO-water flux value of 1.72 LMH while only 0.078 wt.% PNaAA grafted onto MNPs provided a flux value of 5.32 LMH. These results indicate that anchoring polymers onto nanoparticles serves to significantly improve their osmotic performance.

The tremendous enhancement to osmotic pressure and water flux values associated with polymeric solutes anchored to MNPs can be attributed to improved hydration of the polymeric strands. The dense packing of polymer chains around MNPs leads to a more extended, brush-like, conformation due to excluded volume interactions [48, 49]. In addition, Ling et al. [20] ascribe a reduced interaction between PAA-MNPs and the FO-membrane surface as also contributing to the improved performance; carboxyl groups interacting with ester moieties on the membrane surface are not interacting with water and thereby reducing its chemical potential.

3.2. A semiempirical model

While h values can serve as a good assessment of changes in solution ideality, simply using Eqs. (15) and (23) to calculate h requires highly precise measurements of amount and osmotic pressure. Such measurements are likely not practical for osmotic systems incorporating macromolecular species or derivatized MNPs in FO. Fortunately, Fullerton et al. [50] proposed using Eq. (25) to model the osmotic behavior of proteins,

$$\frac{M_w}{M_s} = S \times \frac{1}{\pi} + I \tag{25}$$

where M_w is the mass of water, M_s is the mass of solute, and the two fitting parameters, S and I, are assessments of nonideality. The slope is given by Eq. (26),

$$S = \frac{RT\rho}{A_e} \tag{26}$$

where ρ is the density of water at temperature, T, and A_e is the effective osmotic molecular weight. Parameter I is a measure of solvent/solute interactions and is interpreted as varying directly with solvent-accessible surface area. The model and fitting parameters have been shown to adequately explain the solution properties of macromolecular solutes like BSA [32, 35] and PEG [31]. A free-solvent model proposed by Yousef et al. [51] that uses mole fraction as a measure of composition may also prove useful in analyzing nonidealities and has been shown effective particularly at high solute concentrations.

Figure 2 depicts the application of Eq. (25) to data for TREG [20, 31, 52] both alone in solution and grafted to MNPs. The ungrafted TREG molecules display little deviation from ideality,

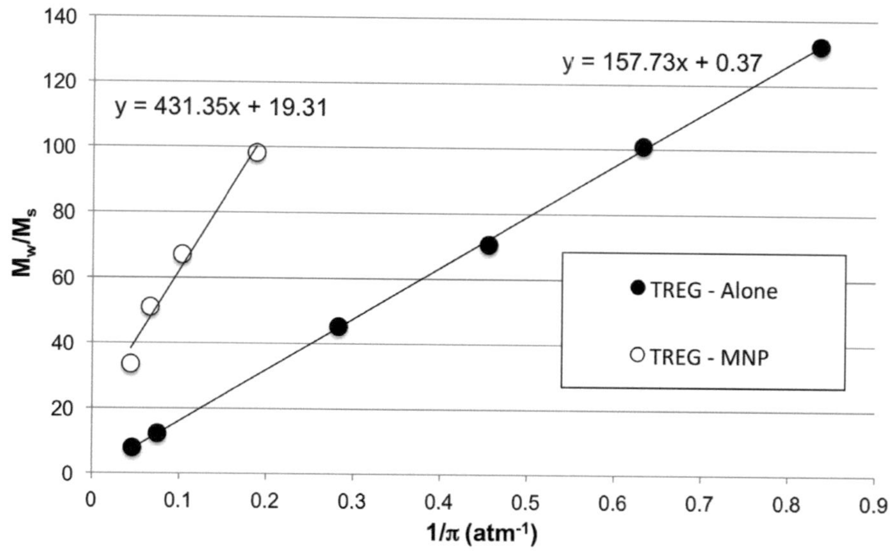

Figure 2. Nonideality analyses for TREG, using data from [20, 31, 52].

with a relatively small I value (0.37) and an effective osmotic molecular weight (153 g mol^{-1}) that is very close to the true molecular weight (150 g mol^{-1}). Though available data is somewhat limited, when grafted, nonideality appears to increase significantly. The value for I (19.3) is quite large when compared to values typically obtained for BSA (~4–12) [35] and for PEG (~1–4) [31], likely resulting from an increase in the amount of water in hydration shells around MNPs when compared to ungrafted TREG. The value for A_e (56.1 g mol^{-1}) is significantly lower than the value for the anchored trimer (149 g mol^{-1}), indicating that the grafted molecule behaves in solution as much smaller molecules.

The application of Eq. (25) to data for which 2-[methoxy-(polyethyleneoxy)$_{6-9}$propyl] trimethoxysilane (MW: 459–591 g mol^{-1}) was used as the grafting agent [47] is provided in **Figure 3**. When compared to TREG data, the greater number of monomers per polymeric strand results in a smaller I value (5.8) and a larger A_e value (101 g mol^{-1}). Although there are differences in particle size and attachment group, these data seem to demonstrate that polymer molar mass affects osmotic performance. Ge et al. [22] found that MNPs coated with PEG^{250}-(COOH)$_2$ provided the best FO performance when compared to similar grafting agents of larger molar mass, observing lower osmotic pressures per monomer concentration as polymer length increased. This difference is perhaps attributable to limited interactions between shorter grafted polymeric strands when compared to longer. Because of the close proximity of individual strands when attached to MNPs, longer strands may be more likely to become intertwined with neighboring strands, thus reducing the surface area available for hydration. Interestingly, the opposite trend has been observed for ungrafted PEGs in the range 200 Da to 10,000 Da, with I values generally increasing with molecular weight before leveling off [31]. Ge et al. [22] also found that MNP-dispersibility increases with polymer length. Optimizing FO performance requires balancing the competing effects of polymer size on dispersibility, osmotic pressure, and viscosity.

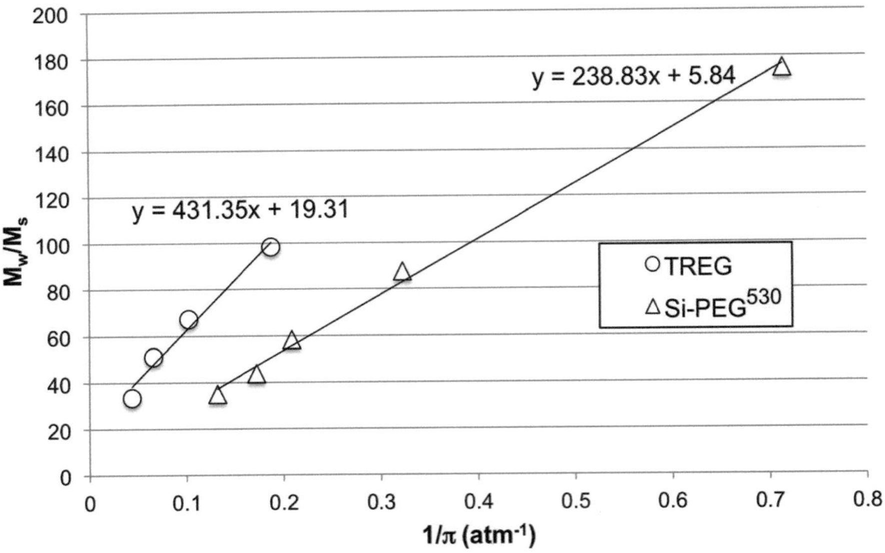

Figure 3. Nonideality analyses for TREG and Si-PEG^{530}, using data from [20, 31, 47, 52].

In **Figure 4**, data for MNPs coated with PAA [20] and HPG [28] are depicted. These results again demonstrate the significant nonideal solution behavior of derivatized MNPs. The large A_e and small I values associated with HPG seem to indicate that the sprawling network of ether linkages may hinder hydration on a per gram of grafting agent basis. By comparison, the long, filamentous PAA^{1800} strands provide an A_e value of 111 g mol^{-1}, which is intermediate between the

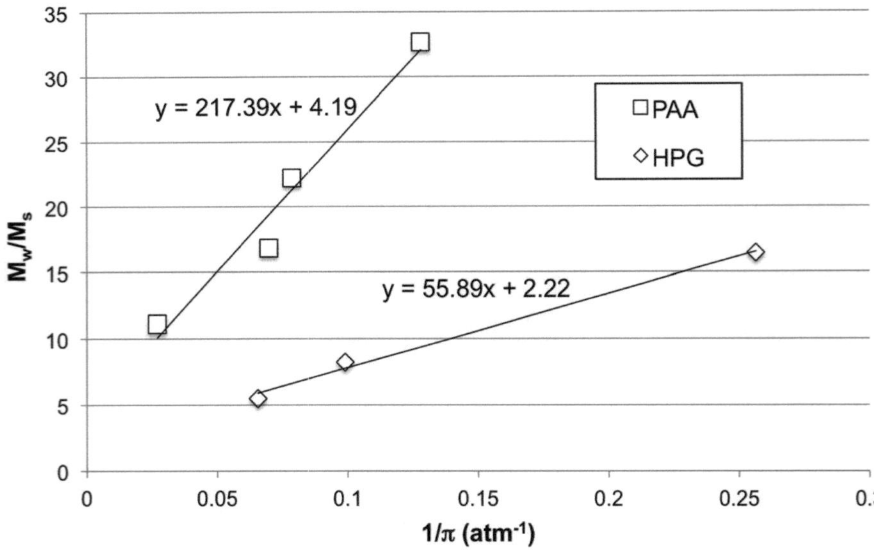

Figure 4. Nonideality analyses for HPG and PAA^{1800}, using data from [20, 28].

Osmotic agent	I	A_e	Ref.
TREG–alone	0.37	153	[20, 31, 52]
TREG–MNP	19.3	56.1	[20]
Si-PEG^{530}–MNP	5.8	101	[47]
PAA^{1800}–MNP	4.2	111	[46]
HPG–MNP	2.2	433	[28]

Table 3. Summary of I and A_e values.

repeating monomer (72 g mol^{-1}) and the full polymer molecular weight (1800 g mol^{-1}). Several researchers [23–25] have also explored PNaAA as an MNP coating agent. Polyelectrolytes exploit greater i values to reduce X_w, however, the extent of ion-pairing between monomer units and counterions greatly influences solution osmotic pressure (**Table 3**).

3.3. Counterion binding

Another significant contributing factor to the osmotic potential of draw solutions incorporating polyelectrolytes is counterion binding. Oosawa was among the first to introduce the concept of counterion condensation around a polyion [53]. His model considers a fraction of counterions that is *bound* to the polyelectrolyte and the remainder is *unbound* in the bulk aqueous phase. Oosawa's expression, provided in Eq.(27), relates the degree of polyelectrolyte dissociation, β; the apparent volume fraction in which counterions are located, ϕ; the absolute value of charge on the counterion, z; and, the intensity of the potential at the polymer surface, Q.

$$\ln\left(\frac{1-\beta}{\beta}\right) = \ln\left(\frac{\phi}{1-\phi}\right) + \beta z Q ln\left(\frac{1}{\phi}\right) \tag{27}$$

Using this model, bound counterions would not contribute to osmotic pressure while unbound ions would. Polymeric structural features that influence the magnitude of Q would therefore significantly impact the osmotic properties of solutions containing that polymer, either alone or grafted onto MNPs. Gwak et al. [54] demonstrated that poly(sodium aspartate) (PNaAsp) provided better osmotic performance than PNaAA, a result attributed to greater polyelectrolyte dissociation (larger β) in the case of PNaAsp. The larger spacing between charged moieties on PNaAsp strands results in a lower surface potential and therefore a higher degree of unbound counterions. Tian et al. [55] investigated the use of ungrafted PNaSS as a draw solute in FO, observing that conductivity and osmotic pressure increase with increasing PNaSS molecular weight, particularly at higher molecular weights. These results indicate that β and Q vary with polymer molecular weight.

3.4. Particle size

Data also indicate that MNP particle size influences their osmotic performance because smaller particles have a larger surface area per volume, thus allowing for more effective grafting-agent coverage and increased nonideality. Ling et al. [20] demonstrated the inverse relationship between nanoparticle size and osmolality using PAA-MNPs. However, Kim et al. [56] found that particles smaller than 11 nm were difficult to separate from solution even with the application of a strong magnetic field, while the removal of particles larger than about 20 nm from the magnetic separator column was problematic. Additionally, the larger the mass percentage of coating material on a Fe_3O_4 core, the lower the saturated magnetization value on a per gram of particle basis. More coating material likely imparts greater osmotic pressure, but it reduces the efficacy of separation. Another significant challenge associated with MNP draw agents is particle aggregation following magnetic separation.

Ge et al. [22] observed a flux decline to approximately 80% of its original value after 9 recycles; this flux decline was accompanied by a particle size increase to 141% of the original value. That study used MNPs with an initial diameter <20 nm. Mino et al. [25] used much larger particles, with diameters of approximately 160 nm, and observed no aggregation even after 10 recycles, though the larger particles achieved only modest osmotic pressures. Park et al. [47] demonstrated that Si-PEG^{530}-MNPs (diameter$_{\text{initial}}$ = 13.6 nm) showed no significant aggregation or FO performance decline after 8 recycles, while Si-COOH-MNPs displayed considerable aggregation after only 5 recycles. Aggregation of the Si-COOH-MNPs was attributed to strong hydrogen bonding between carboxylate groups on adjacent particles when brought into close proximity during magnetic separation and subsequent drying. The oxalic acid- and citric acid-coated MNPs studied by Ge et al. [19] showed no significant particle agglomeration during regeneration, likely the result of strong electrostatic repulsion between particles. Zhao et al. [26] also observed only a slight decline in water flux (<10%) following recycles of their negatively charged PNaSS-PNIPAM-coated particles. In addition, Na et al. [27] demonstrated that small MNPs (3–8 nm) penetrate pores within the FO-membrane support layer (10–40 nm) and become lodged leading to a decline in flux values with time.

4. Summary

While it is now generally accepted that FO processes do not offer an overall energy cost savings when compared to RO for seawater desalination, the prospects of niche applications for FO where RO is unsuitable are numerous. A major challenge for the wider use of FO technology is the development of draw agents that provide high water flux, low reverse solute flux, and facile recovery. Organic-coated superparamagnetic nanoparticles provide properties that address these requirements. The FO performance of MNPs is a function of coating material, particle size, and concentration; with mitigation of particle aggregation during recovery being an essential consideration. The osmotic performance of organic compounds improves significantly when grafted onto MNPs, likely resulting from increased solvent-accessible surface area and enhanced hydration. Application of a simple semiempirical model provides assessments of the nonideality associated with MNPs through calculation of a solvent/solute interaction parameter (I) and the effective osmotic molecular weight (A_e). When attached to MNPs, polymers behave osmotically as much smaller molecules. MNPs derivatized with filamentous, charged molecules (i.e. PNaAA) seem to provide the best results, both in terms of water flux and recoverability. Other significant contributing factors to the overall efficacy of

MNP-based draw solutions are particle size and the extent of counterion binding, with particles in the range 10–20 nm, coated with polyelectrolytes demonstrating high degrees of dissociation, proving most favorable. While the search for the *ideal* draw solute will certainly continue, organic-coated MNPs, because of their enhanced nonideal behavior, offer an encouraging avenue of possibility and opportunity.

Acknowledgements

Support for this work was provided by the Qatar Foundation and was made possible in part by a grant from the Qatar National Research Fund under its Undergraduate Research Experience Program award no. UREP13-018-1-001. Its contents are solely the responsibility of the authors and do not necessarily represent the official views of the Qatar National Research Fund.

Author details

Jimmy D. Roach*, Mandy M. Bondaruk and Zain Burney

*Address all correspondence to: jar2038@qatar-med.cornell.edu

Pre-Medical Education Unit, Weill Cornell Medicine-Qatar, Doha, Qatar

References

[1] Zhao S, Zou L, Tang CY, Mulcahy D. Recent developments in forward osmosis: Opportunities and challenges. Journal of Membrane Science. 2012;**396**:1-21

[2] Klaysom C, Cath TY, Depuydt T, Vankelecom IF. Forward and pressure retarded osmosis: Potential solutions for global challenges in energy and water supply. Chemical Society Reviews. 2013;**42**:6959-6989

[3] Phuntsho S, Sahebi S, Majeed T, Lotfi F, Kim JE, Shon HK. Assessing the major factors affecting the performances of forward osmosis and its implications on the desalination process. Chemical Engineering Journal. 2013;**231**:484-496

[4] Shaffer DL, Werber JR, Jaramillo H, Lin S, Elimelech M. Forward osmosis: Where are we now? Desalination. 2015;**356**:271-284

[5] Akther N, Sodiq A, Giwa A, Daer S, Arafat HA, Hasan SW. Recent advancements in forward osmosis desalination: A review. Chemical Engineering Journal. 2015;**281**:502-522

[6] Qasim M, Darwish NA, Sarp S, Hilal N. Water desalination by forward (direct) osmosis phenomenon: A comprehensive review. Desalination. 2015;**374**:47-69

[7] Cai Y, Hu X. A critical review on draw solutes development for forward osmosis. Desalination. 2016;**391**:16-29

[8] Cath TY, Elimelech M, McCutcheon JR, McGinnis RL, Achilli A, Anastasio D, Brady AR, Childress AE, Farr IV, Hancock NT, Lampi J, Nghiem LD, Xie M, Yip NY. Standard methodology for evaluating membrane performance in osmotically driven membrane processes. Desalination. 2013;**312**:31-38

[9] Wei J, Qiu C, Tang CY, Wang R, Fane AG. Synthesis and characterization of flat-sheet thin film composite forward osmosis membranes. Journal of Membrane Science. 2011;**372**:292-302

[10] Alejo T, Arruebo M, Carcelen V, Monsalvo VM, Sebastian V. Advances in draw solutes for forward osmosis: Hybrid organic-inorganic nanoparticles and conventional solutes. Chemical Engineering Journal. 2017;**309**:738-752

[11] Achilli A, Cath TY, Childress AE. Selection of inorganic-based draw solutions for forward osmosis applications. Journal of Membrane Science. 2010;**364**:233-241

[12] Ge Q, Su J, Amy GL, Chung T-L. Exploration of polyelectrolytes as draw solutes in forward osmosis processes. Water Research. 2012;**46**:1318-1326

[13] Lecaros RLG, Syu Z-C, Chiao Y-H, Wickranasinghe SR, Ji Y-L, An Q-F, Hung W-S, Hu C-C, Lee K-R, Lai J-Y. Characeterization of a thermoresponsive chitosan derivative as a potential draw solute for forward osmosis. Environmental Science & Technology. 2016;**50**:11935-11942

[14] McCutcheon JR, McGinnis RL, Elimelech M. Desalination by ammonia–carbon dioxide forward osmosis: Influence of draw and feed solution concentrations on process performance. Journal of Membrane Science. 2006;**278**:114-123

[15] Wang Y, Yu H, Xie R, Zhao K, Ju X, Wang W, Liu Z, Chu L. An easily recoverable thermosensitive polyelectrolyte as draw agent for forward osmosis process. Chinese Journal of Chemical Engineering. 2016;**24**:86-93

[16] Roach JD, Al-Abdulmalek A, Al-Naama A, Haji M. Use of micellar solutions as draw agents in forward osmosis. Journal of Surfactants and Detergents. 2014;**17**:1241-1248

[17] Li D, Zhang XY, Yao JF, Simon GP, Wang HT. Stimuli-responsive polymer hydrogels as a new class of draw agent for forward osmosis desalination. Chemical Communications. 2011;**47**:1710-1712

[18] Li D, Wang H. Smart draw agents for emerging forward osmosis application. Journal of Materials Chemistry A. 2013;**1**:14049-14060

[19] Ge Q, Yang L, Cai J, Xu W, Chen Q, Liu M. Hydroacid magnetic nanoparticles in forward osmosis for seawater desalination and efficient regeneration via integrated magnetic and membrane separations. Journal of Membrane Science. 2016;**520**:550-559

[20] Ling MM, Wang KY, Chung T-S. Highly water-soluble magnetic nanoparticles as novel draw solutes in forward osmosis for water reuse. Industrial & Engineering Chemistry Research. 2010;**49**:5869-5876

[21] Bai H, Liu Z, Sun DD. Highly water soluble and recovered dextran coated Fe_3O_4 magnetic nanoparticles for brackish water desalination. Separation and Purification Technology. 2011;**81**:392-399

[22] Ge Q, Su J, Chung T-S, Amy G. Hydrophillic supermagnetic nanoparticles: Synthesis, characterization, and performance in forward osmosis. Industrial & Engineering Chemistry Research. 2011;**50**:382-388

[23] Ling MM, Chung T-S. Surface-dissociated nanoparticle draw solutions in forward osmosis and the regeneration in an integrated electric field and nanofiltration system. Industrial & Engineering Chemistry Research. 2012;**51**:15463-15471

[24] Dey P, Izake EL. Magnetic nanoparticles boosting the osmotic efficiency of a polymeric FO draw agent: Effect of polymer conformation. Desalination. 2015;**373**:79-85

[25] Mino Y, Ogawa D, Matsuyama H. Functional magnetic particles providing osmotic pressure as reusable draw solutes in forward osmosis membrane process. Advanced Powder Technology. 2016;**27**:2136-2144

[26] Zhao Q, Chen N, Zhao D, Lu X. Thermoresponsive magnetic nanoparticles for seawater desalination. ACS Applied Materials & Interfaces. 2013;**5**:11453-11461

[27] Na Y, Yang S, Lee S. Evaluation of citrate-coated magnetic nanoparticles as draw solute for forward osmosis. Desalination. 2014;**347**:34-42

[28] Yang H-M, Seo B-K, Lee K-W, Moon J-K. Hyperbranched polyglycerol-coated magnetic nanoparticles as draw solute in forward osmosis. Asian Journal of Chemistry. 2014;**26**: 4031-4034

[29] Beaman DK, Robertson EJ, Richmond GL. From head to tail: Structure, solvation, and hydrogen bonding of carboxylate surfactants at the organic–water interface. Journal of Physical Chemistry C. 2011;**115**:12508-12516

[30] Wills PR, Scott DJ, Winzor DJ. Thermodynamics and thermodynamic nonideality. In: Roberts GK, editor. Encyclopedia of Biophysics. Springer; 2013. pp. 2583-2589

[31] Zimmerman RJ, Chao H, Fullerton GD, Cameron IL. Solute/solvent interaction corrections account for non-ideal freezing point depression. Journal of Biochemical and Biophysical Methods. 1993;**26**:61-70

[32] Kanal KM, Fullerton GD, Cameron IL. A study of the molecular sources of nonideal osmotic pressure of bovine serum albumin solutions as a function of pH. Biophysical Journal. 1994;**66**:153-160

[33] Minton AP. A molecular model for the dependence of the osmotic pressure of bovine serum albumin upon concentration and pH. Biophysical Chemistry. 1995;**57**:65-70

[34] Yousef MA, Datta R, Rodgers VGJ. Understanding nonidealities of the osmotic pressure of concentrated bovine serum albumin. Journal of Colloid and Interface Science. 1998;**207**: 273-282

[35] Zimmerman RJ, Kanal KM, Sanders J, Cameron IL, Fullerton GD. Osmotic pressure method to measure salt induced folding/unfolding of bovine serum albumin. Journal of Biochemical and Biophysical Methods. 1995;**30**:113-131

[36] Yamada S, Tanaka M. Softness of some metal ions. Journal of Inorganic and Nuclear Chemistry. 1975;**37**:587-589

[37] Loeb S, Titelman L, Korngold E, Freiman J. Effect of porous support fabric on osmosis through a Loeb-Sourirajan type asymmetric membrane. Journal of Membrane Science. 1997;**129**:243-249

[38] Wilson AD, Stewart FF. Deriving osmotic pressures of draw solutes used in osmotically driven membrane processes. Journal of Membrane Science. 2013;**431**:205-211

[39] Zavitsas AA. Properties of water solutions of electrolytes and nonelectrolytes. Journal of Physical Chemistry B. 2001;**105**:7805-7817

[40] Marcus Y. Ions in Solution and their Solvation. Wiley; 2015. p. 140

[41] Bathe M, Rutledge GC, Grodzinsky AJ, Tidor B. Osmotic pressure of aqueous chondroitin sulfate solution: A molecular modeling investigation. Biophysical Journal. 2005;**89**:2357-2371

[42] Carrillo J-MY, Dobrynin AV. Salt effect on osmotic pressure of polyelectrolyte solutions: Simulation study. Polymer. 2014;**6**:1897-1913

[43] Tirosh O, Barenholz Y, Katzhendler J, Priev A. Hydration of polyethylene glycol-grafted liposomes. Biophysical Journal. 1998;**74**:1371-1379

[44] Hansen PL, Cohen JA, Podgornik R, Parsegian VA. Osmotic properties of poly(ethylene glycols): Quantitative features of brush and bulk scaling laws. Biophysical Journal. 2003;**84**:350-355

[45] Essafi W, Lafuma F, Baigl D, Williams CE. Anomalous counterion condensation in salt-free hydrophobic polyelectrolyte solutions: Osmotic pressure measurements. Europhysics Letters. 2005;**71**:938-944

[46] Ling MM, Chung TS. Desalination process using super hydrophilic nanoparticles via forward osmosis integrated with ultrafiltration regeneration. Desalination. 2011;**278**:194-202

[47] Park S–Y, Ahn H–W, Chung J–W, Kwak S–Y. Magnetic core-hydrophilic shell nanosphere as stability-enhanced draw solute for forward osmosis (FO) application. Desalination. 2016;**397**:22-29

[48] Achilleos DS, Vamvakaki MV. End-grafted polymer chains onto inorganic nano-objects. Materials. 2010;**3**:1981-2026

[49] Milner ST. Polymer brushes. Science. 1991;**251**:905-914

[50] Fullerton GD, Zimmerman RJ, Cantu C, Cameron IL. New expressions to describe solution nonideality: Osmotic pressure, freezing-point depression and vapor pressure. Journal of. Biochemistry and Cell Biology. 1992;**70**:1325-1331

[51] Yousef MA, Datta R, Rodgers VGJ. Free–solvent model of osmotic pressure revisited: Application to concentrated IgG solution under physiological conditions. Journal of Colloid and Interface Science. 1998;**1997**:108-118

[52] Sun T, Teja A. Density, viscosity, and thermal conductivity of aqueous ethylene, diethylene, and triethylene glycol mixtures between 290 K and 450 K. Journal of Chemical & Engineering Data. 2003;**48**:198-202

[53] Oosawa F. A simple theory of thermodynamic properties of polyelectrolyte solutins. Journal of Polymer Science. 1957;**23**:421-430

[54] Gwak G, Jung B, Han S, Hong S. Evaluation of poly (aspartic acid sodium salt) as a draw solute for forward osmosis. Water Research. 2015;**80**:294-305

[55] Tian E, Hu C, Qin Y, Ren Y, Wang X, Wang X, Xiao P, Yang X. A study of poly (sodium 4–styrenesulfonate) as draw solute in forward osmosis. Desalination. 2015;**360**:130-137

[56] Kim YC, Han S, Hong S. A feasibility study of magnetic separation of magnetic nanoparticle for forward osmosis. Water Science and Technology. 2011;**64**:469-476

Chapter 8

Fouling in Forward Osmosis Membranes: Mechanisms, Control, and Challenges

Amira Abdelrasoul, Huu Doan, Ali Lohi and
Chil-Hung Cheng

Additional information is available at the end of the chapter

http://dx.doi.org/10.5772/intechopen.72644

Abstract

Continuously escalating global water demand places a substantial burden on the available water and energy resources. Forward osmosis (FO) is an evolving membrane desalination technology that has recently raised interest as a promising low-energy process. FO is a unique method since it utilizes natural osmosis as the driving force, and hence, it ensures that the energy consumption is significantly reduced, in comparison to other pressure-driven membrane processes that are constrained by their excessive energy consumption and unsustainable cost. Therefore, the growing interest in FO from various disciplines and industrial sectors calls for a better understanding of the FO process and further advances in the FO technology management. This chapter aims to provide an in-depth assessment of the water transport phenomenon in FO membranes by focusing on the influence of internal concentration polarization, membrane structure/material, and membrane orientation on the permeate flux. This chapter offers critical insight that can lead to the potential development of new FO membranes with reduced internal concentration polarization and higher water permeability. In addition, key strategies for FO membrane development, some of its challenges, and the perspectives for future investigations of FO membrane fouling and effective FO fouling control methods are explored in this chapter.

Keywords: forward osmosis, fouling, concentration polarization, mass transfer, water filtration

1. Introduction

As the fossil fuels are depleted and the world population continues to rapidly increase, energy and water became two of the most vital global resources. Energy emergencies and the lack of water have severely affected communities worldwide [1–3]. Reports indicate that more than 1.2 billion people do not have access to safe and clean drinking water sources, while 2.6 billion do not have adequate levels of sanitation [1, 4, 5]. In fact, the overall annual financial loss in Africa caused by the lack of access to basic sanitation and clean water is valued at $28 billion, or 5% of Africa's gross domestic product [5]. While oceans are covering the majority of the planet's surface, only 0.8% of the world's water can be defined as potable [6]. Moreover, the recent world energy outlook report [2] indicates that the world's marketed energy use is predicted to rise by 49% from 2007 to 2035. Data such as this reflects a dangerous trend, especially since currently 1.5 billion people, or more than 1/5 of the world's population, still do not have access to reliable electricity.

Interdisciplinary research groups need to remain aware of the explicit connection that exists between energy and water. The process of making freshwater accessible is a highly energy-demanding process, while the production of the required power frequently necessitates substantial amounts of water [7, 8]. A relatively new technology, forward osmosis (FO), shows a lot of potential in energy production and water supply, especially for applications in controlled-release–type drug medication, medical product enrichment, and food processing. Over the last decade, FO has incited substantial interest in the areas of seawater/brackish desalination [9–11], food processing [12–15], power generation [16–19], and wastewater treatment [20–22]. In terms of its methodology, FO is an osmotically driven membrane process that relies on the osmotic pressure gradient and that moves water across a semipermeable-type membrane from the feed solution side, with the low osmotic pressure, to the draw solution side, featuring high osmotic pressure. Because of its lower hydraulic pressure demands, FO provides multiple benefits, such as lower fouling tendency, easier fouling removal [20, 22, 23], smaller energy input [24], and greater water recovery [25, 26], if compared to pressure-driven processes such as ultrafiltration (UF), nanofiltration (NF), and reverse osmosis (RO).

2. Advantages of forward osmosis

There are numerous potential benefits offered by FO, especially because of the lower hydraulic pressure values necessary for this osmotically driven–type process. FO's benefits are reflected by its various water treatment applications. First, FO can help obtain smaller energy consumption potentials and as a consequence lower the overall costs and contribute to the production of technically and economically innovative solutes and their respective regeneration methodologies [3, 18, 24]. Arguably, this is one of the key advantages of FO, considering the ongoing global energy crisis. Research studies have shown that membrane fouling in FO is comparatively small [20], somewhat more reversible [23, 27], and may be lowered using hydrodynamics optimization [28]. Furthermore, a number of contaminants may be successfully filtered out with the aid of the FO process [29, 30]. FO can likewise feature greater water recovery and improved water flux because of the higher osmotic pressure gradient occurring across the membrane. Greater water recovery can help reduce the desalination brine volume, especially as it is a substantial environmental concern when it comes to desalination plants and inland desalination facilities [9]. Moreover, in the industries like pharmaceutical and food processing,

FO offers the benefits of preserving the physical properties of the feed, such as color, aroma, nutrition, and taste, without diminishing the overall quality, as it is not heated or pressurized [14, 31, 32]. When it comes to medical uses, FO can help with the release of drugs featuring low oral bioavailability, or poor solubility, in a controlled way and implementing osmotic pumps [33, 34].

3. Modeling of water transport in forward osmosis

The general equation for water flux in forward osmosis (FO), reverse osmosis (RO), or pressure-retarded osmosis (PRO) is [16]:

$$J_w = A(\sigma\Delta\pi - \Delta P) \tag{1}$$

where J_w is the water flux, A is the membrane's water permeability constant, σ is the reflection coefficient, $\Delta\pi$ is the osmotic pressure difference across the membrane, and ΔP is the applied hydraulic pressure variance. In Eq. (1), the term (σ $\Delta\pi$ − ΔP) signifies the effective driving force necessary for the water molecules' transport across the membrane. In the FO desalination process, there is no hydraulic pressure applied and the change in osmotic pressures is the sole driving force; Eq. (1) can be expressed as:

$$J_w = A\sigma\Delta\pi_{Bulk} = A\sigma(\Delta\pi_{Draw} - \Delta\pi_{Feed}) \tag{2}$$

where $\Delta\pi_{Feed}$ stands for the feed solution's bulk osmotic pressure, and $\Delta\pi_{Draw}$ is the draw solution's bulk osmotic pressure. Eq. (2) is restricted by the assumption that the membrane does not permit draw solute permeation [35, 36]. Furthermore, Eq. (2) is applicable for dense symmetric membranes, in which the driving force for water molecules is the difference between the osmotic pressures of the bulk feed and draw solutions, as reflected in **Figure 1**.

If it can be assumed that the difference between the bulk osmotic pressure of the feed and the draw solution is the driving force responsible for water permeation through membranes in FO, then Lee et al. [37] proposed the following model for low water flux cases:

$$J_w = \frac{1}{K} ln\left(\frac{\pi_{Draw}}{\pi_{Feed}}\right) \tag{3}$$

where K stands for the resistance to diffusion of solute within the porous support layer of the FO membrane, and π_{Draw} and π_{Feed} are the respective bulk osmotic pressures of the draw and the feed solution. K can be estimated using Eq. (4) [16]:

$$K = \frac{t\tau}{\varepsilon D} = \frac{S}{D} \tag{4}$$

where t is the membrane's thickness, τ is tortuosity, ε is membrane porosity, S is membrane's structural parameter, and D is the solute's diffusion coefficient.

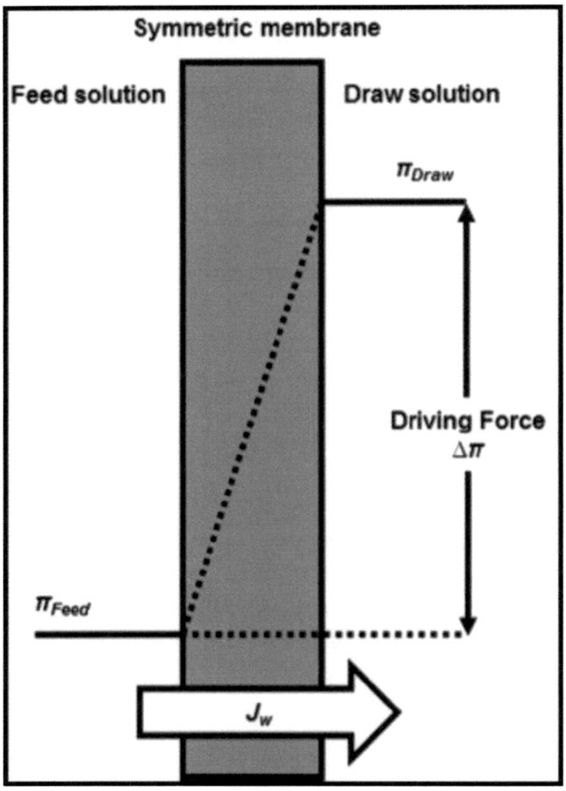

Figure 1. Ideal osmotic pressure driving force in the case of symmetric membranes.

4. Challenges in forward osmosis

FO applications are still facing some critical challenges even though the osmotically driven membrane processes have been extensively researched in relation to a range of applications and environments. FO's primary issues are connected to aspects such as membrane fouling, reverse solute diffusion, further membrane development, concentration polarization, and the improvement of the draw solute design.

4.1. Concentration polarization mechanism in forward osmosis

When it comes to the osmotically driven and pressure-driven membrane processes, the concentration polarization is an inevitable and frequent phenomenon [11, 38–42]. As illustrated in **Figure 2**, in the osmotically driven membrane processes, the concentration polarization is produced by the overall concentration variance occurring between the draw solution and the feed solution through the asymmetric FO membrane. The internal concentration polarization (ICP) and external concentration polarization (ECP) can happen during the FO processes. In general, ICP happens within the membrane's porous support layer, and ECP happens at the surface of the membrane's dense active layer. The sections below further describe both ECP and ICP.

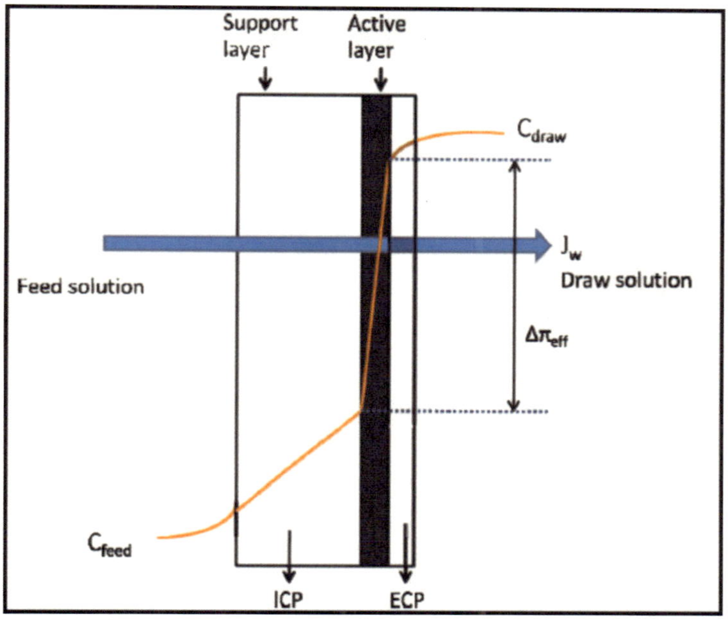

Figure 2. Internal concentration polarization (ICP) and external concentration polarization (ECP) through an asymmetric FO membrane [16].

4.1.1. External concentration polarization: modeling and mechanism

ECP in FO occurs at the surface of the membrane's active layer, similar to the other pressure-driven membrane processes. Their distinction is due to the fact that only concentrative ECP can occur in a pressure-driven membrane process, and both dilutive ECP and concentrative ECP can happen in an osmotically driven membrane process, conditional on the membrane's orientation with regard to the feed and the draw solutions. The dilutive ECP happens when the membrane's support layer is facing the feed solution, while the concentrative ECP occurs in instances where the membrane's support layer is facing the draw solution. ECP lowers the overall driving force due to the higher osmotic pressure at the membrane's active layer interface located on the membrane's feed side, or the lowered osmotic pressure at the membrane's active layer surface located on the draw solution side. The unfavorable effects of ECP on the permeate flux can be alleviated by optimizing the water flux and raising the flow's velocity or turbulence [11]. With the application of the boundary-layer film theory McCutcheon and Elimelech have successfully modeled ECP in FO [38, 43]. The generalized equation for concentration polarization modulus in pressure-driven membrane processes may be expressed in Eq. (5), as follows.

$$\frac{C_m}{C_b} = exp\left(\frac{J_w}{k}\right) \tag{5}$$

where J_w is the water flux, k is the mass transfer coefficient value, and C_m and C_b are the concentrations of the feed solution at the membrane's surface and in the bulk, respectively. The mass transfer coefficient (k) is related to the Sherwood number (Sh) by:

$$k = \frac{ShD}{D_h} \tag{6}$$

where D is the solute diffusion coefficient value, and D_h is the hydraulic characteristic length. When the feed solution concentration is relatively low in FO, the concentrations in Eq. (5) can be substituted by the osmotic pressures. As a result, the concentrative ECP modulus can be expressed as follows:

$$\frac{\pi_{m-feed}}{\pi_{b-feed}} = exp\left(\frac{J_w}{k_{feed}}\right) \tag{7}$$

where k_{feed} is the mass transfer coefficient on the feed side, π_{m-feed} and π_{b-feed} are the osmotic pressures of the feed solution at the membrane's surface and in the bulk, respectively. Similarly, the dilutive ECP modulus in FO can be expressed as:

$$\frac{\pi_{m-draw}}{\pi_{b-draw}} = exp\left(-\frac{J_w}{k_{draw}}\right) \tag{8}$$

where k_{draw} is the mass transfer coefficient on the draw side, and π_{m-draw} and π_{b-draw} are the osmotic pressures of the draw solution at the membrane's surface and in the bulk, respectively. Eqs. (1) and (2) reflect the water transport in RO, FO, and pressure-retarded osmosis (PRO), as indicated in Eqs. (1) and (2), as shown in Section 3. Both π_{draw} and π_{feed} should be the effective osmotic pressures at the membrane's surfaces, specifically.

$$J_w = A(\pi_{m-draw} - \pi_{m-feed}) \tag{9}$$

By substituting Eqs. (7) and (8) into Eq. (9), Eq. (10) can be obtained as below:

$$J_w = A\left[\pi_{b-draw}\ exp\left(-\frac{J_w}{k_{draw}}\right) - \pi_{b-feed}\ exp\left(\frac{J_w}{k_{feed}}\right)\right] \tag{10}$$

Although the dilutive ECP and concentrative ECP have been examined in Eq. (10) [43], there are multiple key points that have to be noted in Eq. (10). First of all, the mass transfer coefficient values on the feed and draw solution sides are not the same because of the varying hydraulic conditions between the draw solution side and the feed side. Next, this model relies on multiple assumptions, including that the solute permeability's coefficient is zero (i.e., the reflection coefficient $\sigma = 1$ [44]) and that the draw and feed solution concentration values are reasonably low, since only in this case can it be accepted that the concentration is equal to the corresponding osmotic pressure values. Finally, it must be noted that this model is adequate only in instances with a dense symmetric film, instead of an asymmetric-type membrane. As a result, the uses for this model can be somewhat limited. It is necessary to examine the dynamic where an asymmetric FO membrane is applied in a manner that would replicate its real-world practical uses and where the ICP effects become more significant.

4.1.2. Internal concentration polarization: modeling and mechanism

ICP is a critical aspect of the osmotically driven membrane-type processes. Research indicates that the water flux decline in FO is primarily produced by ICP [38, 44–46]. The early research projects that looked at FO suggested that ICP might lower the water flux by more than 80% [45, 47]. As indicated in **Figure 3**, there are two types of ICP, concentrative ICP and dilutive ICP, occurring within the membrane's support layer, and they depend on the membrane's orientation [48]. Once the draw solution is situated against the membrane's support layer, dilutive ICP can successfully happen within the membrane's support layer since the water permeates across the membrane, from the feed solution to the draw solution. In a different membrane orientation where the feed solution is opposite the membrane support layer, concentrative ICP happens when the solute properly accumulates within the membrane's support layer located on the feed side. The ICP process is happening in the support layer and, as a result, it cannot be weakened through a change in the hydrodynamic conditions, including higher turbulence or flow rate.

The effects of ICP on FO water flux have been modeled using an adaptation of the classical solution-diffusion theory [38, 43]. The dilutive ICP dominates the water flux (Jw) when the draw solution is placed against the membrane support layer (i.e., FO mode) and can be expressed [49] as follows:

$$J_w = \frac{1}{K} ln \frac{A\pi_{draw} + B}{A\pi_{feed} + B + J_w} \tag{11}$$

where B is the membrane's solute permeability coefficient, and K is the solute resistivity value, a measure of solute transport in the membrane's support layer. K is used to quantify the solute's capacity to diffuse into or out of the membrane's support layer, and it can reflect the degree of ICP available in the support layer. Lower K values indicate less ICP and cause greater pure water flux (J_w). K is defined earlier in Eq. (4). It should be noted that the structural parameter S, in Eq. (4), is an essential membrane quality since it governs ICP in the membrane's support by establishing membrane's tortuosity, porosity, and thickness values. As a

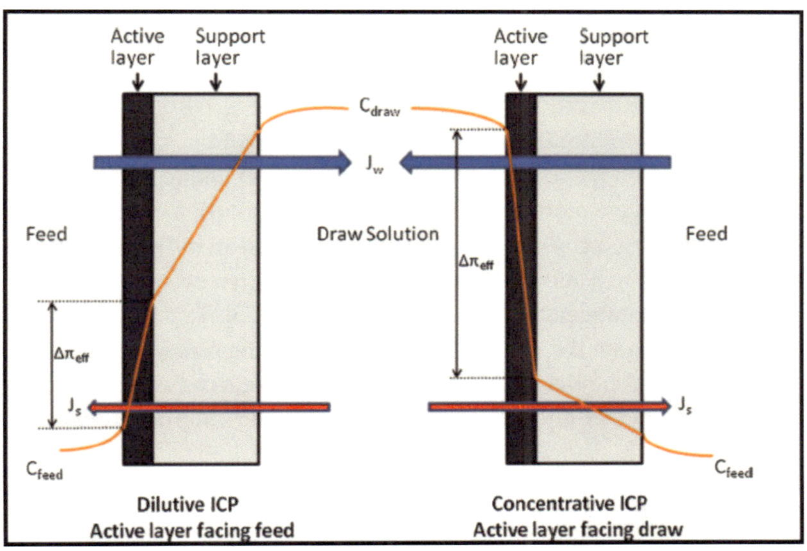

Figure 3. Dilutive ICP and concentrative ICP across an asymmetric FO membrane [48].

result, when it comes to the newly developed membranes, it is crucial to describe the membrane's structural parameter S. The value of S can be calculated based on Eqs. (4) and (11) and by fitting FO test results [50, 51]. On the other hand, in a specific membrane orientation, both ICP and ECP happen concurrently rather than occurring separately. Researchers McCutcheon and Elimelech have designed models that consider the characteristics and influences of both ECP and ICP. For the FO mode, the analytical model capturing the effects of both concentrative ECP and dilutive ICP on permeate flux may be conveyed [38, 43] by:

$$J_w = A\left[\pi_{draw}\ exp\left(-J_w K\right) - \pi_{feed}\ exp\left(\frac{J_w}{k}\right)\right] \tag{12}$$

As seen in Eq. (4), it appears that ICP in the membrane's support layer is formed based on the membrane properties, such as membrane's tortuosity, porosity, and thickness, as well as the diffusion solute properties, like the diffusion coefficient of the solute. A research project by Zhao and Zou has connected ICP to additional properties of the solution, like viscosity and diffusion solute size, by considering the idea of constrictivity [48]. The equation that corresponds to this dynamic is embodied in the following:

$$K = \frac{t\tau}{\delta\varepsilon_{eff} D} \tag{13}$$

In this case, a new parameter δ is expressed as the constrictivity factor, and ε_{eff} is the effective transport through porosity, as it can be lower than the overall membrane porosity if certain small pores are not available to the larger solute. In particular, the constrictivity parameter relies on the ratio of the pore diameter to the solute molecule diameter:

$$\lambda = \frac{molecule\ diameter}{pore\ diameter} < 1 \tag{14}$$

Tang et al. researched the cumulative effect of fouling and ICP on FO flux behavior. [50]. Tang et al. noted two critical phenomena during the experimental runs. The first phenomenon was that the water flux was comparatively stable and its decrease was minor during the FO mode, whereas during the PRO mode, the flux decrease was substantial and especially prominent when membrane fouling happened. The second phenomenon had to do with the fact that the effects of ICP on FO flux were more distinct at greater draw solution concentration values [48]. A number of new modeling techniques have been used to research the concentration polarization (CP) phenomenon, such as the computational fluid dynamics (CFD) [52], numerical simulation [53], and the finite element method (FEM) [54, 55]. A project spearheaded by Li et al. used FEM to interrogate the relationship between the membrane's porous structure and ICP [54]. The mathematical models that came out of this project can serve as a valuable toolkit for improving FO performance and optimizing the membrane's support construction [54].

4.2. Membrane fouling mechanism in forward osmosis

Like concentration polarization, membrane fouling is an unavoidable as well as essential phenomenon influencing all types of membrane processes [28, 56–63]. As a consequence, smaller membrane fouling potential ensures that there is less cleaning, longer membrane life, and more water produced, which effectually decreases capital and operational costs. On the other hand, the membrane-type fouling happening in osmotically driven membrane processes is distinct from the types of fouling present in pressure-driven membrane processes, as a result of the low hydraulic pressure being used in the former case. Initially, Cath et al. researched membrane fouling in FO in relation to systems used in long-term space missions [64, 65]. Cath et al. suggested that FO could have the capacity to reduce membrane fouling, since there was no flux decrease due to fouling detected during the experimental runs [65]. During the last few years, FO has been applied in osmotic membrane bioreactors (OMBR) primarily for wastewater treatment because of its lower energy consumption and lower fouling needs [20, 22], both of which are two challenges for membrane bioreactors [59, 66]. In a research project by In addition, the OMBR system was used to treat activated sludge. The results report that neither irreversible nor reversible fouling was seen whenever the membrane's active layer was positioned in a way facing the activated sludge [22]. An experiment conducted by Achilli et al. relied on a submerged OMBR so as to treat domestic wastewater over the prolonged period of up to 28 days, indicating that the decrease of water flux was primarily due to membrane fouling [20]. On the other hand, the flux of the initial values could be recovered by roughly 90% through the process of osmotic backwashing. This experimental result suggests that membrane fouling in OMBR may in fact be reversible. Similarly, the data reflect that membrane fouling does exist in FO and is apparent during long-term operational runs. Mi and Elimelech interrogated the inorganic and organic fouling in FO [23, 27, 62]. Mi and Elimelech determined that, first of all, the intermolecular adhesion and organic fouling were connected and that foulant-foulant interactions had an important role in organic cleaning and fouling. Second, Mi and Elimelech found out that FO fouling was controlled by the coupled effects of chemical, for example, calcium binding, and hydrodynamic, for instance permeation drag and shear force, interactions. They likewise noted that membrane materials had a key role in organic fouling and cleaning, which was later verified with the help of atomic force microscope (AFM) measurements. Mi and Elimelech also found that both inorganic and organic types of fouling in FO were nearly fully reversible using water rinsing. This could be attributed to the less compact fouling layer created by the applied low hydraulic pressure, which suggests that chemical cleaning could be prevented. Moreover, researchers comparing membrane fouling in FO and RO suggested that it could be diverse from one case to another with respect to water cleaning efficiency and reversibility [23, 27, 28]. Although it was irreversible in RO, Lee et al. observed that membrane fouling in FO was almost entirely reversible [28]. Alternatively, Lee et al. linked the FO fouling to the accelerated cake-enhanced osmotic pressure (CEOP) created by the reverse solute (salt) diffusion process in the draw solution. **Figure 4** outlines the mechanics of this process [31]. Once the draw solution faces the membrane's support layer, using the reverse diffusion, the draw solute collects on the active layer's surface located on the feed side, lowering the net osmotic driving force and improving the concentration polarization layer. The draw solute featuring a less hydrated radius value (e.g., NaCl) is more easily capable of initiating CEOP, when compared to the ones with a greater hydrated radius values, like dextrose. In an experiment by Lay et al., it was noted that the reverse diffusion of the draw solute could worsen the CEOP effect as well as intensify FO fouling [67]. Alternatively, new research suggests that FO fouling could be substantially lowered if the cross flow velocity is increased [28].

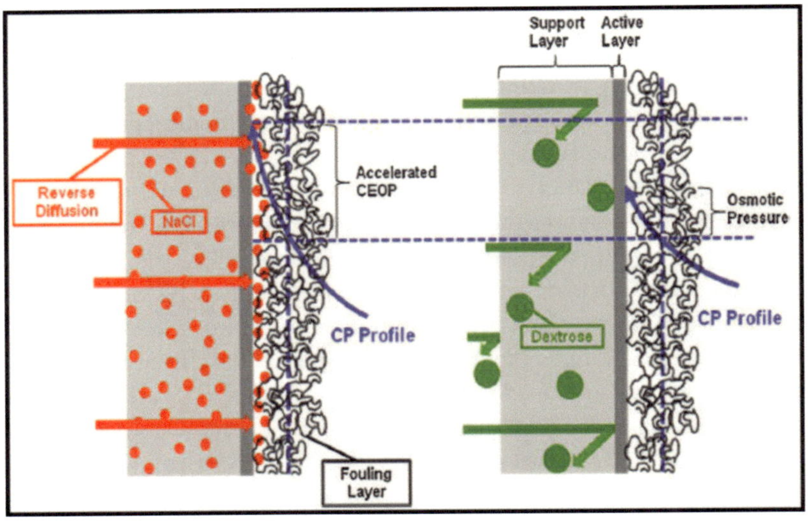

Figure 4. The effects of draw solute reverse diffusion on cake-enhanced osmotic pressure (CEOP) in FO for two different draw solutions: (a) NaCl and (b) dextrose [31].

In a recent experimental attempt, Tang's group used direct microscopic observation to study FO fouling and its mechanisms. They determined that the critical flux concept could also be relevant to osmotically drive the types of processes [63]. Direct microscopic observation has been implemented to research the production of fouling in pressure-driven membrane processes and currently remains one of the primary membrane fouling characterization methodologies [68–71]. Admittedly, direct microscopic observation is relevant only for the cases with large foulants in colloidal or biofouling fouling, such as microbes or particles. Thus, direct microscopic observation can likewise be implemented in the research of membrane fouling if visible fouling layers or large foulants are present in FO.

Usually defined as the level of permeate flux where membrane fouling becomes noticeable, critical flux has been widely used in pressure-driven membrane processes [72–75]. Critical flux can also be applied to osmotically driven membrane processes. A recent study by Zhao et al. confirms its presence in FO [76]. It is necessary to note that the critical flux identified by Zhao et al.'s FO research study, as well as by Tang and coworkers, was detected when the membrane's surface was partially covered with visible foulant [63, 76]. As a consequence, the critical flux in FO could have an implicit connection to the visible fouling layer. This particular connection must be investigated in greater depth. Research suggests that greater working temperatures can have various negative influences on FO cleaning and scaling in brackish water desalination processes, potentially because of the change of HCO_3^- into CO_3^{2-} at high temperature values [26]. The report indicates that, caused by the polymerization of dissolved silica, the silica scaling of FO membranes was the primary inorganic type of fouling in real-case seawater desalination examples [77]. The silica polymerization might likewise quicken the organic fouling, which is removed much easier using water rinsing if compared to the silica scaling [77].

Alternatively, membrane fouling could improve the FO membrane's solute rejection potential. It was also detected that organic foulants located on the membrane's surface, or its active layer, could improve the negative charge property and surface hydrophilicity, and this in turn can raise the hydrophilic compound absorption capacity [78]. Changes like these can increase the critical rejection potential for many new contaminants, including trace organic compounds, as well as hydrophobic neutral compounds and hydrophilic ionic compounds [78]. Once the FO tests were run continuously for prolonged periods of time at the pilot scale, the rejection performance values improved even further as more substantial fouling happened [79]. In a project by Jin et al., it was determined that organic fouling can likewise have substantial consequences for the elimination of inorganic contaminants, like arsenic and boron [80]. In particular, their influences relied on the membrane's orientation. For instance, in the FO mode, where the membrane's support layer is facing the draw solution, the organic fouling on the membrane's active layer can improve the sieving influence and essentially increase the arsenic rejection in the feed. Alternatively, in the PRO mode, where the membrane's support layer is facing the feed, the organic fouling in the membrane's support structure can lower the boron rejection [80]. Membrane fouling and concentration polarization remain critical phenomena in FO processes since they have the capacity to heighten the additional membrane resistance and lower membrane permeability potential. Researchers must continue to further examine their functions and mechanisms if they want to improve the FO process and its performance capacity. Successful application of FO in real settings will remain problematic until a more comprehensive analysis becomes available.

4.3. Reverse solute diffusion

In membrane processes that are osmotically driven, the solute's reverse diffusion, from the draw solution and through the membrane toward the feed solution, is likewise almost certainly due to the concentration variances. Cath et al. (2009) state that the draw solute reverse diffusion has to be carefully studied as it could endanger the success of the process [11, 81]. Some research studies have linked draw solute reverse diffusion with the membrane fouling phenomenon. Lay et al. and Lee et al. have shown that the draw solute reverse diffusion can, on the one hand, improve the CEOP influence and, on the other hand, intensify FO fouling [28, 67]. Thus, multivalent ion solutions featuring smaller diffusion coefficient values are better for certain uses in which higher rejection potentials are required [11]. Alternatively, in other cases, multivalent ions, like Ca^{2+} and Mg^{2+}, could impede the foulants in the feed solution following reverse diffusion, a dynamic that can worsen the overall membrane fouling [82]. Furthermore, multivalent ions could likewise incite a more substantial ICP due to their smaller diffusion coefficients and bigger ion sizes [48]. Defined as the ratio of the reverse solute flux to the forward water flux, specific reverse solute flux has been added as another potential measure of membrane's selectivity [81, 82]. Specific reverse solute flux parameter offers a third dynamic for the proper FO performance evaluation, together with the salt rejection and the permeate flux parameters. A greater specific reverse solute flux suggests reduced membrane selectivity potential, as well as an inferior FO efficiency value. Although the specific reverse solute flux depends on the membrane's active layer selectivity, it is independent of the structure of the membrane support layer and the draw solution concentration values [51]. This key outcome grants another standard for the production of a new type of FO membrane, that is, greater selectivity of the membrane's active layer. Moreover, engaging a multivalent ion solution as the draw solution could reduce membrane fouling [28, 67] and lower the reverse solute diffusion [81], but in this case, there is also a potential to have a higher ICP [48] and a greater risk of fouling [82]. To sum up, reverse solute diffusion remains one of the main challenges in osmotically driven membrane processes and as a result must be reduced during the production and design of draw solutes and FO membranes.

4.4. Membrane development

Based on the available membrane fabrication methodologies, newly produced and designed membranes can be organized into three categories: the thin film composite (TFC) membranes, the chemically modified membranes, and the phase inversion-formed cellulosic membranes. Reverse solute diffusion, membrane fouling, and ICP are three of the crucial concerns that exist with respect to the osmotically driven membrane processes, since they effectually direct the FO performance. As a result, when considering innovative FO membrane development, it is essential to characterize its salt rejection, antifouling, and anti-ICP characteristics. When compared with other types of processes, FO could be viewed as more competitive when treating challenging waters with higher fouling potential or solid content, since ICP and fouling are frequently much more serious. When it comes to FO, the water flux is affected by the water permeability, while the reverse solute flux is shaped by the membrane solute permeability. In this instance, there is a type of trade-off between salt rejection and water permeability [83]. Higher water permeability values are desirable, as well as lower salt rejection potential. In most cases, FO membrane featuring higher water permeability potential likewise offers higher salt flux, and the reverse relationship holds true as well. As a consequence, defined as the ratio of the reverse solute flux to the forward water flux, specific reverse solute flux can be a superior parameter to evaluate when considering the FO performance [81]. In fact, it might be better to assess the FO performance with the aid of the osmotic water flux and specific reverse solute flux when membrane fouling and ICP are present. Thus, the characterization and design of new FO membrane in the forthcoming future must reflect on the antifouling and the anti-ICP properties, as well as salt rejection (solute permeability), structural parameters, and water permeability.

4.4.1. Phase inversion-formed cellulosic membranes

Asymmetric cellulosic osmotically driven membranes developed through phase inversion have been created specifically for osmotic drug delivery before they were used for water treatment purposes [33, 84, 85]. Most of these membranes were created using conventional phase inversion and with the help of cellulose acetate as the dip-coating polymer. A research breakthrough in Loeb and Sourirajan's method occurred when they prepared RO membranes through phase inversion based on cellulose acetate polymer. Cellulose acetate offers a variety of desirable properties, such as a comparatively high hydrophilicity favoring lower fouling propensity and greater water flux, wide availability, improved mechanical strength, as well as enhanced resistance to degradation by chlorine and other types of oxidants [86, 87]. This particular form of cellulosic membrane is implemented in energy generation, such as osmotic power, and through a PRO process [88]. Recently, Chung's research group has produced a number of cellulose ester-based membranes specifically for FO applications and containing flat sheet modules and hollow fiber [89–91]. In this case, the methods for creating these cellulose derivative membranes are relatively similar, in the form of phase inversion that is followed by hot water annealing at 60–95°C. Chung's research group determined that the resulting membrane could have two selective skin layers that are capable of lowering ICP in the membrane support layer [87, 90]. A more recent study modeled this type of double-skinned FO membrane [92]. Chung's research group likewise noted that the relationship between the casting substrate and the polymer had an important role during phase inversion for the development of the membrane's structure [87, 91]. Furthermore, Sairam et al. implemented this phase inversion approach in order to create flat sheet FO membranes using cellulose acetate [93]. Specifically, they applied maleic acid, zinc chloride, and lactic acid as pore-forming agents, while casting the membrane onto nylon fabric at a range of annealing temperature values. Sairam et al. noted that the membrane developed with zinc chloride as the pore-forming agent allowed for a reasonably effective FO performance. On the other hand, the disadvantages of cellulose acetate have to be examined before it is used in FO membranes. While cellulose acetate membranes are more resistant to chloride degradation and more hydrophilic, if compared to the TFC polyamide RO membranes, they have lower resistance potential to biological attachments and hydrolysis [86, 94, 95]. To reduce the hydrolysis of cellulose acetate membranes, it is crucial to modify the pH of the feed and draw solutions within the ranges of 4–6 and to sustain the working temperature that does not rise above 35°C [86, 94].

4.4.2. *Thin film composite membranes*

It has been noted that there is a key trade-off dynamic occurring between salt rejection and water permeability potential. For instance, the raising of the trimesoyl chloride (TMC) concentration or the reduction of the m-phenylenediamine (MPD) concentration caused greater membrane permeability potential but lower salt rejection values [83]. Research likewise indicates that greater hydrophilicity of the support layer may prefer water diffusion across the FO membrane [96, 97]. Wang et al. prepared polyethersulfone (PES)/sulfonated polysulfone (PSF)-alloyed–type membranes as the substrates of interfacial polymerization and produced high-performance FO membranes. On the other hand, Yu et al. developed a nonporous polyethersulfone (PES) FO-type membrane with the aid of phase inversion, however, without using interfacial polymerization [98]. In this case, the polyester nonwoven fabrics were implemented for backing support. This membrane creation approach was comparable to the one used by Elimelech's group, with the exception of the additional interfacial polymerization phase. According to the report, the membrane produced by this method featured an active layer formed on top of the support layer, high water flux value, and low reverse solute flux potential [98].

Song et al. reported the creation of a nanofiber TFC FO-type membrane using electrospinning, which was followed by interfacial polymerization (ES-IP) [99]. Song et al. noted that the nanocomposite FO membrane allowed for an improved FO performance mostly because of high porosity and low tortuosity that significantly decreased the structural parameters of the membrane. If compared to the TFC FO membrane made using phase inversion followed by interfacial polymerization (PI-IP), the electrospinning-formed nanofiber support layer offers a porous structure resembling a scaffold with interlocked pores between individual nanofibers [99]. Due to this structure, the water flux value of the ES-IP–formed FO membrane was found to be three times as high as the water flux potential of the PI-IP–formed membrane. In this instance, the performance of the FO membrane was enhanced with respect to osmotic water flux, while salt rejection was obtained as well as confirmed by Bui et al. research group [100]. The majority of the approaches used for preparing TFC FO membranes and asymmetric cellulose acetate FO membranes are indistinguishable from the original traditional RO membrane methods, like phase inversion followed by interfacial polymerization, or phase inversion and annealing.

When it comes to a TFC FO membrane, the membrane support layer made using phase inversion governs the ICP, water flux in FO, and the membrane's active layer controls reverse solute flux potential and salt rejection values. A high salt rejection can be obtained when the TFC membranes are developed with the help of interfacial polymerization. In fact, FO's performance is shaped by the membrane's support layer. Next-generation FO membrane production must pay attention to the membrane's support layer and its role. All in all, an effective FO membrane has to provide a design that appears sufficiently porous and offers improved hydrophilic support combined with lower tortuosity capable of decreasing ICP, as well as a selective active layer that can lower reverse solute diffusion and augment salt rejection potential.

4.4.3. Chemically modified membranes

Over the course of the last several years, chemical modification methodologies have been implemented during the development of innovative FO membranes. Arena et al. research group (2011) applied polydopamine (PDA) as a new bioinspired hydrophilic polymer for the modification of the support layers in commercial TFC RO membranes catering to engineered osmosis applications [101]. This modified membrane showed improved water flux and lower ICP during the conducted FO tests. Furthermore, Setiawan et al. created a hollow-type fiber FO membrane featuring a positively charged NF-like selective layer using a polyelectrolyte posttreatment of a polyamide-imide (PAI) microporous substrate with polyethylenimine (PEI) [102]. Setiawan et al. indicated that the final FO membrane produced could be applied in heavy metal removal processes due to its unique positively charged characteristic. This research group likewise designed a flat sheet–type membrane offering a positively charged NF-like selective layer on top of a woven fabric–embedded substrate and implementing a similar methodology. The reported results suggest that the overall thickness of the substrate was reduced to 55 μm when the PAI microporous substrate was successfully embedded within a woven fabric. Moreover, Tang and coworkers relied on a creative layer-by-layer assembly approach in order to produce FO membranes with desirable properties [103, 104]. In Tang's research studies, polyacrylonitrile (PAN) substrate was prepared with the aid of phase inversion and then posttreated by sodium hydroxide so as to improve surface negative charge density and hydrophilicity potential. Poly(sodium 4-styrene-sulfonate) (PSS) and poly(allylamine hydrochloride) (PAH) were implemented as the polyanion and polycation, respectively. Arguably, the majority of the present approaches to the FO membrane preparation are established methods that have been applied during the last few decades for the creation of pressure-driven–type membranes, such as RO and NF. The production and design of innovative high-performance FO membranes are still in their early stages. As a result, the process of relying on the older methodologies for RO or NF membrane preparation is a sensible and practical direction. Forthcoming research may expand the recently developed techniques for the production of high-performance FO membranes, including layer-by-layer assembly [103, 105–112], UV-photographing [113–116], and polyelectrolyte dip-coating [117, 118]. In addition, membranes featuring polyelectrolyte multilayers, charged properties, or double selective layers can provide exciting avenues for specific real-life FO applications.

4.5. Draw solute/solution developments

Despite preexisting setbacks, key innovative approaches for draw solute selection have been suggested [119, 120]. Specifically, there are three conditions for successful selection of a suitable draw solution in FO applications. To begin, the draw solution must offer a reasonably high osmotic pressure values [11]. Next, the diluted draw solution needs to be economically and effortlessly reconcentrated and recovered [11, 121]. Finally, the draw solute has to showcase lowered ICP during the FO processes. A research study by Zhao and Zou indicates that ICP in FO is seriously influenced by draw solution viscosity, draw solution's ion/molecule size of the solute, and solute diffusion coefficient values [48]. Increased diffusion coefficients, coupled with lowered ion/molecule sizes and smaller solution viscosities, will minimize ICP and allow for resulting in improved permeate fluxes [48]. Moreover, additional parameters like low reverse solute permeability [11], zero toxicity, low cost, absence of membrane damage, inertness and stability at or near natural pH, and good biofouling-resistance should be considered when the draw solute/solution is selected.

During the last few decades, numerous draw solutes/solutions have been examined during osmotically driven–type processes. The primary benefit of implementing volatile gases as draw solutes in FO is that the final thermolytic draw solution may be separated or recovered with the help of heating/or distillation. In a separate project, sugars were likewise tested as draw solutes since there is no necessity to separate the diluted nutrient solutions further, and the diluted solutions may be reconcentrated in decreased pressures with loose RO membranes. After the 2000s, Elimelech and coworkers suggested a new draw solution for the purposes of FO desalination, that is, a water-soluble mixture of NH_3 and CO_2 including ammonium bicarbonate (NH_4HCO_3) [9, 10, 24]. The proposed draw solution can offer improved water fluxes as a result of the higher driving forces created by the greater solubilities of the solutes. These types of draw solutes may be effortlessly recovered or recycled using moderate heating (~60°C) [9]. As a result, this innovative draw solution could find a potential application in large-scale desalination, even though the removal the ammonia (NH_3) smell from the produced water could be a concern. Furthermore, various other chemicals have been assessed for the role of the draw solutes [119].

For instance, synthetic materials, like organic compounds [121] and magnetic and/or hydrophilic nanoparticles [122–124], have been proposed for the application as the draw solutes. In the case of the laboratory-designed magnetic nanoparticles, data suggest that the particle size and surface hydrophilicity of the particles had critical roles for the FO separation performance [123]. It was also noted that particle agglomeration happened during draw solute recycling process using magnetic separators [124]. Such an accumulation of magnetic nanoparticles may be reduced with the aid of ultrasonication. When using this method, the particles' magnetic characteristics and the recovery efficacy were threatened by ultrasonication as well. In order to transcend the issue of accumulation during draw solute recycling, the thermal-responsive properties were integrated into the magnetic nanoparticles using the one-step thermal decomposition [125]. UF could likewise be used to recover diluted draw solutes featuring big particle or molecule sizes. Wang's research group has produced a stimuli-responsive polymer hydrogel as another draw solute for FO desalination [126]. Polymer hydrogels such as these have the capacity to pull water from the saline feed during swelling and after that release the water while deflating, the latter being caused by heating and hydraulic pressure. So as to enhance the capacity of swelling ratios and drawing water, a type of light-absorbing carbon particles was introduced into the polymer hydrogels, and as a consequence an improved performance was obtained [127]. A new draw solute separation system simulating the "destabilization" phenomenon was suggested for the process of generating drinking water [128]. In this particular system, $Al_2(SO_4)_3$ was chosen as the draw solute, while the diluted $Al_2(SO_4)_3$ solution pH was attuned using CaO, finally resulting in the creation of a white gel-like mixture made out of positively charged $Al(OH)_3$ and $CaSO_4$. In the following step, negatively charged superparamagnetic nanoparticles were added so as to enable the sedimentation. To accelerate the sedimentation process and enhance separation efficiency, an external magnetic field was introduced. Such an innovative draw solution separation technique can make FO a more economical and eco-sustainable process for efficient drinking water production [128].

The selection criteria for the draw solutes and solutions need to be addressed for the process to be effective and sustainable. An effective draw solute option for FO must offer easy and economical recovery, lower tendency to cause ICP, zero toxicity, reasonable costs, higher solubility, and greater osmotic pressure. The diffusion coefficient, viscosity of the draw solution, and the solute particle size need to be examined as they are directly connected to ICP [48] effectually dominating the water flux in FO [38, 46].

4.6. Forward osmosis fouling control

In this chapter, the focus was on discussing and reviewing the primary five issues that exist in FO. Certainly, these challenges do not exist in isolation but are rather interconnected. To sum up some of these issues, the membrane's support layer needs to be as porous as possible for the lower ICP, and the membrane's active layer needs to be more selective for a lower reverse solute diffusion potential. The smaller reverse solute diffusion can then decrease the membrane fouling. When it comes to the draw solute, small ion or molecule sizes could minimize ICP [48]; however, they can likewise increase membrane fouling and reverse solute diffusion. All of these correlations and criteria make the creation of advantageous draw solutes much more problematic. In most cases, higher reverse solute diffusion may lead to substantial membrane fouling, and this correlation holds the other way as well [28, 67]. In addition, ICP and membrane fouling could lead to multiple adverse properties for water flux in FO [50]. Furthermore, reverse solute diffusion, membrane fouling, and ICP are at their core determined by draw solute properties and membrane qualifications.

5. Conclusion

The membrane processes based on osmosis are new technological directions that have exhibited a lot of promise for a range of applications, and especially water purification, food processing, desalination, wastewater treatment, power generation, and pharmaceutical product dehydration. While FO is not likely to fully replace RO as the primary desalination technology in the foreseeable future, it remains an appealing alternative as an effective desalination approach offering many benefits over pressure-driven–type membrane processes. In order to transfer FO from the laboratory stages of research into hands-on industrial applications, a set of advances in terms of FO membrane and draw solute development needs to happen. In fact, the membranes need to offer critical properties of minimizing ICP, higher mechanical strength, stability, improved water permeability, and better selectivity. To sum up, this chapter examined five essential challenges for FO in the form of membrane fouling, reverse solute diffusion, further membrane development, concentration polarization, and enhanced draw solute design. The innovative draw solutes must be capable of producing higher osmotic pressure, remain easily and economically regenerated/or recycled, and provide minimal ICP. Draw solutes must also offer compatibility with the FO membranes and zero toxicity. A successful draw solute has a vital role in the popularization and efficacy of FO applications. The next level of draw solute development will allow for a much wider use of FO in a range of industrial-scale applications and fields.

Nomenclature

A	water permeability constant of the membrane
B	solute permeability coefficient of the membrane
C_m	concentrations of the feed solution at the membrane surface
C_b	concentrations of the feed solution at the bulk
D	diffusion coefficient of the solute
D_h	hydraulic diameter
J_w	water flux
k	mass transfer coefficient
k_{feed}	mass transfer coefficient on the feed side
k_{draw}	mass transfer coefficient on the draw side
t	thickness of the membrane
τ	tortuosity
σ	reflection coefficient
ε	membrane porosity
S	membrane structural parameter
Sh	Sherwood number
ΔP	applied hydraulic pressure difference
$\Delta\pi$	osmotic pressure difference across the membrane
$\Delta\pi_{Feed}$	bulk osmotic pressure of the feed solution
π_{m-draw}	osmotic pressures of the draw solution at the membrane surface
π_{b-draw}	osmotic pressures of the draw solution in the bulk
$\Delta\pi_{Draw}$	bulk osmotic pressure of the draw solution

Abbreviations

AFM	atomic force microscope
CEOP	cake-enhanced osmotic pressure
CFD	computational fluid dynamics
ECP	external concentration polarization
ES-IP	electrospinning followed by interfacial polymerization
FEM	finite element method
FO	forward osmosis
ICP	internal concentration polarization
MPD	m-phenylenediamine
NF	nanofiltration
OMBR	osmotic membrane bioreactor
PAH	poly(allylamine hydrochloride)
PAI	polyamide-imide
PAN	polyacrylonitrile
PDA	polydopamine
PEI	polyethylenimine
PES	polyethersulfone
PRO	pressure-retarded osmosis
PSF	polysulfone
PSS	poly(sodium 4-styrene-sulfonate)
RO	reverse osmosis
TFC	thin film composite
TMC	trimesoyl chloride

Author details

Amira Abdelrasoul[1]*, Huu Doan[2], Ali Lohi[2] and Chil-Hung Cheng[2]

*Address all correspondence to: amira.abdelrasoul@usask.ca

1 Department of Chemical and Biological Engineering, University of Saskatchewan, Saskatoon, Saskatchewan, Canada

2 Department of Chemical Engineering, Ryerson University, Toronto, Ontario, Canada

References

[1] Shannon MA, Bohn PW, Elimelech M, Georgiadis JG, Marinas BJ, Mayes AM. Science and technology for water purification in the coming decades. Nature. 2008;**452**: 301-310

[2] E.I.A. International. World Energy Outlook 2010. The Energy Information Administration. Washington DC; 2010

[3] Elimelech M, Phillip WA. The future of seawater desalination: Energy, technology, and the environment. Science. 2011;**333**:712-717

[4] Montgomery MA, Elimelech M. Water and sanitation in developing countries: Including health in the equation. Environmental Science & Technology. 2007;**41**:17-24

[5] Elimelech M. The global challenge for adequate and safe water. Journal of Water Supply: Research and Technology—AQUA. 2006;**55**:3-10

[6] Gleick PH. Water resources. In: Schneider SH, editor. Encyclopedia of Climate and Weather. New York: Oxford University Press; 1996

[7] King CW, Webber ME. Water intensity of transportation. Environmental Science & Technology. 2008;**42**:7866-7872

[8] Geise GM, Lee H-S, Miller DJ, Freeman BD, McGrath JE, Paul DR. Water purification by membranes: The role of polymer science. Journal of Polymer Science Part B: Polymer Physics. 2010;**48**:1685-1718

[9] McCutcheon JR, McGinnis RL, Elimelech M. A novel ammonia–carbon dioxide forward (direct) osmosis desalination process. Desalination. 2005;**174**:1-11

[10] McCutcheon JR, McGinnis RL, Elimelech M. Desalination by ammonia–carbon dioxide forward osmosis: Influence of draw and feed solution concentrations on process performance. Journal of Membrane Science. 2006;**278**:114-123

[11] Cath TY, Childress AE, Elimelech M. Forward osmosis: Principles, applications, and recent developments. Journal of Membrane Science. 2006;**281**:70-87

[12] Petrotos KB, Quantick P, Petropakis H. A study of the direct osmotic concentration of tomato juice in tubular membrane-module configuration. I. The effect of certain basic process parameters on the process performance. Journal of Membrane Science. 1998;**150**: 99-110

[13] Petrotos KB, Quantick PC, Petropakis H. Direct osmotic concentration of tomato juice in tubular membrane-module configuration. II. The effect of using clarified tomato juice on the process performance. Journal of Membrane Science. 1999;**160**:171-177

[14] Petrotos KB, Lazarides HN. Osmotic concentration of liquid foods. Journal of Food Engineering. 2001;**49**:201-206

[15] Garcia-Castello EM, McCutcheon JR, Elimelech M. Performance evaluation of sucrose concentration using forward osmosis. Journal of Membrane Science. 2009;**338**:61-66

[16] Lee KL, Baker RW, Lonsdale HK. Membranes for power generation by pressure-retarded osmosis. Journal of Membrane Science. 1981;**8**:141-171

[17] Seppälä A, Lampinen MJ. Thermodynamic optimizing of pressure-retarded osmosis power generation systems. Journal of Membrane Science. 1999;**161**:115-138

[18] McGinnis RL, Elimelech M. Global challenges in energy and water supply: The promise of engineered osmosis. Environmental Science & Technology. 2008;**42**:8625-8629

[19] Achilli A, Cath TY, Childress AE. Power generation with pressure retarded osmosis: An experimental and theoretical investigation. Journal of Membrane Science. 2009;**343**:42-52

[20] Achilli A, Cath TY, Marchand EA, Childress AE. The forward osmosis membrane bioreactor: A low fouling alternative to MBR processes. Desalination. 2009;**239**:10-21

[21] Holloway RW, Childress AE, Dennett KE, Cath TY. Forward osmosis for concentration of anaerobic digester centrate. Water Research. 2007;**41**:4005-4014

[22] Liu Y, Mi B. Combined fouling of forward osmosis membranes: Synergistic foulant interaction and direct observation of fouling layer formation. Journal of Membrane Science. 2012;**407-408**:136-144

[23] Mi B, Elimelech M. Organic fouling of forward osmosis membranes: Fouling reversibility and cleaning without chemical reagents. Journal of Membrane Science. 2010;**348**:337-345

[24] McGinnis RL, Elimelech M. Energy requirements of ammonia–carbon dioxide forward osmosis desalination. Desalination. 2007;**207**:370-382

[25] Martinetti CR, Childress AE, Cath TY. High recovery of concentrated RO brines using forward osmosis and membrane distillation. Journal of Membrane Science. 2009;**331**:31-39

[26] Zhao S, Zou L. Effects of working temperature on separation performance, membrane scaling and cleaning in forward osmosis desalination. Desalination. 2011;**278**:157-164

[27] Mi B, Elimelech M. Gypsum scaling and cleaning in forward osmosis: Measurements and mechanisms. Environmental Science & Technology. 2010;**44**:2022-2028

[28] Lee S, Boo C, Elimelech M, Hong S. Comparison of fouling behavior in forward osmosis (FO) and reverse osmosis (RO). Journal of Membrane Science. 2010;**365**:34-39

[29] Cartinella JL, Cath TY, Flynn MT, Miller GC, Hunter KW, Childress AE. Removal of natural steroid hormones from wastewater using membrane contactor processes. Environmental Science & Technology. 2006;**40**:7381-7386

[30] Cath TY, Hancock NT, Lundin CD, Hoppe-Jones C, Drewes JE. A multi-barrier osmotic dilution process for simultaneous desalination and purification of impaired water. Journal of Membrane Science. 2010;**362**:417-426

[31] Yang Q, Wang KY, Chung T-S. A novel dual-layer forward osmosis membrane for protein enrichment and concentration. Separation and Purification Technology. 2009;**69**:269-274

[32] Jiao B, Cassano A, Drioli E. Recent advances on membrane processes for the concentration of fruit juices: A review. Journal of Food Engineering. 2004;**63**:303-324

[33] Wang C-Y, Ho H-O, Lin L-H, Lin Y-K, Sheu M-T. Asymmetric membrane capsules for delivery of poorly water-soluble drugs by osmotic effects. International Journal of Pharmaceutics. 2005;**297**:89-97

[34] Shokri J, Ahmadi P, Rashidi P, Shahsavari M, Rajabi-Siahboomi A, Nokhodchi A. Swellable elementary osmotic pump (SEOP): An effective device for delivery of poorly water-soluble drugs. European Journal of Pharmaceutics and Biopharmaceutics. 2008;**68**:289-297

[35] Bamaga OA, Yokochi A, Zabara B, Babaqi AS, Hybrid FO. RO desalination system: Preliminary assessment of osmotic energy recovery and designs of new FO membrane module configurations. Desalination. 2011;**268**:163-169

[36] Tang W, Ng HY. Concentration of brine by forward osmosis: Performance and influence of membrane structure. Desalination. 2008;**224**:143-153

[37] Mehta GD. Further results on the performance of present-day osmotic membranes in various osmotic regions. Journal of Membrane Science. 1982;**10**:3-19

[38] McCutcheon JR, Elimelech M. Influence of concentrative and dilutive internal concentration polarization on flux behavior in forward osmosis. Journal of Membrane Science. 2006;**284**:237-247

[39] Zydney AL. Stagnant film model for concentration polarization in membrane systems. Journal of Membrane Science. 1997;**130**:275-281

[40] Sablani SS, Goosen MFA, Al-Belushi R, Wilf M. Concentration polarization in ultrafiltration and reverse osmosis: A critical review. Desalination. 2001;**141**:269-289

[41] Thorsen T. Concentration polarisation by natural organic matter (NOM) in NF and UF. Journal of Membrane Science. 2004;**233**:79-91

[42] Kim S, Hoek EMV. Modeling concentration polarization in reverse osmosis processes. Desalination. 2005;**186**:111-128

[43] McCutcheon JR, Elimelech M. Modeling water flux in forward osmosis: Implications for improved membrane design. AICHE Journal. 2007;**53**:1736-1744

[44] Su J, Chung T-S. Sublayer structure and reflection coefficient and their effects on concentration polarization and membrane performance in FO processes. Journal of Membrane Science. 2011;**376**:214-224

[45] Mehta GD, Loeb S. Performance of permasep B-9 and B-10 membranes in various osmotic regions and at high osmotic pressures. Journal of Membrane Science. 1978;**4**:335-349

[46] Gray GT, McCutcheon JR, Elimelech M. Internal concentration polarization in forward osmosis: role of membrane orientation. Desalination. 2006;**197**:1-8

[47] Mehta GD, Loeb S. Internal polarization in the porous substructure of a semipermeable membrane under pressure-retarded osmosis. Journal of Membrane Science. 1978;**4**:261-265

[48] Zhao S, Zou L. Relating solution physicochemical properties to internal concentration polarization in forward osmosis. Journal of Membrane Science. 2011;**379**:459-467

[49] Loeb S, Titelman L, Korngold E, Freiman J. Effect of porous support fabric on osmosis through a Loeb–Sourirajan type asymmetric membrane. Journal of Membrane Science. 1997;**129**:243-249

[50] Tang CY, She Q, Lay WCL, Wang R, Fane AG. Coupled effects of internal concentration polarization and fouling on flux behavior of forward osmosis membranes during humic acid filtration. Journal of Membrane Science. 2010;**354**:123-133

[51] Phillip WA, Yong JS, Elimelech M. Reverse draw solute permeation in forward osmosis: Modeling and experiments. Environmental Science & Technology. 2010;**44**: 5170-5176

[52] Gruber MF, Johnson CJ, Tang CY, Jensen MH, Yde L, Hélix-Nielsen C. Computational fluid dynamics simulations of flow and concentration polarization in forward osmosis membrane systems. Journal of Membrane Science. 2011;**379**:488-495

[53] Jung DH, Lee J, Kim DY, Lee YG, Park M, Lee S, Yang DR, Kim JH. Simulation of forward osmosis membrane process: Effect of membrane orientation and flow direction of feed and draw solutions. Desalination. 2011;**277**:83-91

[54] Li W, Gao Y, Tang CY. Network modeling for studying the effect of support structure on internal concentration polarization during forward osmosis: Model development and theoretical analysis with FEM. Journal of Membrane Science. 2011;**379**:307-321

[55] Sagiv A, Semiat R. Finite element analysis of forward osmosis process using NaCl solutions. Journal of Membrane Science. 2011;**379**:86-96

[56] Jarusutthirak C, Amy G, Croué J-P. Fouling characteristics of wastewater effluent organic matter (EfOM) isolates on NF and UF membranes. Desalination. 2002;**145**:247-255

[57] Seidel A, Elimelech M. Coupling between chemical and physical interactions in natural organic matter (NOM) fouling of nanofiltration membranes: Implications for fouling control. Journal of Membrane Science. 2002;**203**:245-255

[58] Hoek EMV, Elimelech M. Cake-enhanced concentration polarization: A new fouling mechanism for salt-rejecting membranes. Environmental Science & Technology. 2003; **37**:5581-5588

[59] Le-Clech P, Chen V, Fane TAG. Fouling in membrane bioreactors used in wastewater treatment. Journal of Membrane Science. 2006;**284**:17-53

[60] Ang WS, Elimelech M. Protein (BSA) fouling of reverse osmosis membranes: Implications for wastewater reclamation. Journal of Membrane Science. 2007;**296**:83-92

[61] Tang CY, Chong TH, Fane AG. Colloidal interactions and fouling of NF and RO membranes: A review. Advances in Colloid and Interface Science. 2011;**164**:126-143

[62] Mi B, Elimelech M. Chemical and physical aspects of organic fouling of forward osmosis membranes. Journal of Membrane Science. 2008;**320**:292-302

[63] Wang Y, Wicaksana F, Tang CY, Fane AG. Direct microscopic observation of forward osmosis membrane fouling. Environmental Science & Technology. 2010;**44**:7102-7109

[64] Cath TY, Gormly S, Beaudry EG, Flynn MT, Adams VD, Childress AE. Membrane contactor processes for wastewater reclamation in space: Part I. Direct osmotic concentration as pretreatment for reverse osmosis. Journal of Membrane Science. 2005;**257**:85-98

[65] Cath TY, Adams D, Childress AE. Membrane contactor processes for wastewater reclamation in space: II. Combined direct osmosis, osmotic distillation, and membrane distillation for treatment of metabolic wastewater. Journal of Membrane Science. 2005;**257**:111-119

[66] Meng F, Chae S-R, Drews A, Kraume M, Shin H-S, Yang F. Recent advances in membrane bioreactors (MBRs): Membrane fouling and membrane material. Water Research. 2009;**43**:1489-1512

[67] Lay WCL, Chong TH, Tang C, Fane AG, Zhang J, Liu Y. Fouling propensity of forward osmosis: Investigation of the slower flux decline phenomenon. Water Science and Technology. 2010;**61**:927-936

[68] Kang S-T, Subramani A, Hoek EMV, Deshusses MA, Matsumoto MR. Direct observation of biofouling in cross-flow microfiltration: Mechanisms of deposition and release. Journal of Membrane Science. 2004;**244**:151-165

[69] Wang S, Guillen G, Hoek EMV. Direct observation of microbial adhesion to membranes. Environmental Science & Technology. 2005;**39**:6461-6469

[70] Li H, Fane AG, Coster HGL, Vigneswaran S. Direct observation of particle deposition on the membrane surface during crossflow microfiltration. Journal of Membrane Science. 1998;**149**:83-97

[71] Li H, Fane AG, Coster HGL, Vigneswaran S. Observation of deposition and removal behaviour of submicron bacteria on the membrane surface during crossflow microfiltration. Journal of Membrane Science. 2003;**217**:29-41

[72] Bacchin P, Aimar P, Field RW. Critical and sustainable fluxes: Theory, experiments and applications. Journal of Membrane Science. 2006;**281**:42-69

[73] Zhang YP, Fane AG, Law AWK. Critical flux and particle deposition of bidisperse suspensions during crossflow microfiltration. Journal of Membrane Science. 2006;**282**:189-197

[74] Zhang YP, Fane AG, Law AWK. Critical flux and particle deposition of fractal flocs during crossflow microfiltration. Journal of Membrane Science. 2010;**353**:28-35

[75] Field RW, Wu D, Howell JA, Gupta BB. Critical flux concept for microfiltration fouling. Journal of Membrane Science. 1995;**100**:259-272

[76] Zhao S, Zou L, Mulcahy D. Effects of membrane orientation on process performance in forward osmosis applications. Journal of Membrane Science. 2011;**382**:308-315

[77] Li Z-Y, Yangali-Quintanilla V, Valladares-Linares R, Li Q, Zhan T, Amy G. Flux patterns and membrane fouling propensity during desalination of seawater by forward osmosis. Water Research. 2012;**46**:195-204

[78] Valladares Linares R, Yangali-Quintanilla V, Li Z, Amy G. Rejection of micropollutants by clean and fouled forward osmosis membrane. Water Research. 2014;**45**:6737-6744

[79] Hancock NT, Xu P, Heil DM, Bellona C, Cath TY. Comprehensive benchand pilot-scale investigation of trace organic compounds rejection by forward osmosis. Environmental Science & Technology. 2011;**45**:8483-8490

[80] Jin X, She Q, Ang X, Tang CY. Removal of boron and arsenic by forward osmosis membrane: Influence of membrane orientation and organic fouling. Journal of Membrane Science. 2012;**389**:182-187

[81] Hancock NT, Cath TY. Solute coupled diffusion in osmotically driven membrane processes. Environmental Science & Technology. 2009;**43**:6769-6775

[82] Zou S, Gu Y, Xiao D, Tang CY. The role of physical and chemical parameters on forward osmosis membrane fouling during algae separation. Journal of Membrane Science. 2011;**366**:356-362

[83] Wei J, Liu X, Qiu C, Wang R, Tang CY. Influence of monomer concentrations on the performance of polyamide-based thin film composite forward osmosis membranes. Journal of Membrane Science. 2011;**381**:110-117

[84] Lin Y-K, Ho H-O. Investigations on the drug releasing mechanism from an asymmetric membrane-coated capsule with an in situ formed delivery orifice. Journal of Controlled Release. 2003;**89**:57-69

[85] Herbig SM, Cardinal JR, Korsmeyer RW, Smith KL. Asymmetric-membrane tablet coatings for osmotic drug delivery. Journal of Controlled Release. 1995;**35**:127-136

[86] Baker RW. Membrane Technology and Applications. 2nd ed. Etobicoke, Canada: John Wiley & Sons, Ltd.; 2004

[87] Zhang S, Wang KY, Chung T-S, Chen H, Jean YC, Amy G. Well-constructed cellulose acetate membranes for forward osmosis: Minimized internal concentration polarization with an ultra-thin selective layer. Journal of Membrane Science. 2010;**360**:522-535

[88] Gerstandt K, Peinemann KV, Skilhagen SE, Thorsen T, Holt T. Membrane processes in energy supply for an osmotic power plant. Desalination. 2008;**224**:64-70

[89] Su J, Yang Q, Teo JF, Chung T-S. Cellulose acetate nanofiltration hollow fiber membranes for forward osmosis processes. Journal of Membrane Science. 2010;**355**:36-44

[90] Wang KY, Ong RC, Chung T-S. Double-skinned forward osmosis membranes for reducing internal concentration polarization within the porous sublayer. Industrial & Engineering Chemistry Research. 2010;**49**:4824-4831

[91] Zhang S, Wang KY, Chung T-S, Jean YC, Chen H. Molecular design of the cellulose ester-based forward osmosis membranes for desalination. Chemical Engineering Science. 2011;**66**:2008-2018

[92] Tang CY, She Q, Lay WCL, Wang R, Field R, Fane AG. Modeling double skinned FO membranes. Desalination. 2014;**283**:178-186

[93] Sairam M, Sereewatthanawut E, Li K, Bismarck A, Livingston AG. Method for the preparation of cellulose acetate flat sheet composite membranes for forward osmosis–desalination using $MgSO_4$ draw solution. Desalination. 2011;**273**:299-307

[94] Vos KD, Burris FO, Riley RL. Kinetic study of the hydrolysis of cellulose acetate in the pH range of 2–10. Journal of Applied Polymer Science. 1966;**10**:825-832

[95] Merten U. Flow relationships in reverse osmosis. Industrial & Engineering Chemistry Fundamentals. 1963;**2**:229-232

[96] Widjojo N, Chung T-S, Weber M, Maletzko C, Warzelhan V. The role of sulphonated polymer and macrovoid-free structure in the support layer for thin-film composite (TFC) forward osmosis (FO) membranes. Journal of Membrane Science. 2011;**383**:214-223

[97] McCutcheon JR, Elimelech M. Influence of membrane support layer hydrophobicity on water flux in osmotically driven membrane processes. Journal of Membrane Science. 2008;**318**:458-466

[98] Yu Y, Seo S, Kim I-C, Lee S. Nanoporous polyethersulfone (PES) membrane with enhanced flux applied in forward osmosis process. Journal of Membrane Science. 2011;**375**:63-68

[99] Song X, Liu Z, Sun DD. Nano gives the answer: Breaking the bottleneck of internal concentration polarization with a nanofiber composite forward osmosis membrane for a high water production rate. Advanced Materials. 2015;**23**:3256-3260

[100] Bui N, Lind ML, Hoek EMV, McCutcheon JR. Electrospun nanofiber supported thin film composite membranes for engineered osmosis. Journal of Membrane Science. 2011:385, 10-386, 19

[101] Fayyazi F, Feijani E, Mahdavi H. Chemically modified polysulfone membrane containing palladium nanoparticles: Preparation, characterization and application as an efficient catalytic membrane for Suzuki reaction. Chemical Engineering Science. 2015;**134**: 549-554

[102] Setiawan L, Wang R, Li K, Fane AG. Fabrication of novel poly(amide–imide) forward osmosis hollow fiber membranes with a positively charged nanofiltration-like selective layer. Journal of Membrane Science. 2011;**369**:196-205

[103] Saren Q, Qiu CQ, Tang CY. Synthesis and characterization of novel forward osmosis membranes based on layer-by-layer assembly. Environmental Science & Technology. 2011;**45**:5201-5208

[104] Qiu C, Qi S, Tang CY. Synthesis of high flux forward osmosis membranes by chemically crosslinked layer-by-layer polyelectrolytes. Journal of Membrane Science. 2011; **381**:74-80

[105] Hong SU, Bruening ML. Separation of amino acid mixtures using multilayer polyelectrolyte nanofiltration membranes. Journal of Membrane Science. 2006;**280**:1-5

[106] Hong SU, Ouyang L, Bruening ML. Recovery of phosphate using multilayer polyelectrolyte nanofiltration membranes. Journal of Membrane Science. 2009;**327**:2-5

[107] Bruening ML, Dotzauer DM, Jain P, Ouyang L, Baker GL. Creation of functional membranes using polyelectrolyte multilayers and polymer brushes. Langmuir. 2008;**24**:7663-7673

[108] Stanton BW, Harris JJ, Miller MD, Bruening ML. Ultrathin, multilayered polyelectrolyte films as nanofiltration membranes. Langmuir. 2013;**19**:7038-7042

[109] Malaisamy R, Bruening ML. High-flux nanofiltration membranes prepared by adsorption of multilayer polyelectrolyte membranes on polymeric supports. Langmuir. 2005;**21**: 10587-10592

[110] Zhang G, Yan H, Ji S, Liu Z. Self-assembly of polyelectrolyte multilayer pervaporation membranes by a dynamic layer-by-layer technique on a hydrolyzed polyacrylonitrile ultrafiltration membrane. Journal of Membrane Science. 2007;**292**:1-8

[111] Jin W, Toutianoush A, Tieke B. Use of polyelectrolyte layer-by-layer assemblies as nanofiltration and reverse osmosis membranes. Langmuir. 2003;**19**:2550-2553

[112] Lee H, Lee Y, Statz AR, Rho J, Park TG, Messersmith PB. Substrateindependent layer-by-layer assembly by using mussel-adhesive-inspired polymers. Advanced Materials. 2008;**20**:1619-1623

[113] Akbari A, Desclaux S, Rouch JC, Aptel P, Remigy JC. New UV-photografted nanofiltration membranes for the treatment of colored textile dye effluents. Journal of Membrane Science. 2006;**286**:342-350

[114] Akbari A, Desclaux S, Rouch JC, Remigy JC. Application of nanofiltration hollow fibre membranes, developed by photografting, to treatment of anionic dye solutions. Journal of Membrane Science. 2007;**297**:243-252

[115] Li X-L, Zhu L-P, Xu Y-Y, Yi Z, Zhu B-K. A novel positively charged nanofiltration membrane prepared from N,N-dimethylaminoethyl methacrylate by quaternization cross-linking. Journal of Membrane Science. 2011;**374**:33-42

[116] Deng H, Xu Y, Chen Q, Wei X, Zhu B. High flux positively charged nanofiltration membranes prepared by UV-initiated graft polymerization of methacrylatoethyl trimethyl ammonium chloride (DMC) onto polysulfone membranes. Journal of Membrane Science. 2011;**366**:363-372

[117] Miao J, Chen G-H, Gao C-J. A novel kind of amphoteric composite nanofiltration membrane prepared from sulfated chitosan (SCS). Desalination. 2005;**181**:173-183

[118] He T, Frank M, Mulder MHV, Wessling M. Preparation and characterization of nanofiltration membranes by coating polyethersulfone hollow fibers with sulfonated poly (ether ether ketone) (SPEEK). Journal of Membrane Science. 2008;**307**:62-72

[119] Achilli A, Cath TY, Childress AE. Selection of inorganic-based draw solutions for forward osmosis applications. Journal of Membrane Science. 2010;**364**:233-241

[120] Kim T-W, Kim Y, Yun C, Jang H, Kim W, Park S. Systematic approach for draw solute selection and optimal system design for forward osmosis desalination. Desalination. 2012;**284**:253-260

[121] Yen SK, Mehnas Haja F, Su NM, Wang KY, Chung T-S. Study of draw solutes using 2-methylimidazole-based compounds in forward osmosis. Journal of Membrane Science. 2010;**364**:242-252

[122] Ling MM, Chung T-S. Desalination process using super hydrophilic nanoparticles via forward osmosis integrated with ultrafiltration regeneration. Desalination. 2011;**278**:194-202

[123] Ling MM, Wang KY, Chung T-S. Highly water-soluble magnetic nanoparticles as novel draw solutes in forward osmosis for water reuse. Industrial & Engineering Chemistry Research. 2010;**49**:5869-5876

[124] Ge Q, Su J, Chung T-S, Amy G. Hydrophilic superparamagnetic nanoparticles: Synthesis, characterization, and performance in forward osmosis processes. Industrial & Engineering Chemistry Research. 2011;**50**:382-388

[125] Ling MM, Chung T-S, Lu X. Facile synthesis of thermosensitive magnetic nanoparticles as smart draw solutes in forward osmosis. Chemical Communications. 2011;**47**:10788-10790

[126] Li D, Zhang X, Yao J, Simon GP, Wang H. Stimuli-responsive polymer hydrogels as a new class of draw agent for forward osmosis desalination. Chemical Communications. 2011;**47**:1710-1712

[127] Li D, Zhang X, Yao J, Zeng Y, Simon GP, Wang H. Composite polymer hydrogels as draw agents in forward osmosis and solar dewatering. Soft Matter. 2011;**7**:10048-10056

[128] Liu Z, Bai H, Lee J, Sun DD. A low-energy forward osmosis process to produce drinking water. Energy & Environmental Science. 2011;**4**:2582-2585

Chapter 9

Fouling and Cleaning in Osmotically Driven Membranes

Martha Noro Chollom and Sudesh Rathilal

Additional information is available at the end of the chapter

http://dx.doi.org/10.5772/intechopen.73047

Abstract

Fouling is a phenomenon that occurs in all membrane processes. It is a complex problem, which limits the full operation of this technology. Fouling in pressure-driven membranes (PDMs) has been studied extensively, and the occurrence is well understood in that methods of mitigation have been proposed; however, limitations still occur for their full implementation. The use of osmotically driven membranes (ODMs) for water treatment is an emerging technology, which has shown some advantages such as low hydraulic pressure operation, high solute rejection and high recovery over PDMs. However, like in PDMs, fouling still presents a challenge. This chapter is aimed at evaluating the impact of fouling on the ODM performance, exploring the factors and mechanisms governing the fouling behaviour, developing approaches for mitigating fouling, elucidating the effect of membrane fouling and providing mitigation strategies as well as the causes of fouling in ODMs.

Keywords: membrane fouling, fouling mitigation, forward osmosis, pressure retarded osmosis, pretreatment

1. Introduction

The use of osmotically driven membranes (ODMs), such as forward osmosis (FO), pressure retarded osmosis (PRO), direct osmotic concentration (DOC) and osmotic dilution (ODN), for water treatment is an emerging technology that has shown some advantages such as low hydraulic pressure operation and hence low energy consumption, high solute rejection and high recovery over pressure-driven membranes (PDMs) [1–5]. The ODMs are seen to gradually outperform the conventional PDMs. For instance, Mi and Elimelech [6], in their review, noted that forward osmosis is said to consume only about 20% of the electrical energy required by other processes such as desalination. These processes could use low quality heat like the waste heat from power plants as their energy inputs. The advantage associated with it has been the higher recovery, and because of this, less discharge of brine to the environment is noticeable [6].

Until recently, the focus of most studies has been on PDMs; however, a shift in research is being noticed and more research is emerging regarding the application of ODMs. However, the studies on ODMs published has been intensified mainly on issues such as choice of draw solutions, membrane properties and other factors relating to the application of ODMs. Fouling mechanisms in these membranes has, on the contrary, received less attention. On the other hand, fouling in PDMs has been studied extensively and methods of mitigation and control are being adopted for their implementation [6].

Fouling is a phenomenon that occurs in all membrane processes. It is a complex problem that limits the full operation of this technology. Fouling can be caused by the accumulation of suspended particles or colloids, organic molecules and also soluble inorganic compounds, microorganisms, or a combinations of all these on the membrane [7]. Different substances have been identified to cause fouling in membranes and as such, this can result in different fouling mechanisms in the membranes. For example, fouling could occur as a result of the deposition of foulants onto the surface of the membrane thus forming a cake layer. This phenomenon is commonly referred to as external fouling. It could occur within the pores of the membrane. In this instance, the foulant sizes could be relatively smaller than the pores of the membranes, hence penetrates the pores of the membrane thereby leading to pore blocking. This type of fouling is called internal fouling [7].

Fouling occurs in all membrane operations, however, the tendency and its behaviour varies due to the mode of operation, the nature of the membranes and the nature of the foulants. Mi and Elimelech [8] studied the chemical and physical aspects of organic fouling of FO membranes using alginate, bovine serum albumin (BSA), and Aldrich humic acid (AHA) as the exemplary organic foulants. In that study, the effect of chemical and physical interactions such as intermolecular adhesion forces, calcium binding and the membranes initial permeate flux were investigated. Similarly, the membrane orientation on organic fouling of FO membranes was investigated. They observed that there was a relationship between organic fouling and intermolecular adhesion, thus indicating that foulant-foulant interaction is an important aspect that can determine the rate and level of fouling, therefore emphasising that the main factors that control membrane fouling differ from foulant to foulant [8].

Studies on the fouling in ODMs have revealed that fouling propensity within the ODMs is lower as compared to PDMs [2, 3, 9, 10]. The lower fouling propensity is said to be so in the case, whereby the active layer of the membrane is arranged to face the feed solution containing the foulant. In addition, the low flux conditions and lack of applied pressure in the FO process have been highlighted as some of the reasons for this occurrence. However, internal concentration polarisation (ICP) could still occur within the membrane [10]. Therefore, ICP is one of the major drawbacks of ODMs especially in FO [3].

Factors such as draw solutions, hydrodynamics and operating conditions and feed water characteristics could impact fouling in different ways. The effect of these factors, if properly managed, will help mitigate fouling propensity on the membrane. The configuration of the membranes can also affect membrane performance significantly. Tang et al. [3] studied the coupled effects of internal concentration polarisation and fouling on flux behaviour of FO membranes during humic acid filtration. They found that the membrane orientation plays an important role. In their observation, ICP occurred more when the active layer faced the draw solution (AL-facing-DS) as compared to when the membrane active layer faced the feed water (AL-facing-FW). This leads to a dilutive ICP in the FO support layer [3]. However, a more substantial flux stability is said to be achieved by the AL-facing FW as against the dilutions of the bulk draw solution and membrane fouling.

Thus, understanding the phenomenon of fouling in ODMs will provide more information that could lead to the development of new FO membranes with reduced ICP and high water permeability. The aim of this review is to evaluate the impact of fouling on the ODMs performance and to explore the factors and mechanisms governing the fouling behaviour. Further, it aims to develop approaches for mitigating fouling and to further elucidate the effect of membrane fouling and mitigation strategies. The causes of fouling in ODMs will also be described. The performance of FO membranes is defined by three parameters; the pure water permeability coefficient, solute permeability coefficient and the structural parameter. The solute permeability describes mass transport across the membrane active layer while the structural parameter governs the transport phenomena across the membrane support layer. The aforementioned parameters are used to describe the permeate water and solute fluxes of FO processes [11].

2. Fouling in membranes

The fouling phenomenon in PDMs and ODMS differs in some ways. In PDMs, factors that affect membrane fouling can be classified into three categories: membrane properties, operating parameters and the nature of the waste water to be treated. In ODMs, there could be additional factors to the aforementioned, such as the membrane orientation and the type of draw solutions [12–14]. The driving force for PDM systems is pressure. Hence, the relationship between pressure and flux is positive. A rise in pressure causes a rise in flux; however, for a feed mixture there is a point where a further increase in pressure results in a minimal increase in flux. This is because the particles of the component being rejected by the membrane accumulates on the membrane surface and obstructs the passage of the solvent through the membrane. If the process is allowed to continue to run, the rejected layer on the membrane surface grows thicker and becomes more and more resistant to solvent flow and this results in the flux dropping. At this point, it is said that the membrane is fouled and it is no longer economically justifiable to continue to run because the added energy to the system does not recover or even maintain flux [15, 16].

On the other hand, ODM systems use the osmotic pressure gradient, which is the chemical potential difference between the feed water and the concentrated draw solution as its driving force. With this application, the use of external pressure is not needed [3, 10, 17, 18]. As a result of this, the system is said to be more economically viable due to its significantly low energy consumption [19]. This advantage has attracted the application of ODMs in seawater desalination, wastewater reclamation and in liquid food processing. However, like PDMs, a decline in flux always results, due to the severe internal concentration polarisation that always occurs in the porous membrane support [9, 17, 20, 21].

Fouling in membranes could occur internally or externally on the membrane. The extent of fouling in membranes depends on the type of separation and the type of membrane used to carry out the separation. Fouling leads to an overall increase in membrane resistance for mass transport, and hence affects the performance of membranes by a gradual decrease in flux and a decrease in rejection. The effect of this is seen in the deterioration of the membrane properties and as such results in high costs of operation and cleaning of the membranes to restore its initial flux [22].

The orientation of ODMs could be in two ways; active layer-feed solution (AL-FS) and the active layer-draw (AL-DS) solution. Hence, the nature of fouling differs with orientation [17]. The AL-FS mode is reported to be the FO mode (normal), while the AL-DS is the (PRO) reversed AF [23]. AL-FS orientation is when the active layer of the membrane faces the feed solution; and the AL-DS is when the active layer of the membrane faces the draw solution [14]. The type of fouling that will occur in the membrane will depend on the orientation of the membrane. The AL-DS could also be referred to as external fouling because solutes from the feed solution accumulate on the membrane surface thus forming a cake layer with time. This is similar to that formed in PDMs. On the other hand, in the AL-DS orientation, complications arise because other fouling mechanisms, such as pore blocking, could occur in addition to the cake layer formation. The occurrence of this is however dependent on the type and nature of foulants. Smaller sized particles will find their way into the pores of the membranes, thereby causing the membrane pores to be become blocked and therefore, internal fouling occurs. However, if it contains larger particles, these foulants will remain on the surface of the membrane and are thus deposited on the membrane surface hence blocking the pores leading to external fouling. If the feed solution contains a mixture of both sizes of foulants, both types of fouling could be occurring within the membrane [7, 23].

Both orientations affect the performance of the membrane in different ways. The AL-DS orientation has shown to have a lower initial flux, however, a higher fouling resistance, while the AL-FS has a higher initial flux but is less prone to fouling. However, it can be immensely affected by dilutive ICP. Therefore, ODM membranes are faced either with a more severe dilutive ICP in AL-FS or having much greater fouling tendencies in AL-DS, and therefore a balance must be reached in order to obtain optimum performance carrying out mitigating measures [17]. However, Chen et al. [23] reported in their study that the effect of fouling is more enhanced in PRO membranes. The purpose for this is the fact that PRO membranes are composed of a denser or thicker structure than the FO membranes to enable them to withstand the high pressure loading. Therefore, the denser structure contributes to the fouling tendencies.

Mi and Elimelech [6] studied the organic fouling of forward osmosis membranes. The main aim of that study was to examine organic fouling and the cleaning methods that will follow in the FO. Two types of membranes were used; polyamide and cellulose acetate (PA and CA) with alginate as the model foulant. Again, they used atomic force measurement (AFM) to detect the role of membrane materials in determining membrane fouling and cleaning behaviour.

They found that the PA was prone to more fouling propensity. The PA membrane surface were said to contain some adhesive sites even though lower as compared to those from the CA. The higher fouling in the PA membranes were due to the fact that the PA membranes caused more adsorption, thereby leading to a more severe fouling at an early stage [6].

Furthermore, Mi and Elimelech [6], in the same study, considered the flux behaviour in RO and FO and found that similar flux patterns were obtained in membrane types; however, the flux recovery was different. A higher flux was recovered from FO than the RO. The reason for this occurrence was attributed to the fact that the fouling layer formed on the FO membrane was less compact due to a lack of hydraulic pressure application [6].

Xie et al. [24], in their study on the role of pressure in organic fouling in FO and RO, used alginate as the foulant, while varying the contribution pressure in terms of osmotic and hydraulic. From that study, two possible mechanisms of fouling were identified which were permeation drag force and compression of foulants. The fouling thickness that was observed by them was in the decreasing order of FO < PFO < R O. They arrived at the same conclusion that hydraulic pressure plays a significant part in the compression of the fouling layer to a great extent [24]. The drag force was the only applied force in FO; however, this did not necessarily mean that fouling will not occur in the FO membranes.

As stated earlier, different factors are responsible for fouling in membranes. One dominant factor is the nature of contaminants that can be found in the wastewater, for example, colloidal particles or particulate matter, dissolved organics, chemical reactants, micro-organisms and other microbial substances [17]. Foulants are colloidal materials with different properties, which interact with the membrane thereby causing fouling. They can be grouped into four categories: organic precipitates, inorganic precipitates, biological and particulates [13, 25]. Hence, the type of fouling can be grouped based on the foulant type, e.g., inorganic (scaling), organic and biofouling [22].

2.1. Inorganic fouling

Inorganic fouling normally results from the deposition and accumulation of inorganic matter and other precipitates such as metal hydroxides and silica on the surface of the membrane. Inorganic fouling will foul the membrane both on the surface and internally. The precipitates are formed when the concentration of the chemical species is more than their saturation concentrations. This tends to happen on the membrane surface where accumulations of particles occur due to retention on the membranes. The result of this will be a decline in flux [17, 26].

Mi and Elimelech [27], in their study on the gypsum scaling and cleaning in FO, reported a decline in flux in both RO and FO modes. About 96% of the flux was recovered in the FO mode following a water rinse only without the use of any chemical cleaning agent. In the RO mode, however, the flux recovered was 10% lower than that of FO. Similarly, the same authors, Mi and Elimelech [26], reported in their study for silica scaling and scaling reversibility in FO, a decline in flux both in the FO and RO mode. However, 100% flux was recovered in the FO and only 80% in the RO modes. They concluded, after characterising the fouled membrane, that scaling on the membrane originated from the monosilicic acid deposition on the membrane surface, which was followed by polymerisation (the formation of a soft amorphous silica gel layer that hardened with time by a continuous dehydration). Again, on the use of the AFM force measurement, it was revealed that the membrane surface roughness played a crucial part by increasing the adhesion force between the membrane and the silica gel layer, thereby considerably reducing the cleaning efficiency of the membrane [26].

The combination of alginate, which is the main component of polysaccharides with calcium ions in water, could lead to a more pronounced decline in flux due to the formation of a cake layer or gel layer. Chun et al. [28] reported that inorganic scaling, which was caused by calcium and phosphate and the interactions with other organic constituents in the feed solutions used, were the main cause of the reduction in flux of the membrane [28]. The cleaning of the inorganic scaling was, however, poor after using both physical and chemical methods. On further characterisation of the membrane, it was confirmed that gypsum and organic components that were present in the feed solution might have formed a gel layer (calcium bridging), thereby enhancing the fouling layer rigidity [17, 28]. Silica scaling is said to be difficult to be removed physically, while other types of the NOM foulants can be easily removed.

2.2. Organic fouling

The adsorption of organic matter such as humic substances, protein, and grease onto the membrane surfaces is referred to as organic fouling. These organic substances can be hydrophobic, hydrophilic or transphilic in nature. The mechanisms of organic fouling are complicated due to the wide variety of organic foulants existing in natural waters. These organic matters, commonly known as natural organic matter (NOM), are prevalent in most natural water sources such as run-offs, rivers, seawater and ground water [17].

NOM which are terrestrially derived are known as autochthonous NOM. There are also the microbially derived and wastewater NOM. Each of these organic fractions foul membranes differently because of different hydrophobicity, molecular weight size and charge density. However, microbially derived NOM are found to be the worst foulants. Fouling from these fractions is found to be most problematic and severe [29]. A study by Bessiere et al. [30] on the effect of NOM on fouling shows that hydrophilic components of NOM are responsible for the rapid but reversible fouling on the membrane. **Figure 2** elaborates concentration polarisation that could occur in FO membranes. The hydrophobic components were found to be responsible for the slow but irreversible fouling on the membrane. The hydrophilic components were small compared to the hydrophobic components. Both of these components are adsorbed on the membrane material.

The size of NOM plays a great role on the fouling of the membrane. Because NOM adsorbs onto the membrane, small NOM enter the pore of the membrane and get adsorbed on the wall of the pore channel resulting in pore narrowing. Larger NOM components get trapped at the entrance of the membrane pores and block the entrance to the pore channel resulting in cake layer formation as filtration progresses [31]. Fan et al. [32] found that the fouling order of hydrophobic membranes by NOM material is as follows; hydrophilic neutrals > hydrophobic acids > transphilic acids. Again, Chun et al. [17], in their study, noted that hydrophilic, H-bond acceptor, non-H-bond-donor and neutrally charged membranes are said to be resistant to organic fouling; however, hydrophobic and rougher membranes are more prone to fouling by NOM [17].

For ODM systems, a strong correlation has been established between organic fouling and intermolecular adhesion forces. For example, Mi and Elimelech [8] studied the chemical and physical aspects of organic fouling of FO membranes and found a strong correlation between organic fouling and intermolecular adhesion forces, which indicated that foulant-foulant interaction played an important role in determining the extent of the fouling [8]. They used the AFM. Adhesion force measurement was used to elucidate the impact of membrane material fouling. They found that the small adhesive sites on the membrane played a significant role in organic fouling formation [8]. They concluded that permeation drag, hydrodynamic shear force and calcium binding were the main contributing factors that govern organic fouling development [8, 17].

Colloidal matter in a suspension can be charged and depending on the charge of both the membrane and the particle, adhesion or repulsion will occur. The charge of the particles can be altered by adjusting the pH of the suspension. pH adjustment changes the electrostatic interaction between the membrane and particle from attractive to repulsive or from repulsive to attractive [33]. The effect of the ionic strength of colloidal particles on fouling was also studied by Singh and Song [34]. The study found that increasing the ionic strength of colloidal matter and its concentration in the feed solution increases the fouling potential of the water linearly.

2.3. Biofouling

Biofouling in simple terms can be defined as biological fouling. It is a net resultant of microbial attachment to the membranes and the consequent growth and discharge of biopolymers that are connected with this microbial activity. The foulants in biofouling include proteins, organics, organic acids, polysaccharide fats, etc. [14]. Biofoulants in this section will be divided into humic materials and micro-organisms (bacteria) [35]. The attached communities of bacteria in aquatic systems are encased in a glycocalyx matrix that is polysaccharide in nature. This matrix material mediates adhesion. The biofilm is made up of single cells and micro colonies that are enclosed in a hydrated, predominantly anionic exopolymer matrix. The attachment of bacteria to surfaces is irreversible and it results from a secretion by the bacteria itself which is a matrix of extracellular polymeric substances in which the bacteria cells are embedded upon [36].

This adhesion of microbial cells to the membrane surface is the beginning of membrane biofouling. Subsequent to attachment of microbial, a biofilm layer is formed, which has a composition that is vast in diversity of different micro-organisms which could be bacteria, algae, protozoa, fungi, etc. [36]. Basically, three steps are involved in the formation of the biofilm; adsorption of the organic species and other suspended species on the wet membrane, transportation of microbial cells to the formed film and finally, the microbial cells then attach themselves on the membrane surface. The growth and metabolism as well as the biofilm of the attached organisms are then developed [37].

Extracellular polymeric substances (EPS) are high molecular weight secretions of microorganisms that are made up of organic substances such as polysaccharide, protein, nucleic acids and lipids. EPS offer a binding base for biofilm to the membrane surface. They contribute to the mechanical stability of the biofilm and to the organisation of the biofilm community. Once the biofouling has been established, other organic and inorganic materials contribute to the building of the fouling mass [38]. EPS promotes the adhesion of microbial to surfaces by changing the physicochemical characteristics of the biofilm fouled membrane surface such as its charge, hydrophobicity and roughness. The EPS offers building blocks between the membrane pore and microbial cells. High concentration of EPS contributes greater binding capacity. EPS aggregates are comprised of charged groups, and they therefore have both wetting and cross-linking characteristics which contains both hydrophobic and hydrophilic sites on their structure. This enables them to be able to adhere to both hydrophilic and hydrophobic surfaces. The factor that makes biofouling very complex to understand and plan against is the fact that EPS have flexibility and rearrangement characteristics. This means EPS structure will rearrange themselves so that they are able to stick to any surface [37].

Biofouling is one of the most difficult fouling to control as there is a large range of biofoulants that could be present in a particular aqueous system at a particular time for a particular feed solution [17]. Other types of fouling are easily mitigated by the use of chemical and physical pre-treatment. Like other fouling mechanisms, it causes significant losses in flux it is reported that it can cause a 10–15% decline in the membrane performance of the start-up values under the applied operational conditions [17]. Further impacts of biofouling are observed in membrane biodegradation which can lead to an increased salt passage as well as raising energy requirements. Consequently, a higher operating and maintenance cost and possibly shortening of membrane lifetime will be the overall impact [17].

Bogler et al. [14] reported that biofouling in FO has a lower influence on permeate water flux than that in PDM membranes. The same hydrodynamic conditions, feed concentration, membrane type and bacterial concentration were used to test for biofouling in RO and FO membranes, and it was found that there was a 10% decrease in flux as compared to the 30% in RO membranes after 24 h. However, the influence of biofilm in FO has been considered to be more complicated than in RO. This is due to the fact that the additional phenomenon that occurs is unique to membrane systems, which are driven by osmotic force [14]. According to Bogler et al. [14], there is an interaction between the reverse solute and the biofilm by the draw solution especially when it contains divalent cations as calcium [14]. Again, the biofilm formed on the FO membrane is more loosed and thicker than that formed in RO membrane. This was said to enhance CP instead of the additional hydraulic resistance as the main reason for permeate water flux reduction [14].

3. Concentration polarisation in ODMs

The major challenge in bringing about a deep knowledge that will aid to understand membrane fouling is the difficulties in the identification of the actual foulants, and distinction between the indicators of fouling and effect of CP. CP is the occurrence in membrane processes, whereby the concentration of solute near the membrane surface is very different from that of the bulk solution [7, 39, 40]. In membrane systems, using hydraulic pressure, the liquid is passed through the membrane and the particles accumulates near the membrane surface thereby forming a thin layer. In the layer, the particles get stuck in the transverse direction which is close to the membrane surface such that retained particles on the stationary layer provides an added resistance to the permeate flow. The resistance therefore depends on the total number of particles formed on the layer and on as well as the spatial distribution. The stationary layer containing retained particles is called concentration polarisation (CP) and is inherent to all cross flow filtration processes [41]. The retained solutes/particles diffuse back to the bulk solution. However, the rate of permeation in membrane systems is higher than the rate at which the rejected solutes diffuse back to the bulk solution. This results in a higher solute concentration at the membrane surface than in the bulk solution. As filtration progresses, the concentration of the particles on the membrane surface becomes so high that a gel layer is formed which acts as a secondary barrier to permeate flux [42].

The effect of the CP is noticeable during membrane operation by the reduction of permeate flux as well as decline in the effective driving force across the membrane, leading to further fouling of the membrane. This influence occurs both in PDMs and ODMs. A similar scenario is observed with the ODM membranes; however, because the driving force here is osmotic pressure, a difference in CP mechanism is noticed. It has been emphasised that in ODMs, CP could occur as internal concentration polarisation (ICP) or external concentration polarisation (ECP) of the membrane (see **Figure 1**) [7, 17, 21, 39]. **Figure 1** shows that the solutes on the draw side decreases while those on the feed side increases, as a result a concentration gradient is formed and as such, a reduction in the osmotic pressure difference between the two solutions is enriched as shown in **Figure 1B**. The result of this is seen in the build-up of CP, which thus reduces flux flow.

ICP occurs within the membrane porous support layer, while ECP occurs on both sides of the membrane surfaces [17, 39]. The orientation of the membranes plays a vital role with regards to the type of CP that will occur on the membrane and it should be noted that CP is contributed by both convective and reverse solute diffusion (RSD) [21, 39]. The effect of ICP is more pronounced on the membranes than that of ECP. The reason is attributed to the fact that there is an axial flow of salt solution within the asymmetric FO membrane, which is the solute that enters and exits the porous support layer. To further validate and understand the nature of CP, both ECP and ICP have been elucidated and categorised as concentrative external concentration polarisation (CECP) and diluted external concentration polarisation (DECP) for ECP and diluted ICP (DCIP) (**Figure 2**).

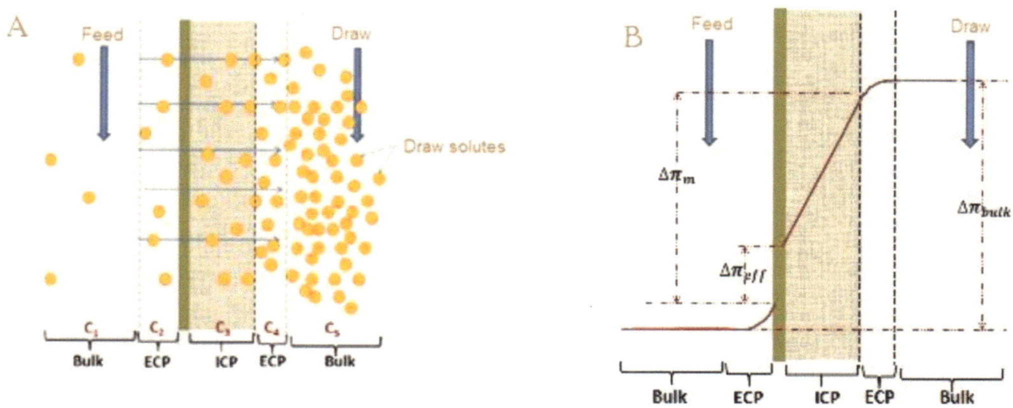

Figure 1. Schematic representation of (A) concentration polarisation on FO membranes (B) osmotic pressure difference due to effects of CP [43].

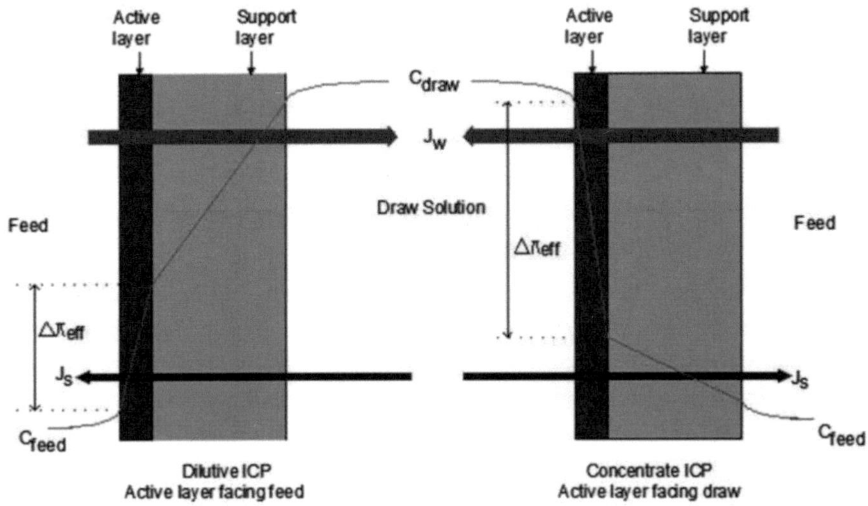

Figure 2. Schematic representation of DECP and DICP on a porous support layer [17].

The logical explanation for this is the fact that the drawn solution is greatly diluted by the permeate water within the porous support of the membrane [39]. Therefore, CECP occurs when the active layer of the membrane faces the feed solution and as such, there is accumulation of the solutes thereby increasing the feed concentration while DECP occurs when the active layer of the membrane faces the draw solution, hence dilution. The outcome of this is seen in the effective osmotic pressure of the feed solution increasing from the bulk solution to the membrane surface and that of the draw solution decreasing from the draw bulk solution to the to the membrane surface. This can be seen in the equation of FO which describes the permeate flux as seen in (Eq. (1)).

$$Jw = A * \Delta\pi = (\pi_{D,m} - \pi_{F,m}) \quad (1)$$

where Jw is the permeate flux; A is the pure water permeability coefficient; $\pi_{F,m}$ is the osmotic pressure of feed solution on the membrane surface; $\pi_{D,m}$ is the osmotic pressure of draw solution on the membrane surface.

Equation (1), above which describes the flux in FO, was first modified by McCutcheon and Elimelech [44].

$$\frac{\pi f, m}{\pi f, b} = \exp\frac{Jw}{K} \quad (2)$$

$$\frac{\pi D, m}{\pi D, b} = \exp\frac{Jw}{K} \quad (3)$$

Where Jw is the permeate flux and k is the mass transfer coefficient. k is related to the Sherwood number (Sh), solute diffusion coefficient and hydraulic diameter of the flow channel

$$k = Sh * \frac{D}{d_h} \quad (4)$$

Depending on the flow regime, *Sh* is calculated using either Eqs. (5) and (6).

$$sh = 1.85 * \left(Re * Sc * \frac{d}{L}\right)^{0.33} \text{ (laminar flow } Re \leq 2100) \quad (5)$$

$$sh = 0.04 * Re^{0.75} Sc^{0.33} \text{ (turbulent flow } Re > 2100) \quad (6)$$

However, if the salt back diffusion across the membrane does not take place, then the permeate flux [Eq. (1)] is modified by taking CECP and DECP into consideration the equation can be transformed to Eq. (7).

$$Jw = A * \left(\pi D, b * \exp\left(-\frac{Jw}{Kd}\right) - \pi F, b * \exp\left(\frac{Jw}{Kf}\right)\right) \quad (7)$$

Equation (7) describes ECP in FO; however, ECP effect on flux decline is not as pronounced as that of ICP. The impact of ICP on the membrane is more prominent on the membrane. Therefore, to account for ICP that occurs in the membrane, the equation is modified [1].

Apparently, due to the nature of most membranes being asymmetric and comprising of a thin selective layer and a thick, non-elective layer, Eq. (7) cannot be used to describe ICP porous support layer. This being due to the fact that the osmotic pressure of a solution can be established only at the interface with the selective layer. Noted also is the fact that asymmetric structure of the membrane.

The asymmetric structure of the membrane is made such that one of the boundary layers is within the support layer which then results in ICP [1]. Therefore, to justify for the porous layer, an effective mass transfer coefficient (K_{eff}) is defined as shown in Eq. (8) [1].

$$K_{eff} = \frac{D_s \epsilon}{\tau \delta} = \frac{D_s \epsilon}{\tau t} \quad (8)$$

Where D_s is the diffusivity of the solute, δ is the thickness of the boundary layer ε, τ, and t are the porosity, tortuosity, and thickness of the porous support layer of the membrane.

In normal mode of FO, Eq. (7) is modified to:

$$Jw = A * \left[\pi D, b * \exp\left(-\frac{Jw}{Kdeff}\right) - \pi F, b * \exp\left(\frac{Jw}{Kf}\right)\right] \quad (9)$$

According to Chun et al. [17], the effect of ECP is suffered on all membrane processes. The effect of CP is experienced more on the interface because it is more in contact with the bulk solution. This is due to the fact that the layer interface becomes polarised. Transport of water and other solutes within this interface is merely on advection and molecular diffusion [17]. Because, it is only a minimal amount of the solute that is able to penetrate through the dense selective layer, back diffusion occurs with an accumulation of solute within the porous layer which leads to the formation of ICP effect [17]. Like PDMs, enhanced cake layer concentration

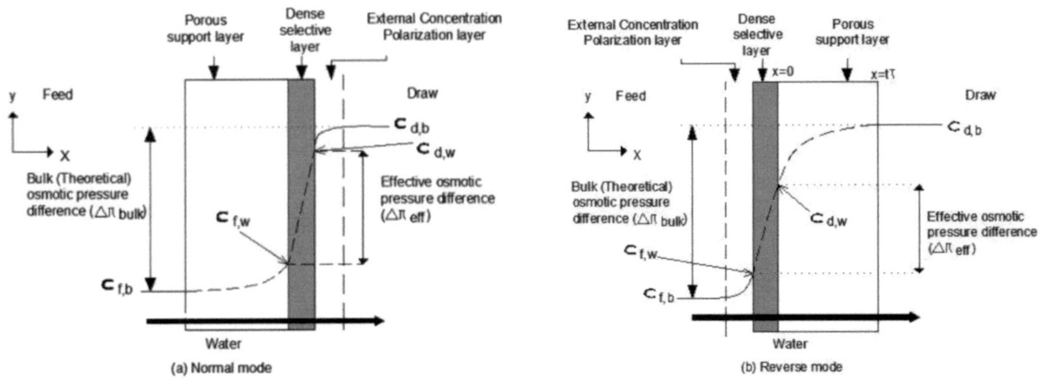

Figure 3. Schematic representation of (a) and (b) shows the different membrane orientation. (a) The normal mode and (b) reverse mode adapted from [24].

polarisation could be formed on the membrane surface. This can happen when the flux is significantly high and thus leads to the formation of a porous fouling layer on the membrane surface such that solute diffusion inside this layer becomes seriously hindered [21, 45]. The effects of Enhanced CP can be expressed through a mass transfer coefficient as shown in Eq. (10) [21].

$$\mathrm{KECP} = \frac{\mathrm{Dml}.\varepsilon \mathrm{la}}{\delta \mathrm{la}.\tau \mathrm{la}} \tag{10}$$

where D_{ml} is diffusion coefficient of the solutes inside the fouling layer; εla is porosity and δla the thickness and τla is the tortuosity of the fouling layer, respectively.

She et al. [39], in his review, outlined the main equations that described both ICP and ECP in ODMs. He noted that the actual solute concentration at the support-active layer interface and that on the active layer surface were not the same with that of the bulk solution.

Xie et al. [24] modified the film models to predict flux behaviour in FO considering its external and internal concentration polarisation. They tested the membranes in two modes; the normal and the reverse. In the normal mode, the dense selective layer faced the DS while the porous layer faced the feed solution; while in the reverse mode, the dense selective layer faced the feed solution while the porous layer faced the draw solution [24]. This is illustrated in **Figure 3**. From their study, they expected the FO to have a greater flux, considering the fact that the influence of ICP in the FS was lesser than in the DS. According to them, the FO process should be preferably operated in the normal mode, this is also the mode implemented which is obtainable commercially as FO membrane for FO processes [24].

4. Membrane materials

One factor that is now increasingly being considered in membrane materials is the material's susceptibility to fouling. Some of the properties of the membrane that affect fouling are charge, roughness and pore size. Membrane material and its properties play an important role in the type of fouling that will occur in ODMs [8]. Knowledge of the nature of the membrane helps in the identification and understanding the fouling mechanism occurring in the membrane.

The development of membranes that can be used for FO has remained a challenge till date. ODMs can be made either by modifying an existing NF or RO membranes or by the development of new membranes with specific design for FO applications [22]. The latter is said to be simple, effective to some extent and cost efficient. Membranes used in RO or NF are made up of a non-porous active layer and a porous support layer and are made from thin-film polymerisation on a polysulfone layer supported by nonwoven fabrics However, they suffer from ICP and thus reduces the effective driving force [7, 22, 46].

The phase inversion and membrane process formation is one way to modify the membranes to improve its properties. These membranes are fabricated with a thin and porous support layer that can reduce ICP effect, while at the same time maintaining a thin and dense selective layer for adequate water flux and salt rejection [46]. Loeb and Sourirajan [47] were the first to use the phase inversion method to fabricate asymmetric polymeric membranes viz. cellulose triacetate (CTA). Ever since, cellulose acetate (CA) has become a popular material for different separation applications [46]. Relatively high hydrophilicity that favours flux and low fouling propensity has been associated with the use of CA. In addition, CA has shown high mechanical strength and availability. The other commonly available membrane type is the polyamide (PA). This is also referred to as the thin film composite (TFC) membrane. This membrane has an asymmetric structure with a dense thin film as well a thick porous support layer. This membrane is said to offer a higher flux and salt rejection and can be operated over a wide range of temperature. Early attempts in using the RO membrane as FO, however, failed due to CP that occurred in the membranes hence reducing flux. Wang et al. [48] further defined asymmetric membranes as consisting of a 0.1–1 μm thick dense layer supported by a highly porous, 100–200 μm thick support layer. The dense layer provides the selectivity of membrane. Hence, the separation properties chemical nature, thickness of the skin layer and pore sizes that are normally between 0.4 and 1 nm [48].

To attain optimum performance of ODM membranes, their selection and fabrication should be based on the following characteristics:

1. The membrane should be dense, ultrathin, have uniform active surface layers, high solute rejection and high permeate flux rate.
2. It should have a thin, porous supporting layer as well as be strong enough to provide mechanical strength to the membrane. The thin layer should help curb ICP and hence increase the membrane flux.
3. Finally, the membrane material should have high hydrophilicity tendency to enhance water flux and reduced membrane fouling [49].

The hydrophobicity of the membrane material plays a major role in membrane fouling. Hydrophobic interaction can be described as "like attracts like." The similar chemical structures owned by both the membranes and the solutes tend to have a natural tendency to be attracted to each other. Hydrophobic attraction is a result of the van der Waals forces, which occur between molecules [50]. Hydrophobic adhesion is a crucial mechanism for fouling which dominated by NOM due to the fact that high molecular weight NOM offers a higher potential for hydrophobic adhesion because of their charge density. Other factors that affect the strength of the adhesion to membrane surfaces are membrane surface roughness and membrane pore size [51]. A study by Bendinger et al. [52] showed that most foulants that are hydrophobic and slightly hydrophilic adhere better on hydrophobic surfaces than on hydrophilic surfaces. Only highly hydrophilic foulants attach stronger on hydrophilic material. Extremely hydrophobic materials do not adhere too well on the hydrophobic and hydrophilic material.

Hydrophilic membranes have higher fouling resistance than hydrophobic membranes. This means that hydrophobic membranes can be impregnated with water-soluble materials such as poly-vinyl pyrrolidone or poly-vinyl methyl ether. However, this is mostly at the polymer formulation stage [53]. The FO membrane surface roughness does not vary significantly from those of a typical RO and nanofiltration membrane (NF) [6]. The rough and large pore size membranes are shown to be more prone to fouling than the smooth, small pore membranes. This is because the bigger pores are more accessible to foulants. The nature and the extent of the fouling are determined by the specific physical and chemical characteristics of the each component as well as the membrane [6].

In RO membranes, it is expected that the porous support layer material should be thick enough to be able to withstand the high pressures involved, but for FO membranes, which uses osmotic pressures, the thickness of the support layer could be reduced since mechanical strength is not an issue here. Therefore, modifications can be made to reduce the thickness and adjust the structure of the support layer to mitigate the CP phenomenon [19]. The modification of membranes is potentially one of the suitable ways to mitigate and prevent fouling. Therefore, attempts have been made to modify the singly skinned asymmetric FO membranes into a double-skinned membrane structure. This is made such that it contains a porous support, which can be sandwiched between the two rejection skins [10, 46]. The single skinned asymmetric FO membranes face a dilemma of either experiencing more severe dilutive ICP in AL-FS or having much higher fouling propensity in AL-DS [10].

Also, some FO membranes are modified from RO/NF membranes. Hence, they are composed of asymmetric structures which are characterised by a dense active layer on top of a porous support layer. This main separation and structural properties of the active support layer governs both the water and solute transportation. This further enhances the membrane fouling behaviour.

Membranes made up of superior separation properties and structural properties such as the higher water permeability, selectivity and smaller structural parameter could provide much higher water flux [49]. However, an increase in the membrane fouling could be observed due to the enhanced hydrodynamic drag force. Therefore, a balance between mechanical strength and porosity of the membrane is needed. The mechanical strength of the membranes should be reduced so as to increase the porosity and tortuosity [22]. McCutcheon and Elimelech [49], in their study, removed the backing fabric support layer (thickness of 80–120 μm) of commercial RO membranes (overall thickness of 200 μm) and the FO water flux of the modified membranes was improved by a factor of 5.

Pore wettability of the membrane is tied to its ease to wet easily with water. Therefore, for FO membranes, it is important that the pore wettability is improved because the presence of un-wetted pore regions may block the water flux and significantly intensify ICP [22]. The use of a highly hydrophilic polymer, like polydopamine (PDA), to coat the membranes has been demonstrated to be an effective technique in the improvement of the wettability. It has been reported that the wettability can be increased ten folds after coating with FDA [22, 54, 55].

5. Fouling mitigation

Membrane fouling mitigation deals mainly with the management or minimization of the effect of membrane fouling since fouling itself cannot be completely avoided in membrane filtration [56]. Membrane fouling can be controlled and managed at different stages. These include feed pre-treatment to reduce the fouling tendencies, and improve on its antifouling properties. Others such as membrane cleaning and optimisation of operating conditions could further be of benefit also [50].

5.1. Hydrodynamic/operating conditions

Hydrodynamic condition controls the rate of particle deposition on the membrane. According to She et al. [7], most of the conclusions drawn regarding fouling mechanisms in PDMs can also be drawn on ODMs. The operating conditions and properties of the membrane play an important role in the mass transport of the ODMs. Cath et al. [1], in their study, supported the fact that the effect of operating conditions is more noticeable in ODMs than in PDMs. They reiterated that newly developed ODMs are tested under varying temperatures, draw solution compositions and as well the concentrations, flow rates and pressure [1, 48]. Hence, optimum operating conditions should be established to serve as a basis of comparison. Like in PDMs, severe fouling could occur at a higher water flux and lower cross flow velocity. Cross flow velocity has been the most common and widely used method to control fouling at the membrane surface; however, it cannot certainly prevent internal fouling. High cross flow velocity influences membrane fouling through CP and mass transfer near the membrane surface [7, 50].

High cross-flow velocity creates mixing on the membrane surface thereby improving the mass transfer coefficient, but the increase in mass transfer coefficient is different for different feed solutions [57]. At the membrane surface for any filtration system, rejected particles accumulate in a boundary layer. According to Fick's law, particles in the boundary layer collide with each other more frequently thereby improving particle diffusion from the boundary layer to the bulk solution. This diffusion can be improved by what is called shear-induced diffusion. This is achieved by causing movement of the liquid close to the boundary layer. When the movement of the liquid is increased, the particle collision becomes vigorous and the particle diffusivity is increased. Shear-induced diffusion of particles is highest at the membrane surface or at the boundary layer because of the high particle density in that region [57]. The membrane orientation should also be considered, because AL-FS is preferred due to low fouling propensity, however, the ICP is more prone in this orientation.

Hydrodynamic conditions in PDMs mostly refers to initial permeate flux, transmembrane pressure and turbulence at the membrane surface. Initial flux is the flux at the beginning of filtration and is usually high because at this stage the membrane is clean. Due to high initial flux, particles in a suspension are dragged towards the membrane surface faster than they are diffused/dispersed back into the bulk solution. Therefore, more particles are deposited on the membrane during high initial flux [58].

The study by Hwang et al. [59] showed that high initial flux, results in a large number of particles being simultaneously transported towards the membrane surface. The simultaneous arrival of these particles on the membrane surface was found to be the factor that benefits flux because entry to the membrane pore is hindered and particles only deposit on the membrane surface rather than the membrane internals. The opposite was found for a low initial flux. The finding of Hwang et al. [59] was also confirmed by Wang and Tarabara [60].

For most PDM systems, the effect of aeration as a means to mitigate fouling has been extensively studied especially for membrane bioreactors [61]. The introduction of aeration to cross flow velocity helps to reduce fouling on the membrane surface. This concept has not been widely researched on ODMs. Therefore, there is the need to investigate the effect of aeration on fouling in ODMs.

5.2. Temperature

Temperature of the solution is one of the parameters that can be altered to reduce the effects of fouling. However, this parameter is not often used for fouling control particularly in water treatment [62]. For FO processes, factors such as osmotic pressure, fluid viscosity, mass transfer and mineral solubility depends on temperature, hence it needs to be maintained so that the membrane performance is not altered [62]. Zhao and Zou [62] elaborated that at a higher temperature there is a higher initial permeate flux, higher water recovery and higher concentration factors, and since temperature effect can significantly impact on the membrane, it is important that this parameter is optimised.

Salahi et al. [63] found that when the temperature of the feed water (oily wastewater) used in their study was increased by 20°C, there was an increase in flux of about 60%. This was attributed to an increase in the diffusion rate as the temperature was raised. The flux increase was attributed to the combined effects as listed by She et al. [7] to be (1) a decrease in solution viscosity which can reduce the membrane resistance and as such can cause an increase in the water permeability, (2) an increment in the solute diffusivity which also can increase the mass transfer around the boundary layer and thus leading to a reduction in CP most importantly, ICP and (3) finally an increment in the osmotic pressure thereby increasing the effective driving force. The effect of temperature on ODM fouling was outlined to be through the influence of hydrodynamic conditions such as mass transfer of foulants and initial flux thermodynamic conditions such as osmotic pressure of the solution, solubility and stability of the foulant and finally the interaction of the foulants and the membrane [7].

Kim et al. [64], in the study of the fouling types and mechanisms in a FO membrane processes, under raised temperature, found that flux due to organic fouling was more pronounced when the draw solution was increased. This increase was attributed to the increased permeation drag at increased initial flux level. However, on increasing the feed solution, less fouling was observed because of the organic back diffusion from membrane surface and the increase in the organic solubility [64]. The same authors, Kim et al. [64], in their study observed that membrane fouling became more enhanced when the initial flux was increased to a certain critical flux as temperature for both the feed and DS was increased. This was because organic convection by permeation drag dominated the fouling mechanism. At critical flux, only localised deposition on the membrane occurs, because the rate at which particles deposit on the membrane surface is almost equal to the rate at which they are diffused back into the solution [65]. However, if the process is operated above the critical flux, enhanced fouling is observed on the membrane.

5.3. Feed pretreatment

The feed water to be treated, in most cases, are made up of various components which might include divalent ions, humic substance, alginate, silica and a host of others. These particles could accumulate on the porous membrane structure thereby causing severe decline in membrane permeate flux [23]. The extent to which the feed water is pre-treated depends on the quality of the water; hence, this factor is also dependent on the sources of the water. Pretreatment of feed can be divided into two: physical and chemical. Physical pretreatment involves the use of mechanical filtration such screening, cartridge filters, sand filters or membrane filtration while chemical pretreatment involves the addition of scale inhibitors, coagulants, disinfectants and polyelectrolytes [26].

Extensive studies regarding feed pretreatment in PDMs especially for NF and RO membranes have been investigated extensively, basically for removal of particulate matter [27]. Ultrafiltration (UF) and microfiltration (MF) membranes are used as feed pretreatment to most NF and RO processes due to their porosity. At other times even NF membranes can be used as pretreatment method. The permeates from these membranes have been presented to have low turbidity and silt density index thus increasing recovery in the RO process. For instance, Mi and Elimelech [27] compared three pretreatment technologies; powdered activated carbon (PAC), addition of coagulants such ferric chloride and pretreatment using UF before RO desalination.

The use of NF as pre-treatment to ODM systems however has not been comprehensively studied and remains a crucial aspect for further investigations. Chen et al. [23] studied the first systematic investigation on the use of a loose NF to pretreat feed wastewater in practical PRO practice. They found that the low pressure NF was able to mitigate the fouling potential from multivalent ions and organic matters. Thus, they found the NF method of pretreatment as cost effective. However, the low-pressure NF was able to mitigate the fouling potential from multivalent ions and organic matters, but silica scaling was still predominant, hence, they recommended further investigation. This comparison was made based on a previous study of theirs. That study made use of retentate from a RO unit of a municipal water recycling plant as the main feed stream for an osmotic power generation. Two pre-treatment methods were used: anti-scaling and pH adjustment. The pH adjustment was accompanied by water flushing and 100% by air bubbling thereby resulting in an increased flux [43].

Chemical pre-treatment, on the other hand, involves the addition of chemicals to the feed water. The addition of chemicals for pH adjustment, prevention of scaling and fouling is also used for the pre-treatment of feed to RO processes. This however, in most cases still requires a physical method to be used alongside. For example, a membrane filtration process could be used to pre-treat the feed water and thereafter the chemicals could be added. The advantage presented in following this path is the reduction in chemical consumption. Chlorination, however, should be added to the feed water independent of the pre-treatment method that is being employed. This is to prevent biofouling of the membrane [26]. Nonetheless, after the chlorination, a dechlorination of the feed has to be considered since most of the membranes are susceptible to chlorine attacks.

The addition of coagulants and flocculants causes the dissolve matter to adsorb on the hydroxides and also to cause the agglomeration of colloidal matter. The use of the coagulants aids in reducing the fouling potential on the membrane and also provides a better quality feed water to the RO [26]. The addition of antiscaling agents is considered as one of the pre-treatment methods as well. The precipitation of salts on the membrane surface is referred to as scaling and it is caused by super saturation. It reduces membranes productivity and as well the recovering of water. Different scale inhibitors can be used as antiscalant. These inhibitors control the scaling caused by sulphates, carbonates and calcium fluoride [26].

5.4. Selection of draw solutes

One of the key factors in ODMs is the selection of the right draw solution (DS). The knowledge on the various types of DS used is needful to understand the crucial issues that are related with FO such as CP and mass transport [11]. The following factors should be considered in the selection of DS in ODMs; the solution should produce a substantial amount of osmotic pressure, it should not be expensive ant toxic to the environment and easily regenerated [5, 11, 17, 66]. The commonly used DS is NaCl, because of its high water solubility and it is relatively easy to reconcentrate using desalination processes [17, 66]. Other low molecular weight salts used as DS in recent times include; $MgCl_{2,}$ $CaCl_2$, KCl, Mg_2SO_4. Others such as sucrose, glucose, 2-methylimidazole-based compounds have also been used. Further still, magnetic particles, thermolytic inorganic salts for example ammonia-carbon dioxide and hydrogels have all been tried as DS. It is expected that these solutions should provide a high osmotic pressure and at the same time be easily regenerated and recovered [17].

Cai and Hu [5] reviewed draw solutes used in FO, where they categorise DS into two, namely responsive and non-responsive. The non-responsive solutes were defined as those which when a stimuli such as temperature, pH and others were added to them, no significant change was observed in their water affinity. While on the contrary, the responsive DS were those that, upon exposure to a stimulus, underwent a significant change in their water affinity and thereby accompanied by phase transitions between two states with different water affinities [5].

There is a general perspective regarding the increase in the concentration of the draw solution. Increasing the DS concentration leads to an increase in initial water flux and as such an increase in membrane fouling. The occurrence has been attributed to the effect of increase in hydraulic drag force which is a result of the higher flux that promotes foulant deposition on the membrane [3, 8, 11]. The effect of increasing the DS concentration also influences the RSD by elevating it from the FS to the DS thereby increasing fouling also [67]. She et al. [68] reiterated in their study how RSD influences the deposition of solutes on the membrane surface. The result of this is a change in the feed water chemistry and thus may cause more severe fouling. In that study, they observed that greater alginate fouling occurred on the FO membrane when the DS contained higher concentration of divalent ions of Ca^{2+} and Mg^{2+}. They attributed that the RSD enhanced organic fouling relates to the nature of the DS and to the rate of its diffusion into the feed solution and its ability to interact with the foulant [68]. Therefore, the type and nature of the DS can affect the membrane fouling and the water chemistry too. It was observed that divalent ions in DS, as mentioned above, could influence an additional fouling which is more than the DS even without the specific ions at the same initial water flux level. This occurs as a result of the strong attraction between the ions (foulants) in the solution and the specific ions after they reversely diffuse from DS into FS.

She et al. [39] studied the relationship between reverse and forward solute diffusion to membrane fouling in ODMs. The types of DS used were; NaCl, $MgCl_2$, $CaCl_2$ and $Ca(NO_3)_2$ to reiterate the connection that exists between RSD and forward solute diffusion (FSD). They found that the extent of fouling for the chosen DS was in the order of $Ca(NO_3)_2 > CaCl_2 > MgCl_2 > NaCl$. They concluded that NaCl DS had the highest RSD, this was followed by $Ca(NO_3)_2$ DS, then $CaCl_2$ and finally the least was $MgCl_2$ DS. According to them, the order of the RSD was consistent with the order of their solute permeability. Therefore, the RSD of divalent ions impacted more on the feed solution thus leading to an alginate membrane fouling. Fouling propensity was in the order $Ca(NO_3)_2 > CaCl_2 >> MgCl_2 > NaCl$. Even though a greater amount of NaCl was reversing, the effect of fouling was limited using the NaCl, reason been that the Na^+ did not interact with the alginate. This was related to the cation and anion of the DS and rate of its reverse diffusion.

6. Membrane cleaning

Membrane cleaning is an integral and an important part of membrane processes [16]. Cleaning could be done either hydraulically or chemically. Membrane cleaning becomes necessary when avoiding irreversible fouling of the membrane. The longer the membrane is allowed to operate in its fouled state, the harder it becomes to remove the foulants from the membrane. It, therefore, becomes necessary to use chemicals or greater force to recover a highly fouled membrane.

Physical and chemical methods of cleaning can be employed for fouled membranes. Physical method is also referred to as the hydraulic method. It employs the use of mechanical forces to displace and remove the fouling agents from the membrane surface [69]. These methods of cleaning are typically used in the cleaning in place (CIP) situations. Series of studies have been carried out for the cleaning of ODM membranes using physical methods such as membrane surface flushing and membrane backwashing [7, 16]. The surface washing (forward washing) is achieved when the cross flow velocity is increased on the membrane surface to remove the deposited foulants [13]. Backwashing involves pumping permeate water at a high cross-flow velocity in the opposite direction from which the feed comes in. It is a reversed filtration process in which the permeate of backwashing solution is flushed through the membrane back to the concentrate side. These methods have both shown to be effective against the membrane fouling under different of conditions.

Mi and Elimelech [6] determined the efficiency of surface flushing to investigate the reversibility of FO and RO membranes fouled with organic foulants. Their findings indicate that fouling in FO was more easily reversible than in RO. The reason was due to the hydraulic compaction imposed on the RO membrane which was absent in the FO membranes. It is recommended that for higher recovery of flux, backwashing should be combined with surface flushing. Both surface flushing and backwashing are limited to the fact that only the surface foulants are removed. The internal foulants within the membrane remains after the whole procedure; however, backwashing is moderately successful in removing internal clogging material from the membrane internals.

For FO and PRO membranes, osmotic backwashing has been developed for these processes. The process employs the use of high salinity water to replace the feed solution while a lower salinity water is used to replace the draw solution. Just like in PDMs, the water permeation direction is in the reverse form, thereby creating a negative water flux. The action of this results in the breaking of the foulants away from the membrane [7]. Even though success of osmotic backwashing has been reported by many researchers on recovering of flux, a few others have contrary views where efficiency of osmotic backwashing for water flux recovery was low [7].

When a fouled membrane can no longer be completely removed by physical cleaning, the membrane is irreversibly fouled and therefore, chemical cleaning is required. Caution is however to be employed when cleaning the membranes chemically because the membranes can also be damaged by the chemicals used for membrane cleaning [26, 70]. The choice of chemicals for membrane cleaning must be able to completely dissolve the foulants on the membrane but not damage the membrane itself [69].

Chemical cleaning is a reaction between the chemicals and the foulants on the membrane surface. The process involves mass transfer of the chemicals to the fouling layer and the products of the reaction are dispensed back to the bulk liquid phase. Effectiveness of the chemical cleaning is improved by hydrodynamic conditions that promote contact between the cleaning chemicals and the fouling layer on the membrane surface [50].

The recovery of flux through cleaning has been enumerated to be more in FO than RO membranes. The reason is due to the fact that most fouling in FO is more reversible than that in PDMs [28]. This has extensively been studied by Mi and Elimelech [6], Mi and Elimelech [27] where they carried out chemical and physical cleaning on alginate, bovine serum albumin (BSA) and Aldrich humic acid (AHA) as model organic foulants. They reported a fouling reversibility in the FO and attributed that to the less compacted organic fouling layer formed due to the absence of hydraulic pressure. Another study on the efficiency of physical cleaning in inorganic scaling experiments was also carried out by Zhao and Zou [62] under different temperatures of 25, 35 and 45°C. Membranes were cleaned by the use of water at a cross flow velocity of 33.3 cm/s for 20 min, thus no chemicals were used. Their findings revealed that the higher temperature resulted in higher initial permeate fluxes, higher water recoveries and higher concentration factors. However, more compressed solutes were deposited on the membrane surface and thus the membrane cleaning efficiency was affected [62].

Air scouring induces shear force at the membrane surface as the air bubbles rise travelling adjacent to the membrane surface. The mechanisms responsible for the shear force in the membrane surface are fall film effect and wake effect. These are a net result of the quick rise of air bubbles and the feed solution. Air scouring generates localised cross-flow conditions along the membrane surface thereby reducing the deposition of particles and the development of a cake layer on the membrane surface [61].

7. Conclusion

Fouling in ODM membranes was the main objective of this book chapter. Despite the recognition that ODMs have received in applications in various industries, the use of this this technology is still limited by fouling, thus hindering its overall performance. The information on the fouling mechanisms is still limited and thus needs to be examined critically. This book chapter provides vital information on the impact of fouling on ODMs performance and it explored the factors and mechanisms governing fouling in ODMs. Further still, the effects of membrane fouling were expounded and approaches on the mitigation and cleaning of the membranes were outlined.

Acknowledgements

The authors wish to thank the Durban University of Technology and the National research foundation for providing PhD scholarship.

Author details

Martha Noro Chollom* and Sudesh Rathilal

*Address all correspondence to: mnchollom@gmail.com

Faculty of Engineering and the Built Environment, Department of Chemical Engineering, Durban University of Technology, Durban, South Africa

References

[1] Cath TY, Elimelech M, McCutcheon JR, McGinnis RL, Achilli A, Anastasio D, Brady AR, Childress AE, Farr IV, Hancock NT, Lampi J, Nghiem LD, Xie M, Yip NY. Standard methodology for evaluating membrane performance in osmotically driven membrane processes. Desalination. 2013;**312**:31-38. DOI: 10.1016/j.desal.2012.07.005

[2] Zou S, Gu Y, Xiao D, Tang CY. The role of physical and chemical parameters on forward osmosis membrane fouling during algae separation. Journal of Membrane Science. 2011;**366**:356-362. DOI: 10.1016/j.memsci.2010.10.030

[3] Tang CY, She Q, Lay WCL, Wang R, Fane AG. Coupled effects of internal concentration polarization and fouling on flux behavior of forward osmosis membranes during humic acid filtration. Journal of Membrane Science. 2010;**354**:123-133. DOI: 10.1016/j.memsci.2010.02.059

[4] Shirazi S, Lin C-J, Chen D. Inorganic fouling of pressure-driven membrane processes—A critical review. Desalination. 2010;**250**:236-248. DOI: 10.1016/j.desal.2009.02.056

[5] Cai Y, Hu XM. A critical review on draw solutes development for forward osmosis. Desalination. 2016;**391**:16-29. DOI: 10.1016/j.desal.2016.03.021

[6] Mi B, Elimelech M. Organic fouling of forward osmosis membranes: Fouling reversibility and cleaning without chemical reagents. Journal of Membrane Science. 2010;**348**:337-345. DOI: 10.1016/j.memsci.2009.11.021

[7] She Q, Wang R, Fane AG, Tang CY. Membrane fouling in osmotically driven membrane processes: A review. Journal of Membrane Science. 2016;**499**:201-233. DOI: 10.1016/j.memsci.2015.10.040

[8] Mi B, Elimelech M. Chemical and physical aspects of organic fouling of forward osmosis membranes. Journal of Membrane Science. 2008;**320**:292-302. DOI: 10.1016/j.memsci.2008.04.036

[9] Achilli A, Cath TY, Marchand EA, Childress AE. The forward osmosis membrane bioreactor: A low fouling alternative to MBR processes. Desalination. 2009;**239**:10-21. DOI: 10.1016/j.desal.2008.02.022

[10] Ong CS, Al-anzi B, Lau WJ, Goh PS, Lai GS, Ismail AF, Ong YS. Anti-fouling double-skinned forward osmosis membrane with Zwitterionic brush for oily wastewater treatment. Scientific Reports. 2017;**7**:6904. DOI: 10.1038/s41598-017-07369-4

[11] Kim B, Lee S, Hong SA. Novel analysis of reverse draw and feed solute fluxes in forward osmosis membrane process. Desalination. 2014;**352**:128-135. DOI: 10.1016/j.desal.2014.08.012

[12] Zahrim AY, Tizaoui C, Hilal N. Coagulation with polymers for nanofiltration pre-treatment of highly concentrated dyes: A review. Desalination. 2011;**266**:1-16. DOI: 10.1016/j.desal.2010.08.012

[13] Chollom MN, Rathilal S, Pillay VL, Alfa D. The applicability of nanofiltration for the treatment and reuse of textile reactive dye effluent. Water SA. 2015;**41**:398. DOI: 10.4314/wsa.v41i3.12

[14] Bogler A, Lin S, Bar-Zeev E. Biofouling of membrane distillation, forward osmosis and pressure retarded osmosis: Principles, impacts and future directions. Journal of Membrane Science. 2017. DOI: 10.1016/j.memsci.2017.08.001

[15] Miller DJ, Paul DR, Freeman BD. A crossflow filtration system for constant permeate flux membrane fouling characterization. Review of Scientific Instruments. 2013;**84**:035003. DOI: 10.1063/1.4794909

[16] Chollom MN, Pikwa K, Rathilal S, Pillay VL. Fouling mitigation on a woven fibre microfiltration membrane for the treatment of raw water. South African Journal of Chemical Engineering. 2017;**23**:1-9. DOI: 10.1016/j.sajce.2016.12.003

[17] Chun Y, Mulcahy D, Zou L, Kim IS. A short review of membrane fouling in forward osmosis processes. Membranes. 2017;**7**:30. DOI: 10.3390/membranes7020030

[18] Chekli L, Phuntsho S, Kim JE, Kim J, Choi JY, Choi J-S, Kim S, Kim JH, Hong S, Sohn J, Shon HK. A comprehensive review of hybrid forward osmosis systems: Performance, applications and future prospects. Journal of Membrane Science. 2016;**497**:430-449. DOI: 10.1016/j.memsci.2015.09.041

[19] Xu W, Chen Q, Ge Q. Recent advances in forward osmosis (FO) membrane: Chemical modifications on membranes for FO processes. Desalination. 2017;**419**:101-116. DOI: 10.1016/j.desal.2017.06.007

[20] Akther N, Sodiq A, Giwa A, Daer S, Arafat H, Hasan S. Recent advancements in forward osmosis desalination: A review. Chemical Engineering Journal. 2015;**281**:502-522. DOI: 10.1016/j.cej.2015.05.080

[21] Lay WCL, Zhang J, Tang C, Wang R, Liu Y, Fane AG. Factors affecting flux performance of forward osmosis systems. Journal of Membrane Science. 2012;**394-395**:151-168. DOI: 10.1016/j.memsci.2011.12.035

[22] Le NL, Nunes SP. Materials and membrane technologies for water and energy sustainability. Sustainable Materials and Technologies. 2016;**7**:1-28. DOI: 10.1016/j.susmat.2016.02.001

[23] Chen Y C, Liu C, Setiawan L, Wang Y-N, Hu X, Wang R. Enhancing pressure retarded osmosis performance with low-pressure nanofiltration pretreatment: Membrane fouling analysis and mitigation. Journal of Membrane Science. 2017;114-122. DOI: 10.1016/j.memsci.2017.08.051

[24] Xie M, Lee J, Nghiem LD, Elimelech M. Role of pressure in organic fouling in forward osmosis and reverse osmosis. Journal of Membrane Science. 2015;**493**:748-754. DOI: 10.1016/j.memsci.2015.07.033

[25] Field R, Fundamentals of fouling. In: Peinemann K-V, Nunes SP, Editors. Membrane Technology, Membranes for Water Treatment. 2010, Wiley-VCH: Weinheim. p. 1-23. ISBN: 978-3-527-31483-6

[26] Mi B, Elimelech E. Silica scaling and scaling reversibility in forward osmosis. Desalination. 2013;**312**:75-81. DOI: 10.1016/j.desal.2012.08.034

[27] Mi B, Elimelech M. Gypsum scaling and cleaning in forward osmosis: Measurements and mechanisms. Environmental Science & Technology. 2010;**44**:2022-2028. DOI: 10.1021/es903623r

[28] Chun Y, Zaviska F, Kim S-J, Mulcahy D, Yang E, Kim IS, Zou L. Fouling characteristics and their implications on cleaning of a FO-RO pilot process for treating brackish surface water. Desalination. 2016;**394**:91-100. DOI: 10.1016/j.desal.2016.04.026

[29] Zhou J, Gao N, Peng G, Deng Y. Pilot study of ultrafiltration-nanofiltration process for the treatment of raw water from Huangpu river in China. Journal of Water Resource and Protection. 2009;**1**:203-209. DOI: 10.4236/jwarp.2009

[30] Bessiere Y, Jefferson B, Goslan E, Bacchin P. Effect of hydrophilic/hydrophobic fractions of natural organic matter on irreversible fouling of membranes. Desalination. 2009; **249**:182-187. DOI: 10.1016/j.desal.2008.12.047

[31] Kim H-C, Hong J-H, Lee S. Fouling of microfiltration membranes by natural organic matter after coagulation treatment: A comparison of different initial mixing conditions. Journal of Membrane Science. 2006;**283**:266-272. DOI: 10.1016/j.memsci.2006.06.041

[32] Fan X, Tao Y, Wang L, Zhang X, Lei Y, Wang Z, Noguchi H. Performance of an integrated process combining ozonation with ceramic membrane ultra-filtration for advanced treatment of drinking water. Desalination. 2014;**335**:47-54. DOI: 10.1016/j.desal.2013.12.014

[33] Madaeni SS. The application of membrane technology for water disinfection. Water Research. 1999;**33**:301-308. DOI: 10.1016/S0043-1354(98)00212-7

[34] Singh G, Song L. Quantifying the effect of ionic strength on colloidal fouling potential in membrane filtration. Journal of Colloid and Interface Science. 2005;**284**:630-638. DOI: 10.1016/j.jcis.2004.10.030

[35] Characklis WG, Cooksey KE. Biofilms and microbial fouling. Advances in Applied Microbiology. 1983;**29**:93-138. DOI: 10.1016/S0065-2164(08)70355-1

[36] Nguyen T, Roddick F, Fan L. Biofouling of water treatment membranes: A review of the underlying causes, monitoring techniques and control measures. Membranes. 2012; **2**:804. DOI: 10.3390/membranes2040804

[37] Al-Ahmad M, Aleem FA, Mutiri A, Ubaisy A. Biofuoling in RO membrane systems part 1: Fundamentals and control. Desalination. 2000;**132**:173-179. DOI: 10.1016/S0011-9164(00)00146-6

[38] Komlenic R. Rethinking the causes of membrane biofouling. Filtration & Separation. 2010;**47**:26-28. DOI: 10.1016/S0015-1882(10)70211-1

[39] She Q, Jin X, Li Q, Tang CY. Relating reverse and forward solute diffusion to membrane fouling in osmotically driven membrane processes. Water Research. 2012;**46**:2478-2486. DOI: 10.1016/j.watres.2012.02.024

[40] Tian E, Wang X, Zhao Y, Ren Y. Middle support layer formation and structure in relation to performance of three-tier thin film composite forward osmosis membrane. Desalination. 2017;190-201. DOI: 10.1016/j.desal.2017.02.014

[41] Song L, Elimelech M. Theory of concentration polarization in crossflow filtration. Journal of the Chemical Society, Faraday Transactions. 1995;**91**:3389-3398. DOI: 10.1039/FT9959103389

[42] Baker RW. Membrane Technology and Applications. 2nd ed., England: John Wiley and Sons, Ltd; 2004. 545. ISBN: 978-0-470-02038-8

[43] Liu Y. Fouling in forward osmosis membrane processes: Chracterization, mechanisms, and mitigation [thesis]. University of Maryland; 2013

[44] McCutcheon JR, Elimelech M. Influence of concentrative and dilutive internal concentration polarization on flux behavior in forward osmosis. Journal of Membrane Science. 2006;**284**:237-247. DOI: 10.1016/j.memsci.2006.07.049

[45] Duan J, Litwiller E, Pinnau I. Solution-diffusion with defects model for pressure-assisted forward osmosis. Journal of Membrane Science. 2014;**470**:323-333. DOI: 10.1016/j.memsci.2014.07.018

[46] Zhang S, Wang KY, Chung T-S, Chen H, Jean YC, Amy G. Well-constructed cellulose acetate membranes for forward osmosis: Minimized internal concentration polarization with an ultra-thin selective layer. Journal of Membrane Science. 2010;**360**:522-535. DOI: 10.1016/j.memsci.2010.05.056

[47] Loeb S, Sourirajan S. Sea Water Demineralization by Means of an Osmotic Membrane. 1962, Ameriacn Chemical Society. DOI: 10.1021/ba-1963-0038.ch009

[48] Wang J, Dlamini DS, Mishra AK, Pendergast MTM, Wong MCY, Mamba BB, Freger V, Verliefde ARD, Hoek EMV. A critical review of transport through osmotic membranes. Journal of Membrane Science. 2014;**454**:516-537. DOI: 10.1016/j.memsci.2013.12.034

[49] McCutcheon JR, Elimelech M. Influence of membrane support layer hydrophobicity on water flux in osmotically driven membrane processes. Journal of Membrane Science. 2008;**318**:458-466. DOI: 10.1016/j.memsci.2008.03.021

[50] Abdelrasoul A, Doan H, Lohi A. 2013. Mass Transfer-Advances in Sustainable Energy and Environment Oriented Numerical Modeling, Chapter 8: Fouling in Membrane Filtration and Remediation, Methods 2013. ISBN 978-953-51-1170-2

[51] Liu C, Caothien S, Hayes J, Caothuy T, Otoyo T, Ogawa T. Membrane Chemical Cleaning: From Art to Science. Port Washington, NY: Pall Corporation; 2001. p. 11050

[52] Bendinger B, Rijnaarts HH, Altendorf K, Zehnder AJ. Physicochemical cell surface and adhesive properties of coryneform bacteria related to the presence and chain length of mycolic acids. Applied and Environmental Microbiology. 1993;**59**:3973-3977

[53] Chan R, Chen V. Characterization of protein fouling on membranes: Opportunities and challenges. Journal of Membrane Science. 2004;**242**:169-188. DOI: 10.1016/j.memsci.2004.01.029

[54] Arena JT, McCloskey B, Freeman BD, McCutcheon JR. Surface modification of thin film composite membrane support layers with polydopamine: Enabling use of reverse osmosis membranes in pressure retarded osmosis. Journal of Membrane Science. 2011;**375**:55-62. DOI: 10.1016/j.memsci.2011.01.060

[55] McCloskey BD, Park HB, Ju H, Rowe BW, Miller DJ, Chun BJ, Kin K, Freeman BD. Influence of polydopamine deposition conditions on pure water flux and foulant adhesion resistance of reverse osmosis, ultrafiltration, and microfiltration membranes. Polymer. 2010;**51**:3472-3485. DOI: 10.1016/j.polymer.2010.05.008

[56] Franklin ACM. Prevention and Control of Membrane Fouling: Practical Implications and Examining Recent Innovations. Membraan Applicatie Centrum Twente b.v.; 2009

[57] Scott K, Handbook of Industrial Membranes. Access Online via Elsevier; 1995. ISBN-10: 1856172333

[58] Yoon S-H. Membrane Bioreactor Processes: Principles and Applications. CRC Press; 2015. ISBN-10: 1482255839

[59] Hwang K-J, Liao CY, Tung K-L. Effect of membrane pore size on the particle fouling in membrane filtration. Desalination. 2008;**234**:16-23. DOI: 10.1016/j.desal.2007.09.065

[60] Wang F, Tarabara VV. Pore blocking mechanisms during early stages of membrane fouling by colloids. Journal of Colloid and Interface Science. 2008;**328**:464-469. DOI: 10.1016/j.jcis.2008.09.028

[61] Judd S, Le-Clech P, Taha T, Cui Z. Theoretical and experimental representation of a submerged membrane bio-reactor system. Membrane Technology. 2001;**2001**:4-9. DOI: 10.1016/S0958-2118(01)80232-9

[62] Zhao S, Zou L. Effects of working temperature on separation performance, membrane scaling and cleaning in forward osmosis desalination. Desalination. 2011;**278**:157-164. DOI: 10.1016/j.desal.2011.05.018

[63] Salahi A, Abbasi M, Mohammadi T. Permeate flux decline during UF of oily wastewater: Experimental and modeling. Desalination. 2010;**251**:153-160. DOI: 10.1016/j.desal.2009.08.006

[64] Kim Y, Lee S, Shon HK, Hong S. Organic fouling mechanisms in forward osmosis membrane process under elevated feed and draw solution temperatures. Desalination. 2015;**355**:169-177. DOI: 10.1016/j.desal.2014.10.041

[65] Bacchin P. A possible link between critical and limiting flux for colloidal systems: Consideration of critical deposit formation along a membrane. Journal of Membrane Science. 2004;**228**:237-241. DOI: 10.1016/j.memsci.2003.10.012

[66] Cath TY, Childress AE, Elimelech M. Forward osmosis: Principles, applications, and recent developments. Journal of Membrane Science. 2006;**281**:70-87. DOI: 10.1016/j.memsci.2006.05.048

[67] She Q, Jin X, Tang CY. Osmotic power production from salinity gradient resource by pressure retarded osmosis: Effects of operating conditions and reverse solute diffusion. Journal of Membrane Science. 2012;**401**:262-273. DOI: 10.1016/j.memsci.2012.02.014

[68] She Q, Wong YKW, Zhao S, Tang CY. Organic fouling in pressure retarded osmosis: Experiments, mechanisms and implications. Journal of Membrane Science. 2013;**428**:181-189. DOI: 10.1016/j.memsci.2012.10.045

[69] Garcia-Fayos B, Arnal J, Gimenez A, Alvarez-Blanco S, Sancho M. Static cleaning tests as the first step to optimize RO membranes cleaning procedure. Desalination and Water Treatment. 2015;**55**:3380-3390. DOI: 10.1080/19443994.2014.957924

[70] Gao W, Liang H, Ma J, Han M, Chen Z-L, Han Z-S, Li G-B. Membrane fouling control in ultrafiltration technology for drinking water production: A review. Desalination. 2011;**272**:1-8. DOI: 10.1016/j.desal.2017.04.016

Chapter 10

Forward Osmosis as a Pre-Treatment Step for Seawater Dilution and Wastewater Reclamation

Machawe M. Motsa and Bhekie B. Mamba

Additional information is available at the end of the chapter

http://dx.doi.org/10.5772/intechopen.72289

Abstract

This chapter presents the exploration of the combined process of wastewater reclamation and seawater dilution using forward osmosis (FO). Wastewater and seawater are the two most abundant water sources that are free of the hydrological cycle and could serve as an alternative potable water source. Forward osmosis was chosen as the an ideal pre-treatment step to dilute seawater prior to desalination at relatively lower energy demand and low fouling propensity. Membrane fouling behavior was studied and investigated using different feed compositions bearing fractions of effluent organic matter (EfOM). The negative surface charge of all organic foulants was reduced by the adsorption of calcium ions. Filtration of feed streams containing single, simple organic foulants revealed that alginate (polysaccharides) and bovine serum albumin (BSA) resulted in significant loss in process performance as a result of permeate flux reduction. The complex mixture of alginate, BSA and humic acid caused severe loss in membrane performance due to dominant favorable synergistic interactions between foulants and between foulants and membrane surface. The forward osmosis process presents a viable alternative for a simple and effective seawater dilution step using wastewater as the feed solution. Process performance can be improved by selecting a foulant resistant membrane with matching flux.

Keywords: desalination, fouling, forward osmosis, membrane, seawater, wastewater

1. Introduction

Water forms part of the fundamentals of human existence, however; growth in human population and current extreme climatic conditions have resulted in many parts of the world (particularly arid areas) faced with minimal or no access to water supply. Statistics and research have predicted that over the next decade the impact of water crisis will increase fourfold.

It has been shown that developing countries are the most affected and about 80–90% of all diseases and 30% of all deaths result from exposure to poor quality drinking water [1, 2]. The lack of good quality water has adverse impacts on essential factors of human survival such as food and energy supply. Adequate supply of good quality water and affordable energy sources are vital to sustaining good public health and growing economic rate. Thus, there is a growing awareness among governments and corporations that the future prosperity of societies is intimately tied to the availability of fresh and safe drinking water [3, 4]. The possibility of wastewater reuse instead of disposing it has received increasing attention over the past decades as a viable solution towards minimizing the effect of water scarcity. Past studies have provided a baseline information that wastewater, brackish water, and seawater have great potential to augment shortage water supply, however; the energy expenditure and equipment required for purification of such water streams has limited their potential in many parts of the world [2, 5]. The reuse of wastewater for other applications rather than drinking purpose is already established and examples include the irrigation of golf courses or industrial cooling [6].

Thus water reuse and desalination technologies have been identified as promising strategies to provide safe drinking water to water-stressed communities [2]. Desalination and wastewater reclamation using pressure-driven membrane processes such as nanofiltration (NF) and reverse osmosis (RO) processes have been elaborately applied to produce potable water from brackish and seawater as well as treated wastewater effluent [7]. Pressure-driven membrane processes such as RO and NF rely on the use external hydraulic pressure to overcome the osmotic pressure of the feed solution and produce purified permeate water [8]. The applied pressure is the driving force for mass transport through the membrane. The over-arching advantage of RO is that it produces high quality permeate water that in most cases ready for use. However, there are several inherent drawbacks such as its heavy reliance on hydraulic pressure, large concentrate volumes, and high membrane fouling propensity have greatly restricted its sustainable development in recent times, especially in developing countries, due to the soaring oil and electricity prices [9].

Normally wastewater is composed of a wide range of pollutants and substances which could negatively affect human and aquatic life. The nature of the compounds found in reclaimed water may be of concern in drinking water, but not in water intended for landscape irrigation and other peripheral uses. Among the constituents of wastewater is effluent organic matter (EfOM) which comprises of a range of low- to high-molecular-weight organic compounds such as polysaccharides, proteins, humic and fulvic acids, organic acids and lipids [1, 10]. And it has been repeatedly reported that among the different EfOM components; humic acids, polysaccharides and proteins were responsible for extensive membrane fouling [11]. The chemical complexity and heterogeneous nature of wastewater present a challenge to developing a proper understanding on the key role of the interactions between the different kinds of organic compounds in permeate flux decline as well as fouling layer formation. And numerous findings have attributed the observed difficulty in treating wastewater to the synergistic effects between co-existing organic species [1, 12].

Thus, major efforts have been made to design water treatment technologies that are environmentally friendly, energy-saving and have greater permeate water recoveries with high produced water quality [13]. Innovative membrane separation processes such as forward osmosis (FO) have shown great potential for application in seawater dilution, wastewater treatment and reclamation [14]. Several advantages make the forward osmosis process a more attractive alternative compared to other techniques and they include low energy utilization, lower membrane fouling propensity, simplicity as well as the good rejection of a wide variety of foulants compared to pressure-driven membrane processes [13, 15]. The forward osmosis can also be fitted as an additional step to pressure-driven processes resulting in hybrid processes with potentially improved water recovery and energy savings [16, 17].

Thus, this work seeks to develop insight into the performance of a forward osmosis process as a pre-treatment step for seawater dilution. Significant focus was directed to developing a proper mechanistic understanding of forward osmosis membrane fouling behavior during seawater dilution and wastewater reclamation; where the fouling processes are more complex compared to simple feed and draw solutions. Combined wastewater reclamation and seawater dilution have the potential of fouling both sides of the membrane and thus hugely impacting the process performance. This is due to altered foulant-membrane and foulant-foulant interactions as well as more severe internal concentration polarization effects.

1.1. Forward osmosis membrane processes for water treatment

The main driving force in a forward osmosis membrane separation process is the chemical potential difference between the two solutions across a semi-permeable membrane; that is: pure water diffuses from an impaired source (feed solution) through a semi-permeable membrane to a solution of higher solute concentration (draw solution) induced by osmotic pressure difference. Forward osmosis has inherent potential advantages that highlight it as a promising alternative to pressure-driven membrane separation technologies [16]. These advantages include low energy consumption due to minimal or non-use of external hydraulic pressure. As a result of utilizing low external hydraulic pressure, the subsequent fouling cake layer is much less compressed and can be easily detached by simple physical cleaning methods. Thus, many of the possible forward osmosis applications can be performed with low-quality feed water, including domestic and industrial wastewater/wastewater effluent. Osmotic driven processes can also be integrated to pressure-driven membrane counter-parts such as reverse osmosis to form FO-RO hybrids aimed at improving process performance and lowering energy utilization. However, energy expenditure can only be reduced when forward osmosis is situated before reverse osmosis, as a pre-treatment step to reduce reverse osmosis fouling and scaling; subsequently minimizing the intensity of hydraulic pressure applied to treat water. Thus, in pure thermodynamic terms energy saving is not possible in a closed-loop FO-RO unit. Forward osmosis also has a high rejection of a wider range of inorganic and organic contaminants. In addition, the claimed lower membrane fouling propensity when compared to pressure-driven membrane processes is yet to be proven at high fluxes in real practice. Its process further presents the ability to recover and reuse the osmotic agent [17].

Forward osmosis has found application in a variety of fields such as the production of nutrient-rich drinks that are used as part of life-saving equipment in life boats. The process has also been applied in food processing, in emergencies such as natural disasters as an osmotic concentration of liquid foods [18–26]. As previously stated, it is a robust membrane separation technique that boasts of good rejection of a broad range of pollutants and foulants and dissolved ions. It is therefore for these reasons that it's being researched and improved for water treatment applications such as seawater desalination [27–29], wastewater reclamation [30–33], industrial wastewater treatment [34], brine concentration [35], osmotic membrane bioreactors [36] and the use of the salinity gradient for power generation or osmotic dilution prior to reverse osmosis seawater desalination (using impaired water as feed and seawater as draw solution) [37].

Some of the recent performed research studies in water treatment include comparing the impacts of membrane surfaces (such as the asymmetric polyamide thin-film composite and cellulose triacetate) and system operating conditions on the performance of forward osmosis membranes for membrane desalination of produced water using a standard method and system operating conditions similar to those applied in the operation of industrial-scale spiral wound reverse osmosis membranes conducted by Coday et al. [1, 38]. They found that rejection of inorganic solutes was greater than 94% and dissolved organic carbon was higher than 93%. However, the rejection of total nitrogen (TN) was poor, recording a moderate 63%. Duong and co-workers, [39] evaluated the performance of the forward osmosis process in treating stable oil–water emulsions. Their study demonstrated that FO was successful in the treatment of a wide range of oil–water emulsions from low to very high concentrations of up to 200,000 ppm. Water was separated from oily feeds containing 500 ppm or 200,000 ppm emulsified oil at a relatively high flux of 16.5 ± 1.2 Lm^{-2} h^{-1} or 11.8 ± 1.6 Lm^{-2} h^{-1}, respectively, using a thin film composite membrane at a draw solution concentration of 1 M NaCl. The membrane managed to achieve an oil rejection of 99.88% and producing permeate water with negligible oil concentrations.

The forward osmosis process was used for the dilution of concentrated fertilizer solution which was then applied for fertigation purposes [35, 40]. Furthermore, the idea of combining wastewater treatment and desalination using FO-RO hybrid system was also proposed and investigated [41–43]. Hancock et al. [44] piloted a forward osmosis process scale during simultaneous seawater desalination and wastewater reclamation and subsequently evaluating its performance. A commercial spiral wound forward osmosis membrane element was tested continuously for 1300 h of processing 900,000 L of wastewater effluent and producing 10,000 L of treated water through a hybrid FO-RO process. Water flux was maintained at a relatively constant rate of 5.7 ± 0.2 Lm^{-2} h^{-1} with membrane bioreactor permeate feed and seawater draw solution. Test of sample fluorescence showed that the forward osmosis membrane and the hybrid process provided a strong rejection of protein-like species associated with wastewater effluent. There was also 99.9% removal of orthophosphate and dissolved organic carbon and more than 96% rejection of nitrate. However, as briefly stated, most forward osmosis applications are still restricted to small-scale laboratory experiments. More elaborate studies and conceptual proofs are required to turn its promising performance into a fully-fledged water treatment process.

1.2. Challenges and progress in water reuse and desalination technologies

The process of water desalination requires high electrical power input to achieve high water recoveries, which implies high capital and overall operational costs. It is believed that the cost of saline water desalination including infrastructure, maintenance and energy are very exceeded those needed for other common alternatives such as treating surface water and or ground water. The heavy energy demand of this process remains the hindrance to its extensive application. The theoretical value of about 0.86 kWh of energy is required to desalinate 1 m^3 of salt water (34,500 ppm) which is equal to 3 kJ kg^{-1}. However, in reality this value is normally inflated 5 to 26 times depending on the type of process used. Thus, clearly; it is necessary to make attempts to reduce the energy demand of process as much as possible [45].

The other aspects of saline water desalination include environmental impacts that need consideration. Thus, the disposal of saline concentrate into the water bodies also represents a huge environmental issue when using RO desalination technology. Recent years have seen stricter regulation being established in to protect receiving water bodies, aquatic life, and public drinking water sources by reducing total dissolved solids in brine that could be discharged into waterways. So it can be concluded that the combined treatment of wastewater and seawater could lead to double (heavy) membrane fouling, but; eliminating the use of pressured membrane process where the cake layer can be easily compacted eases the fouling irreversibility [46, 47]. This provides more motivation to explore forward osmosis processes that inherently have low membrane fouling propensity due to the absence of applied hydraulic pressure.

1.3. Determining factors of the forward osmosis membrane process

Permeate flux rate is commonly used as one of the primary performance indicators for membrane-based processes and is influenced by several factors that can be generally categorized as membrane properties, reverse salt diffusion and concentration polarization, feed water quality (and fouling) and operating conditions [48].

1.3.1. Membrane properties

The efficiency of an FO processes is directly linked to its membrane which in –turn is defined by its intrinsic separation properties stemming from the material used in its synthesis or preparation. The most used membrane performance parameters include the pure water permeability (A), solute rejection (R), solute permeability coefficient (B) and structural parameter (S). The membrane water permeability (A) is defined as the transport/passage of water through the membrane per unit driving force. The ability of a membrane to partially or completely retain solutes while allowing free passage of water molecules is referred as solute rejection (R), whilst the solute permeability coefficient (B) is described as the transport of a particular solute through the membrane per unit driving force at given water flux. The structural parameter (S) is a factor that defines the influence of membrane support thickness, porosity and tortuosity on mass transfer in the support layer [49, 50]. Membranes commonly used for osmotically driven filtration processes are characterized by an asymmetric structure defined by a dense thin top selective layer usually followed by a porous sub-layer. Ideally, a membrane needs to be freely permeable to the solvent (water) and completely retain the solute. Therefore, water permeability describes the extent to which water is able to percolate through the membrane's structure (Eq. 1), which is usually determined using hydraulic pressure.

$$A = \frac{J_w}{\Delta P} = \frac{V_{perm}}{A_m \Delta t \Delta P} \tag{1}$$

Where A_m represents the membrane's effective surface area (m^2), V_{perm} is the volume of the permeated water (L), Δt is the time elapsed during the permeation (h) and ΔP is the pressure difference across the membrane (bar).

The water transport across an osmotic membrane is generally described according to:

$$J_w = A(\Delta P - \Delta \pi) \tag{2}$$

Where A is the membrane water permeability ($L\ h^{-1}\ bar^{-1}$), ΔP is the pressure difference across the membrane (bar), $\Delta \pi$ is the osmotic pressure differential across the membrane (bar). The osmotic pressure of a solution can be calculated from the Morse equation. This equation is derived from the van't Hoff work (Eq. 3) on osmotic pressure and only applies to solutions with dilute concentrations (i.e. < 0.5 M). This equation indicates that osmotic pressure is linearly proportional to the solute concentration, (i.e. the higher the solute concentration, the higher the osmotic pressure of the solution).

$$\pi = imRT = i\left(\tfrac{n}{v}\right)RT \tag{3}$$

Where: i is the van't Hoff factor, (defines the number of ions produced during dissociation of a solute in an aqueous solution), m is the molarity of the solute which is equal to the ratio of the number of solute moles (mol) to the volume of the solution (L), R is the universal gas constant (8.3145 $J\ K^{-1}\ mol^{-1}$), T is the absolute temperature (K).

However, this equation does not hold for solutions with higher concentrations (usually >0.5 M). When computing the osmotic pressure of concentrated draw solutions other factors such as solution viscosity come into play [51]. In addition to water permeability property, a membrane has to have selectivity for solutes and is expressed by a rejection coefficient (R):

$$R = \frac{C_f - C_p}{C_f} = 1 - \frac{C_p}{C_p} \tag{4}$$

Where the solute concentrations on the feed and permeate, are represented by C_f and C_p respectively. Water permeability (A) and solute rejection (R) are membrane characteristics

http://dx.doi.org/10.5772/intechopen.72289

which are mainly linked to the active layer, that is; the active layer should permit water molecules to diffuse across while retaining solutes and other unwanted substances.

1.3.2. Draw solution

A draw solution is described as the solution of higher solute concentration and osmotic potential, relative to the feed solution, such that net pure water is induced through the membrane from the less concentrated impaired water to the draw solution to dilute it [52–55]. Different varieties of draw solutions have been evaluated for forward osmosis processes with the aim to achieve a low energy separation method for clean water production. And currently reverse osmosis is the best option for post-treatment of FO treated water, it can be used to separate the draw solution to produce clean potable water. However, there are still concerns about its reliance on hydraulic pressure which translates to high energy demand. Thus, FO draw solute regeneration can be made less energy intensive through the use of low grade energy sources but there can be some gains in energy depending on the type of energy used. In a closed loop FO-RO hybrid set-up, the energy utilization will always be higher than that of just reverse osmosis. But, when fouling in reverse osmosis is reduced then the practical energy consumption of FO-RO hybrid might be lower than just reverse osmosis.

Several fundamental criteria are considered when selecting draw solutions and are as follows: (i) the solute must have a high osmotic efficiency which results from high solubility in water and relatively low molecular weight; (ii) the osmotic agent must also be easily and economically separated from the diluted draw solution to yield potable water without being used up in the process; and (iii) the osmotic agents should ideally be inert, stable, neutral or near neutral pH, and nontoxic. Furthermore, the draw solutions should not be toxic to the membrane's physical structural integrity [52, 54]. Therefore, this makes finding the appropriate draw solution a systematic task. The solutes used to generate osmotic pressure for osmotic processes can be put into four major categories: inorganic solutes, thermolytic/volatile solutes, organic solutes, and polymer-based macro-solutes [56–59].

Inorganic salts are by far the most used draw solutes in FO and PRO research and this is down to abundant availability, affordability, and the ability to generate high osmotic pressures that induce high membrane flux [57, 58].

Thermolytic salts, on the other hand, are considered a unique kind of draw solutes, constituting of highly soluble gases and or volatile solutes that can produce high osmotic pressures and can be easily recovered [59]. This presents the opportunity to evaporate and regenerate the draw solute via the use low temperatures from poor quality heat sources (e.g., power plants) [60, 61]. The NH_3–CO_2 mixture has received extensive attention as a potential thermolytic draw solution. In the case of high draw solution concentrations can be created through manipulating the ratio of NH_3 to CO_2 [59, 62]. Application of thermal processes, heating to around 58°C is required to boil away the NH_3 and CO_2 and produce clean permeate water. These gases (NH_3 and CO_2) are then are re-combined to produce thermolytic salt and utilized again to generate osmotic pressure. However, the small hydration ions of NH_4^+ compared to those of divalent cations (Ca^{2+} and Mg^{2+}) lead to high reverse salt diffusion rate from the draw solution side to the feed water which greatly reduces the effective driving force. The need to apply a significant amount of thermal energy to boil away NH_3 and CO_2 stands as a major hindrance to the development of this draw solution.

It is for these concerns that recent studies have emphasized on polymer-based macro-solutes as potential osmotic agents, which allow easier recovery using low-pressure filtration processes such as ultrafiltration [63, 64]. However, the efficient use of ultrafiltration in the separation can have counter-effects, the accumulation of osmotic agents on the membrane can lead to a build-up of osmotic pressure that can lower the efficiency of the separation process. One outstanding advantage is that the polymer's molecular configuration and size can be tailored to produce draw solutions that give high osmotic pressure and desirable performance.

The lack of ideal draw solutes in forward osmosis is just but one drawback, the non-existent of cheap and simple draw solute separation strategies from the diluted draw solution to produce clean usable water is an area of paramount importance. Thus, attempts have been made recently towards the design of negatively charged super-paramagnetic nanoparticles that can be used in the recovery of an $Al_2(SO4)_3$ draw solute through coagulation [63, 65]. These previous research work have given an insight that future studies should combine the synthesis of novel, easily separable draw solutes, with novel and effective draw solute recovery technologies.

1.3.3. Feed water quality and osmotic gradient

The performance of the FO process is highly linked to the feed water composition. The targeted feed streams for the FO process include brackish water, seawater, treated wastewater effluent and industrial wastewater [20, 26, 27, 66, 67]. These are impaired water types composed dissolved substances or compounds that can induce membrane fouling and cause a severe decline in permeate flow [23, 24, 68–71]. Therefore, the sustainability of membrane permeate flux during FO operation is hugely influenced by feed water composition (foulant type, concentration and physicochemical properties) as well as the feed solution chemistry (i.e. solution pH, ionic strength and cationic species concentration) [65, 69–72]. The high osmotic pressure of seawater can lower the effective osmotic gradient or driving force, subsequently lowering water recovery which subsequently implies that the direct use of seawater as a feed stream in pressure-driven membrane processes such as RO amounts to huge energy consumptions.

Permeate flux is one performance indicator for a membrane-based process and is primarily dependent on the applied osmotic gradient. Therefore, the use of ideal draw solution that can generate high osmotic pressure ($\Delta\pi$) is critical for advancing FO technology [73]. The osmotic pressure difference is a result of the solute content of both the feed and draw solutions. A higher draw solution concentration gives a large osmotic pressure potential which in turn induces high permeate rates. The relationship between draw solution concentration and permeate flux is not linear mostly due to reverse diffusion of the osmotic agent and concentration polarization which are inherent phenomena of forward osmosis [37].

1.3.4. Operational conditions

Conducting a forward osmosis filtration tests involves the optimization of external operating parameters which have a huge role on the overall performance of the system. They include hydrodynamic parameters such as initial flux and cross-flow velocity as well as temperature. The aforementioned conditions strongly impact the output of an FO process, for example; it has been revealed that a higher cross-flow velocity minimizes the boundary layer thickness and concentration polarization, thus; subsequently lowering membrane fouling rate and enhancing water recovery [74]. Feed water composition and operational temperature can also hugely impact the performance of an FO membrane process. Operating temperature is closely linked to mass transfer, salt solubility, membrane fouling and concentration polarization, regardless of being a difficult parameter to monitor in practice, temperature is one critical operating condition [75–77]. Zhao and Zou, [40] observed increased water fluxes and recovery when higher operating temperatures were applied during filtration which they attributed to the decrease in permeate viscosity and an increase in osmotic pressure (and thus driving force), water permeability and mass transfer. Similar observations were made by Xie et al. [72] they found that the water permeability (A) values for different forward osmosis membranes increased with increasing temperature. However, in addition to increased water fluxes, the solute permeability coefficient (B) value was enhanced as well leading to higher reverse salt diffusion rates. The membrane structure factor, S was found to be unaffected by elevation in operating temperatures.

1.3.5. Membrane fouling

Membrane fouling is a broad term that describes the deposition and eventual accumulation of all kinds of substances on the membrane surface resulting from complex physical and chemical interactions between foulants and membrane. Fouling occurs when unwanted substances in the feed water block membrane pores or form an extra cake layer that generates resistance towards permeate flow through the membrane [75, 76, 78, 79]. Any membrane process using impaired water sources are subject to fouling. The fouling process in forward osmosis is said to be reversible due to the lack of pressure on the feed side, as a result foulants loosely bind to the membrane surface; however, the coupled occurrence of membrane fouling and concentration polarization lead to a severe decline in permeate flux [71]. There are four reported types of fouling often encountered in osmotic membrane filtration: inorganic fouling (scaling), biological fouling, colloidal fouling and organic fouling. Large quantities of research have been dedicated to understanding the subject of membrane fouling in osmotic membranes [77, 78, 80]. As partially highlighted, membrane fouling is linked to membrane and foulant's physicochemical properties [81]. Therefore, in summary, it can be stated that the eventual deposition of foulants on the membrane surface depends on the interplay of many factors that can be grouped into feed-water characteristics, hydrodynamic conditions and membrane properties. Attempts to investigate the fundamentals of membrane fouling have shown that the general mechanisms of fouling in osmotic membranes include adsorption, chemical interactions between solutes and membrane, gel formation and bacterial formation [75, 76, 79, 82, 83].

1.4. Challenges of the forward osmosis membrane process

Despite the various potentially attractive advantages of the FO process, it is still yards away from matching reverse osmosis mainly due to the number of obstacles that need to be resolved before its practical real-world implementation [53, 54]. Some of the efforts directed to advancing the forward osmosis technology include conducting systematic experimental research to solve challenging problems including identification of new draw solutes that will be capable of generating higher osmotic pressure, but are still easily separated from the diluted bulk draw solution at lower energy consumption [61, 64]; in addition to this there is the need of tailoring membranes that will decrease the effect of internal concentration polarization (ICP) which mostly occur in the porous support layer of current forward osmosis membranes and significantly reduces water flux because the diffusion of solutes is hindered and hydrodynamic force cannot effectively mix solutions inside the porous support layer [84].

More strategic progress in membrane and draw solute design need to be made for practical up-scaling of the FO technology. However, the subject of membrane fouling has not been fully understood and developed, but is essential to the significant improvement and viability of osmotically driven membrane processes in water treatment. Investigation of FO membrane fouling needs to be emphasized particularly at sufficiently high fluxes that allow economic sustainability. Even though the fouling propensity in forward osmosis is often stated to be relatively mild compared to reverse osmosis [85–88], there remains a need to effectively minimize fouling in order to increase process performance and prolong membrane lifespan. Membrane fouling does not only lead to a decline in permeate water flux, but also deteriorates the permeate water quality and consequently inflates the operating and membrane replacement costs. This is also the subject of interest in this work; therefore, the next sections will be expanding the discussion on the effect of membrane fouling on membrane flux loss in a forward osmosis processes, as well as potential alleviation remedies.

2. Combined wastewater reclamation and seawater dilution

The forward osmosis membrane process was then used to dilute seawater using simulated secondary treated wastewater effluent as feed solution in a laboratory scale setting. The system performance in recovering water was evaluated. Membrane fouling and fouling behavior were investigated.

2.1. Materials and methods

Sodium alginic acid salt, humic acid, bovine serum albumin and octanoic acid were used to as model organic foulants representing common polysaccharides, part natural organic matter (humic substances), proteins and fatty acids respectively in wastewater. These organic macromolecules have been reported to be the major components of organic fouling during membrane-based filtration systems [89, 90]. Alginate had an average molecular weight of 12–80 kDa. Stock solutions of 2 g/L were prepared by dissolving alginate powder in deionized (DI) water by mixing vigorously for 24 hours then kept at 4°C. The stock solution was stored for a maximum of 12 h before use. The molecular weight of humic acid ranged from 12 to 80 kDa as indicated by the supplier, and was prepared by dissolving 2 g/L in deionized water and vigorously stirred for 24 h after which, it was diluted to the desired concentration. Bovine serum albumin received in a powder form with a molecular weight of approximately 66 kDa. It was stored at 4°C upon delivery and was prepared by dissolving 1 g/L in deionized water over 24 h. Octanoic acid was received in a liquid form and was stored at room temperature. Stock solutions were prepared by mixing 1 g/L with deionized water and its pH was adjusted to around 6.7 using 0.05 M NaOH prior to addition to the feed solution. These model organic compounds were all supplied by Sigma-Aldrich (St Louis, MO) and were used as received. They were selected for this particular work because they are functionally similar to the organic foulants in wastewater effluent, so the observed fouling behavior and mechanisms might be comparable to real water effluent treatment using the FO process. However, the simulated wastewater used in this work does not contain particles, nor all the mentioned organic fractions, so real one on one translations could be difficult.

Three types of forward osmosis membranes were used in this work; cellulose triacetate membrane, thin film composite and Porifera membrane. The first two were supplied by Hydration Technologies, Inc. (Albany, OR) while the Porifera membrane was supplied under a confidentiality agreement. The cellulose triacetate membrane possesses an asymmetric structure made of a cellulose acetate skin layer supported by embedded polyester mesh. The thin film composite had a polyamide surface modification while the Porifera membrane was modified to be resilient to fouling (anti-fouling modification). Both the cellulose triacetate and Porifera membranes were stored in ultrapure water at 4°C prior to use. While the thin film membrane was stored in special packaging away from direct light and was soaked in ultrapure water for a minimum of 3 h before use.

The pure water permeability coefficient (A) of the forward osmosis membranes was determined in a laboratory-scale cross-flow reverse osmosis set-up. The effective membrane area was 49 cm^2 and the cross-flow velocity was fixed at 10 cm s^{-1}. Initially, the membrane permeate flux was equilibrated with deionized water at an applied pressure, ΔP, of 8 bar (116 psi), until the permeate flux reached a steady value. After equilibration, the volumetric permeate flux was measured at applied pressures ranging from 2 to 12 bar (29 to 174 psi) in increments of 2 bar (29 psi). The membrane's water permeability coefficient (A) is given by the slope of water flux plotted against applied pressure [65], using Eq.1.1.

The membranes' intrinsic separation parameters determined using equations Eq. 1–4 are presented in **Table 1**. And it is shown that the traditional flat sheet CTA membrane had the lowest pure water permeability (A) and highest salt permeability coefficient (B), with corresponding with a rather lower salt rejection. The thin-film composite membrane (TFC) had significantly increased pure water permeability compared to cellulose triacetate. It also recorded the lowest salt permeability coefficient (B) which translated to a high salt rejection (R). The novel Porifera membrane had the highest pure water permeability (A) and a high salt rejection almost similar to that of the thin film composite membrane. There was no clear correlation between the membrane structural factors and the other parameters. However, the superior

	A	**B**	**R**	**S**
	L/m² h bar	**×10⁻⁷ m/s**	**%**	**μm**
CTA	0.61	1.5	88.5	663
TFC	1.17	0.2	98.2	1227
POR	1.89	1.3	96.0	344

Table 1. Forward osmosis membrane intrinsic separation properties.

performance of thin film composite and Porifera membranes compared to the cellulose triacetate membrane was demonstrated, based on their respective *A*, *B* and *R* values (**Table 1**). These values also confirm the improvement made in water permeability and solute rejection of thin film composite membranes [86, 87, 91].

The different simulated fractions of effluent organic matter were fixed into the following concentrations: 200 mg/L, 100 mg/L, 80 mg/L and 20 mg/L for humic acid (HA), bovine serum albumin (BSA), alginate (Alg), and octanoic acid (OA) respectively. The total feed ionic strength was fixed at 20 mM using 17 mM NaCl and 1 mM $CaCl_2$. The fouling characteristics and potential of the model organic foulants were determined by conducting single foulant experiments for all four model foulants. Possible synergistic effects between foulants were investigated by preparing mixtures of two or more foulants that were then used to conduct fouling tests. The different feed solutions used to investigate thin film membrane fouling behavior are listed in **Table 2**.

2.2. Laboratory test unit

Laboratory filtration tests were performed using a self-assembled forward osmosis cross-flow set-up. It consisted of two closed loops dedicated to the feed and draw solution streams. These solutions were pumped past the cross-flow membrane cell and circulated using variable speed pumps (Cole-Palmer, USA). The cross-flow membrane cell was custom built with equally structured channels on both sides of the membrane. Each channel had the dimensions of 250, 50 and 1 mm for length, width and depth respectively. A polypropylene diamond spacer mesh was added on either side of the TFC membrane to create turbulence and mimic real membrane filtration processes. The change in feed solution weight was monitored over time through a weighing balance (Ohaus, USA) connected to a computer for data logging. These changes in feed water weight over time were used to calculate the water flux during membrane filtration tests.

During filtration, the permeate water gradually dilutes draw solution which decreases its concentration and in-turns reduces the osmotic drive force across the membrane. To prevent this effect, the concentration of the draw solution was maintained at a constant value using a real-time conductivity based program using a Consort conductivity meter (C3310 model, Turnhout, Belgium). Varying amounts of dry salt were dosed into the draw solution triggered by a decline in conductivity [92]. The schematic of the laboratory scale FO cross-flow unit is illustrated in **Figure 1**.

A program-controlled (LabVIEW software, National Instruments, UK) 3-way valve was installed on the draw solution return tube just before it enters the draw solution tank (**Figure 1**).

Feed solution composition	Ionic strength (mm)	Draw solution concentration (m)
100 mg/L BSA	20 mM (20 mM NaCl)	0.52 M NaCl
80 mg/L Alg		
20 mg/L OA		
200 mg/L HA		
BSA + Alg*	20 (17 mM NaCl +1 mM $CaCl_2$)	
Alg + HA + OA*		
Alg + OA + BSA*		
Alg + HA + BSA*		
Alg + HA + OA*		

*The concentrations of the single foulants were kept the same in their mixtures (100, 80, 20 and 200 mg/L for bovine serum albumin; BSA, alginate; ALG, octanoic acid; OA and humic acid; HA, respectively.

Table 2. Feed solution composition, ionic strength and draw solution concentration.

The valve temporally directs (at set intervals) the draw solution into a filter funnel containing dry solid salt (NaCl) after being triggered by a decline in draw solution conductivity. The dissolved salt then dripped into the bulk draw solution to correct the dropping solution conductivity and keep the draw solute concentration constant [50, 92].

2.3. Seawater dilution testing protocols

Membrane filtration tests were performed with the high-performance polyamide modified thin film composite forward osmosis membrane characterized by a hydrophilic surface using synthetic seawater as a draw solution. Particular emphasis was made on studying the effect of foulant synergy on permeate flux loss during wastewater effluent treatment. Furthermore, the effect of different membrane surfaces on fouling behavior was investigated using two additional forward osmosis membranes.

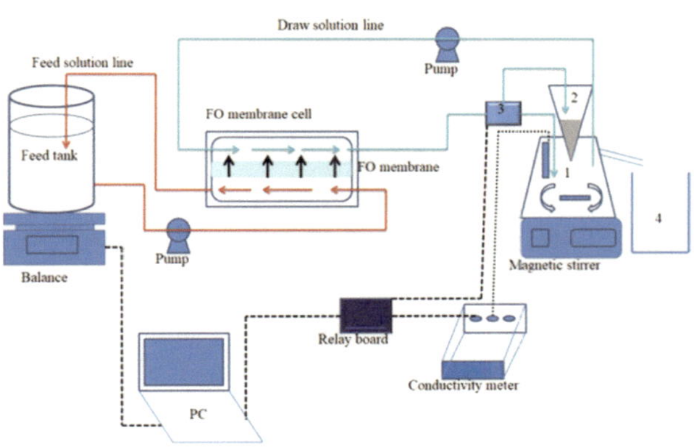

1. Draw solution tank, 2. Filter funnel with salt, 3. 3-way valve, 4. Draw over-flow vessel

Figure 1. Schematic of the laboratory scale forward osmosis cross-flow test unit.

In all fouling tests, sodium chloride (0.5 M) was used to induce permeation across the membrane (as a draw solution). Before conducting each filtration test, a baseline experiment was conducted, where a feed solution containing only the salt solution was filtered for the same duration as the foulant-bearing feed streams. This was performed to isolate flux decline due to due to foulant deposition and cake layer formation from that caused by internal concentration polarization. After the baseline test the membrane was flushed with large amounts of deionized water to wash-off the salt on its surface. Filtration tests were then performed with feed solutions bearing the different single compounds (alginate, humic acid, bovine serum albumin and octanoic acid) or their combinations. After each fouling test; the forward osmosis system was flushed with deionized water at high cross-flow velocity to clean the remnants of the previous test from the tubing system. The feed solution volume was fixed at 10 L while the re-concentrated draw solution volume was 1 L.

The thin film composite membrane was used as the primary membrane for all the filtration tests and its performance and fouling behavior was compared to that of cellulose triacetate and Porifera using the feed solutions that resulted in the most severe permeate flux loss. The concentration of the draw solution was fixed at 0.5 M NaCl for all experimental tests and was adjusted accordingly for the other membranes (cellulose triacetate and Porifera) to achieve an initial permeate flux of 13.5 (± 0.87) $Lm^{-2}\ h^{-1}$. Filtration tests were conducted for 24 h. The active layer-facing-feed solution (FO mode) configuration was used during tests. The cross-flow velocity was fixed at 10 cm s^{-1}.

2.3.1. Characterization techniques

The membrane's electrokinetic properties were investigated via streaming potential measurements which were performed using a SurPASS Electrokinetic Analyzer, (Anton Paar GmbH, Austria). This enabled the determination of membrane surface charge. Measurements were conducted using 0.01 mol/L KCl aqueous solutions as an electrolyte solution at 23°C and a pH of about 5.9. Surface zeta potentials were then derived from the measured streaming potentials according to the Helmholtz–Smoluchowski equation (Eq. 5) [93]. The presented data are average values of three different samples of each membrane type.

$$\zeta = \frac{\Delta V \eta \sigma}{\Delta P \varepsilon\ \varepsilon_0} \tag{5}$$

Where ΔV is the measured streaming potential, η is the electrolyte viscosity (Pa s), electrolyte's electrical conductivity (s/m), ΔP is the applied pressure and $\mathcal{E}$ is the permittivity of water ($C^2N^{-1}\ m^{-2}$). The permittivity is defined as $\mathcal{E} = \mathcal{E}_0.D$, where $\mathcal{E}_0$ is permittivity of vacuum = 8.85×10^{-12} ($C^2N^{-1}\ m^{-2}$) and D the dielectric constant of water = 78.55 at 25°C.

Membrane surface morphology as well as the structural arrangement of fouling layers was assessed using scanning electron microscopy (SEM); using a JEOL IT300 scanning electron microscope (Tokyo, Japan.). Small dried membrane pieces were cut and attached to sample holders using a carbon tape. The sample holder with the membrane pieces was coated with either gold or carbon to provide electrical conductivity and prevent charging during imaging. Analysis was performed at different desired magnifications and working distances.

The topology and roughness of clean and used membranes were studied using an atomic force microscope (AFM: Alpha300, Germany). The average hydrodynamic diameter of the organic aggregates in the different aqueous solutions was measured using the dynamic light scattering (DLS) technique (Malvern nanosizer, Malvern Instruments, UK).

3. Results and discussions

3.1. Feed solution properties

Dynamic light scattering (DLS) measurements provide information on the particle size distribution of a suspension. And it was used in this study to monitor macro-aggregate formation during mixed foulant fouling to gain more insight into foulant-foulant interactions. The intensity of the scattered light is a strong function of the particle size and bigger aggregates cause more scattering which is translated to a larger intensity peak. The role of divalent cations (particularly Ca^{2+}) on organic fouling has been well studied and widely reported using surrogate organic compounds [94]. Their presence has been associated with intense fouling caused by organic foulants via charge neutralization, complexation and forming calcium bridges [95, 96]. In this study, the concentration of Na^{+} was 17 mM and that of Ca^{2+} was fixed to 1 mM. **Table 3** presents hydrodynamic diameters for single foulants and their different combinations. And according to the recorded values, it demonstrated that the cations had a significant influence on the physicochemical properties of the individual and combined foulants. The changes in particle sizes were conspicuous, there was clear aggregation of macromolecules when calcium ions were introduced. This trend

Feed sample	Zeta potential (mv)	Hydrodynamic diameter (nm)
Alg*	−54 ± 3	66 ± 4
HA*	−48 ± 3	213 ± 10
BSA*	−10 ± 1	4
Alg	−14 ± 1	261 ± 8
HA	−27 ± 1	199 ± 2
BSA	−2	8
OA	—	—
Alg + BSA	−20 ± 3	349 ± 15
Alg + HA + OA	−19 ± 2	603 ± 19
HA + BSA + OA	−13 ± 1	377 ± 11
HA + BSA + Alg	−19 ± 5	—
HA + BSA + Alg + OA	—	—

*Measured zeta potential and hydrodynamic diameters in the absence of cations.

Table 3. Measured foulant zeta potentials and average hydrodynamic diameters in the different feed solutions.

was further supported by the surface charge reduction of the aggregates upon exposure to electrolyte solutions. The foulant-cation complexation was more prominent in humic acid and alginate because of the abundant presence of carboxylic acid groups; ionization of carboxylic acids gives carboxylate anions that in turn complex with the positive Na^+ and Ca^{2+} to form aggregates.

The influence of these ions on the fouling potential of each foulant was found to be different. Alginate fouling was consistent with the observed physicochemical (charge and particle size) changes; however, a noticeable deviation was observed with humic acid which resulted in less fouling even in the presence of calcium ions. A possible explanation for this anomalous observation lies on the HA-Ca^{2+} ratio used for the purposes of this study; there were insufficient calcium ions to complex with humic acid macromolecules. Also, Na^+ competed with the Ca^{2+} for the negatively charged HA carboxylate ions.

All three primary foulants were found to exhibit a negative surface charge. Alginate and humic acid had the highest negative charges in solution which can be attributed mainly due to the abundant presence of negative carboxylate groups. Therefore, they had prominent interaction with the cations as evidenced by the large reduction in negative charges in the presence of cations. The determination of both surface charge and hydrodynamic diameter of octanoic acid was unsuccessful. The BSA molecules had the lowest zeta potential values (**Table 3**) and were least influenced by the cations.

The reported zeta potential and hydrodynamic sizes for mixed foulants cannot be tied down to a single factor but rather a combined influence of cationic species' concentrations, molecular size and shape as well as organic–organic interactions. Therefore, the values presented here are averages of a range of sizes and they should be viewed with some reservation. Thus, the discussion is based on qualitative observations rather than on quantitative data. However, the changes in the measured hydrodynamic diameters are in accordance with earlier reported studies on the influence of Na^+ and Ca^{2+} on organic foulants [16, 97], and it was found that the aggregate size followed this order: BSA < humic acid < alginate.

3.2. Membrane surface morphology

Scanning electron and atomic force microscopy analysis of the membrane's feed side gave the micrographs presented in **Figure 2**. The membrane surface appeared to be covered by a thick, loose and flexible fouling layer after filtration of the mixtures of alginate and BSA (**Figure 2(b)**) and that of alginate, BSA and humic acid (**Figure 2(c)**). An indication of heavy foulant deposition during seawater dilution. The fouling layer appears loose and detached which is typical of FO membrane fouling due to the lack of external hydraulic pressure. AFM images show a relatively rough thin film composite membrane in **Figure 2(d)**. The images in **Figure 2(e)** and **(f)** show completely different topologies which suggest the presence of a cake layer on the surface of the membrane.

3.3. Fouling characteristics of single foulants

Filtration tests using feed streams containing single, simple organic compounds yielded varying membrane performances as shown in **Figure 3**. The feed streams containing humic acid,

http://dx.doi.org/10.5772/intechopen.72289

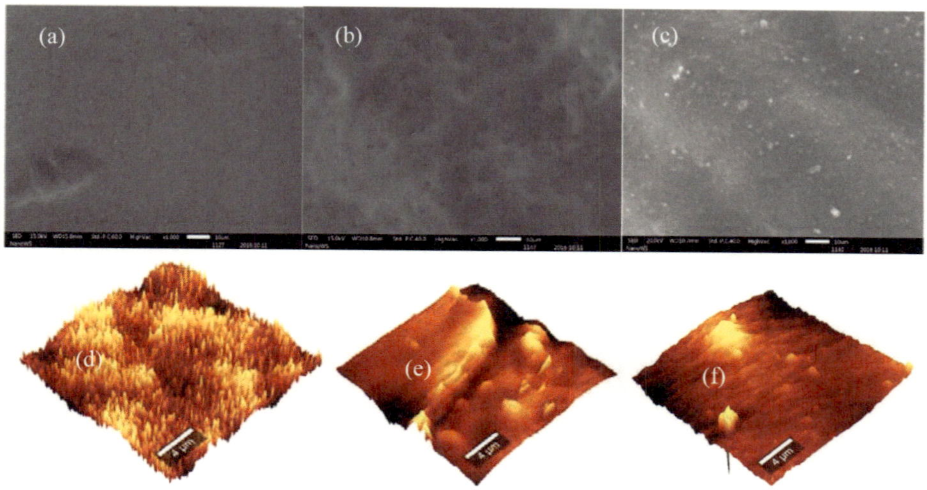

Figure 2. Clean and fouled membrane surface morphology and topology: (a) SEM image of clean TFC membrane, (b) image of membrane fouled with Alg + BSA, (c) image of membrane fouled with Alg + BSA + HA, (d)–(f) corresponding AFM micrographs.

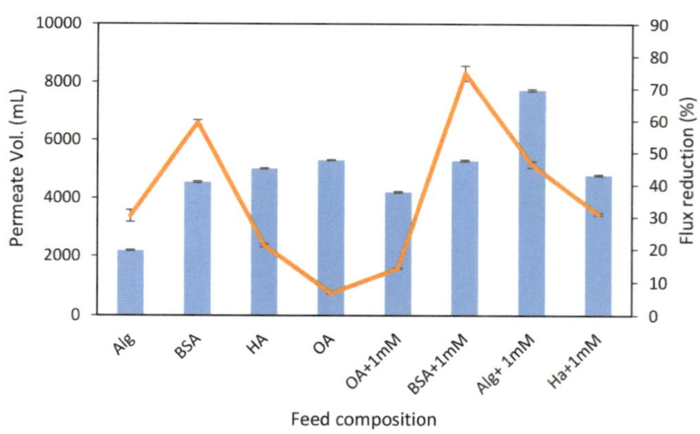

Figure 3. Recorded permeate volumes and flux declines during seawater dilution suing simple feed streams.

alginate and octanoic acid recorded the highest water recovery of above 50%. A slight decline in water recovery was observed when the membrane was used to filtrate the feed solution containing bovine serum albumin recording a 40% recovery. The addition of 1 mM $CaCl_2$ to the feed solutions had a significant influence on membrane performance; particularly, on the feed solutions containing BSA and alginate which correlated to their flux reduction. That is, the calcium ions improved cross-linking of ionized alginate macromolecules forming an organized gel layer that was easily deposited on the membrane surface, creating an extra resistance layer to water permeation. This observation is supported the measured alginate aggregate sizes in **Table 3**, which showed an increase in aggregate size in the presence of Ca^{2+}. The same phenomenon is expected for humic acid, however, the resulting cake layer is porous and offered little resistance to permeate flow, so permeate flux remained the same.

There was a further loss in permeate recovery for the BSA bearing feed solution, the presence of Ca^{2+} enhanced its affinity for the membrane surface. The macromolecular size was reduced to almost neutral values leading to subsequent weakened electrostatic repulsions from the negative membrane surface leading to its multilayer adsorption. Permeate flux reduction patterns correlated to the recorded permeate water recovery rates. The feed solutions containing humic and octanoic acids had the lowest permeate flux reduction with and without calcium ions. The BSA feed exhibited the highest permeate flux loss reduction of 60% before the addition of calcium ions. Permeate flux reduction increased by 16% when Ca^{2+} was added to the alginate feed solution, rising from 30 to 46%. These results revealed that the FO process had an average performance for simple, single foulants bearing feed streams. And humic and octanoic acid had no significant influence on permeate flux unlike, polysaccharides (alginate) and proteins (BSA) that dominated permeate flux loss [95]. The next section investigates the interactions between co-foulants when they co-exist in the same feed solution in an attempt to underpin foulant-foulant interactions.

3.4. Influence of co-foulants on flux

The two fractions that caused the most significant permeate flux decline (alginate and BSA) in the previous section were combined and tested for their impact on permeate flux loss. The resulting fouling trend was compared to those observed during filtration of single foulants as depicted by **Figure 4**. And it can be seen that the co-existence of alginate and BSA resulted in more permeate flux loss. The flux decline curve is similar to that of BSA alone, characterized by the first stable flux region followed by a rapid flux loss rate until a semi-steady flux point was reached. This indicates that BSA macromolecules had a dominant effect on the formation of the combined fouling layer. According to the measured surface charge results the two foulants should electrostatically repel each other due to the negative surface charges; however, hydrophobic interactions among the foulants appear to be dominant in the formation of alginate-BSA aggregates as supported by the increase in sizes when the two foulants are present in the same feed solution (**Table 3**). It is though that the BSA molecules became integrated into the alginate-calcium complexes [98, 99], and since there were favorable interactions that promoted BSA attachment onto the membrane surface; the alginate aggregates were sort of "functionalized" and easily deposited to form the fouling layer shown in **Figure 2(b)**. It can therefore be concluded that the addition of alginate to BSA enhanced permeate flux loss (fouling), which implies the dominant presence of synergistic interactions between the proteins and polysaccharides.

3.5. Filtration tests with complex feed solutions

The feed streams were made more complex by mixing three organic compounds in one feed solution. When the feed solution containing alginate, humic and octanoic acid was filtrated using the TFC membrane, a 51% water recovery was recorded and initial permeate flux was reduced by 30% after 24 h (**Figure 5**). This result was beyond expectations since alginate and humic acid in the presence Ca^{2+} have been reported to worsen fouling due to the formation of HA-Ca^{2+}, Alg-Ca^{2+} and Alg-HA complexes, as the formed fouling layers act to increase

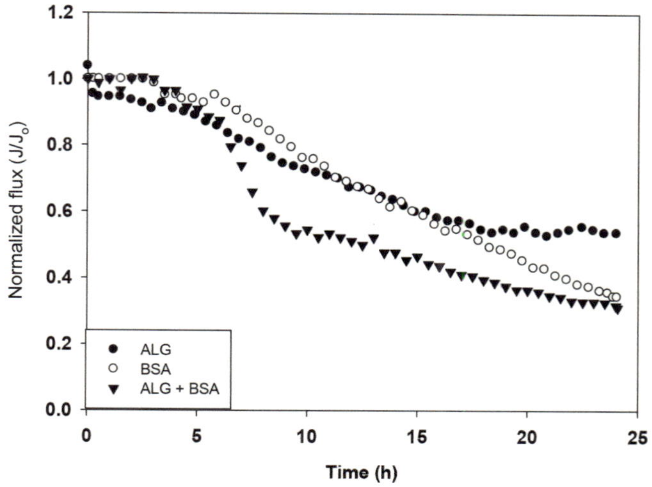

Figure 4. Membrane permeate flux decline pattern during co-foulant (Alg + BSA) feed stream filtration.

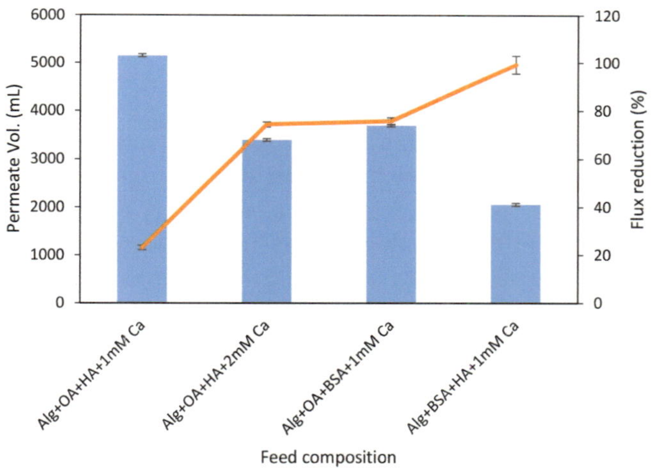

Figure 5. Permeate volumes and permeate flux loss during seawater dilution using complex feed streams.

resistance to permeate flow [100]. The explanation for this could be that the Ca^{2+} concentration was ineffective in causing complete complexation of the foulants (80 mg/L Alg and 200 mg/L HA), thus aggregate formation was in significant and the foulants remained in the bulk solution rather than being deposited. And it also suggests that the cake layer formation during foulant deposition was dominated by humic acid macromolecules which formed a loose porous layer such that permeate flow rate was not significantly lowered. This explanation is backed by the flux reduction and water recovery in the presence of 2 mM Ca^{2+} which shows a 15% reduction in water recovery and a 74% permeate flux loss. The calcium ions interacted with the alginate and humic acid macromolecules to form a thick compact cake layer that offered resistance to permeate flow.

The presence of proteins (BSA) in the feed solution containing alginate and octanoic acid reduced water recovery (37%) and increased flux reduction by 75%. Interestingly, the mixture of alginate, BSA and humic acid resulted in poor process performance with a water recovery of 20% and almost no permeation after 16 hours of filtration. This suggests that there were favorable interactions between the foulants that led to excessive deposition rate onto the membrane surface, resulting in a thick and resistant cake layer which enhanced reverse solute diffusion contributing into flux loss. The differences observed in the permeate flux reductions can be attributed to the various foulant-foulant and organics-membrane interactions during filtration, which then leads to different fouling layer properties.

These results demonstrated that the performance of the FO membrane in treating heavily impaired water using seawater as a draw solution. There was severe flux loss when polysaccharides, humic substances and proteins co-existed in the same feed solution. This is the most likely, occurrence in secondary treated wastewater. However, the organic foulants exist in lower concentrations than what was used in this experiments (worst case scenario). Thus, the combined wastewater-seawater dilution process promises to be a simple and effective water recovery process that might be hindered by membrane fouling. But the resulting fouling layer can be easily washed-off using physical cleaning methods [101, 102].

3.6. Influence of membrane surface

The performance of the commercial thin film composite membrane was compared to that of the low flux cellulose triacetate membrane and two custom-made Porifera membranes using the most complex feed solutions. Average water recovery for the three membranes was above 50% (**Figure 6**). The Porifera membranes had superior performance at the same operating conditions and initial permeate flux, followed by the cellulose triacetate membrane. The observed varying performances are due to differences in surface properties and functionalities. The rough polyamide layer of the TFC membrane was highly susceptible to protein deposition and foulant adhesion. Whilst, the smooth cellulose triacetate surface is resilient

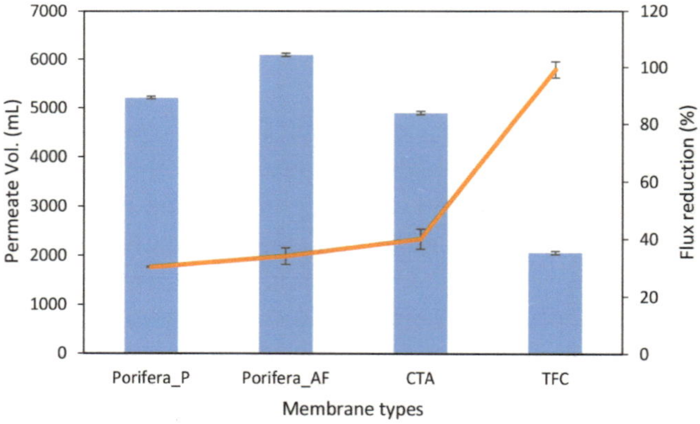

Figure 6. Performance of FO membranes used to filter complex feed streams. Porifera P represents the plain membrane while Porifera AF was modified to induce anti-fouling properties.

to foulant deposition [103]. Therefore, seawater dilution using wastewater can be further improved by choosing a foulant resistant membrane with a matching flux.

4. Summary

The on-going water shortage has opened an opportunity for wastewater and seawater to be explored as alternative water sources to supplement water supply due to the diminishing natural fresh water sources. However, extensive treatment procedures are required to make these water streams suitable for either domestic, industrial or even agricultural purposes, thus forward osmosis membrane process was identified as an ideal candidate to lower the osmotic pressure of seawater prior to desalination using wastewater as a feed source. The fouling behavior of the membrane process was studied. And the results revealed that proteins and polysaccharides had a dominant role in governing permeate flux loss. The presence of divalent cations, especially Ca^{2+} exacerbated the fouling process. Filtration tests demonstrated that there were favorable electrostatic and hydrophobic interactions among foulants and membrane surface that promoted foulant deposition and cake layer formation. The forward osmosis process had an average performance in treating heavily impaired feed water streams under exaggerated conditions. This implies that an even better performance can be expected for real water samples where foulant content is lower. It was also found that the process performance can be improved by selecting/using foulant resistant membranes.

Author details

Machawe M. Motsa* and Bhekie B. Mamba

*Address all correspondence to: machawemmvulane@gmail.com

Nanotechnology and Water Sustainability Research Unit, College of Science, Engineering and Technology, University of South Africa, Pretoria, South Africa

References

[1] Ang WS, Elimelech M. Fatty acid fouling of reverse osmosis membranes: Implications for wastewater reclamation. Water Research. 2008;**42**:198-210

[2] Baker RW. Reverse Osmosis in Membrane Technology and Applications. 2nd ed. John Wiley and Sons Ltd: Chichester; 2004. p. 191

[3] Macedonio F, Drioli E, Gusev AA, Bardow A, Semiat R, Kurihara M. Efficient technologies for worldwide clean water supply. Chemical Engineering and Processing. 2012;**51**:2-17

[4] Shannon MA, Bohn PW, Elimelech M, JGB G, Marinas J, Mayes AM. Science and technology for water purification in the coming decades. Nature. 2008;**452**:301-310

[5] Wang Z, Tang J, Zhu C, Dong Y, Wang Q, Wu Z. Chemical cleaning protocols for thin film composite (TFC) polyamide forward osmosis membranes used for municipal wastewater treatment. Journal of Membrane Science. 2015;**475**:184-192

[6] Zhang X, Ning Z, Wang DK, Diniz da Costa JC. Processing municipal wastewater by forward osmosis using CTA membrane. Journal of Membrane Science. 2014;**468**: 269-275

[7] Lutchmiah K, Verliefde ARD, Roest K, Rietveld LC, Cornelissen ER. Forward osmosis for application in wastewater treatment: A review. Water Research. 2014;**58**:179-197

[8] Lee S, Boo C, Elimelech M, Hong S. Comparison of fouling behavior in forward osmosis (FO) and reverse osmosis (RO). Journal of Membrane Science. 2010;**365**:34-39

[9] Li L, Dong J, Nenoff TM, Lee R. Desalination by reverse osmosis using MFI zeolite membranes. Journal of Membrane Science. 2004;**243**:401-404

[10] Hancock NT, Xu P, Roby MJ, Gomez JD, Cath TY. Towards direct potable reuse with forward osmosis: Technical assessment of long-term process performance at the pilot scale. Journal of Membrane Science. 2013;**445**:34-46

[11] Zang X, Yang H, Wang X, Fu J, Xie YF. Formation of disinfection by products: Effect of temperature and kinetic modeling. Chemosphere. 2013;**90**:634-639

[12] Yangshuo G, Wang Y-N, Wei J, Tang CY. Organic fouling of thin film composite polyamide and cellulose triacetate forward osmosis membranes by oppositely charged macromolecules. Water Research. 2013;**10**:3812-3818

[13] Cath TY, Childress AE, Elimelech M. Forward osmosis: Principles, applications, and recent developments. Journal of Membrane Science. 2006;**281**:70-87

[14] Shaffer DL, Yin Yip N, Gilron N, Elimelech M. Seawater desalination for agriculture by integrated forward and reverse osmosis: Improved product water quality for potentially less energy. Journal of Membrane Science. 2012;**415**:1-8

[15] Ruprakobkit T, Ruprakobkit L, Ratanatamskul C. Dynamic modelling of carboxylic acid filtration in forward osmosis process: The role of membrane CO_2 permeability. Computers and Chemical Engineering. 2017;**98**:100-112

[16] Alsvik IL, Hagg M-B. Pressure retarded osmosis and forward osmosis membranes: Materials and methods. Polymer. 2013;**5**:303-327

[17] Cath TY, Gormly S, Beaudry EG, Flynn MT, Adams VD, Childress AE. Membranes contactor process for wastewater reclamation in space: Part I. Direct osmotic concentration as pre-treatment for reverse osmosis. Journal of Membrane Science. 2005;**257**:85-98

[18] Zhao S, Zou L, Tang CY, Mulcahy D. Recent developments in FO: Opportunities and challenges. Journal of Membrane Science. 2012;**396**:1-12

[19] Park M, Lee J, Boo C, Hong S, Snyder SA, Kim JA. Modeling of colloidal fouling in forward osmosis membrane: Effects of reverse draw solution permeation. Desalination. 2013;**314**:115-123

[20] Ge Q, Su J, Chung T-S, Amy G. Hydrophilic superparamagnetic nanoparticles: Synthesis, characterization, and performance in forward osmosis processes. Industrial and Engineering Chemistry Research. 2011;**50**:82-88

[21] Peñate B, García-Rodríguez L. Current trends and future prospects in the design of seawater reverse osmosis desalination technology. Desalination. 2012;**284**(1-8)

[22] Chung TS, Zhang S, Wang KY, Su JC, Ling MM. Forward osmosis processes: Yesterday, today and tomorrow. Desalination. 2012;**278**:78-81

[23] Beaudry EG, Lampi KA. Membrane technology for direct osmosis concentration of fruit juices. Food Technology. 1990;**44**:121-131

[24] Jiao B, Cassano A, Drioli E. Recent advances on membrane processes for the concentration of fruit juices: A review. Journal of Food Engineering. 2004;**63**:303-324

[25] Dova MI, Petrotos KB, Lazarides HN. On the direct osmotic concentration of liquid foods. Part II. Development of a generalized model. Journal of Food Engineering. 2007; **78**:431-437

[26] Dova MI, Petrotos KB, Lazarides HN. On the direct osmotic concentration of liquid foods. Part I. Impact of process parameters on process performance. Journal of Food Engineering. 2007;**78**:422-430

[27] Kravath RE, Davis JA. Desalination of seawater by direct osmosis. Desalination. 1975; **16**:151-155

[28] Petrotos KB, Quantick PC, Petropakis H. A study of the direct osmotic concentration of tomato juice in tubular membrane-module configuration. I. The effect of certain basic process parameters on the process performance. Journal of Membrane Science. 1998; **150**:99-110

[29] Yangali-Quintanilla V, Li Z, Valladares R, Li Q, Amy G. Indirect desalination of Red Sea water with forward osmosis and low pressure reverse osmosis for water reuse. Desalination. 2011;**280**:160-166

[30] Tang WL, Ng HY. Concentration of brine by forward osmosis: Performance and influence of membrane structure. Desalination. 2008;**224**:143-153

[31] Van Houtte E, Verbauwhede J. Operational experience with indirect potable reuse at the Flemish coast. Desalination. 2008;**218**:198-207

[32] Votta F, Barnett SM, Anderson DK. Concentration of Industrial Waste by Direct Osmosis: Completion Report. Providence, Rhodes Island; 1974

[33] York RJ, Thiel RS, Beaudry EG. Full-scale experience of direct osmosis concentration applied to leachate management. In: Proceedings of the Seventh International Waste Management and Landfill Symposium. 1999. S. Margherita di Pula, Cagliari, Sardinia, Italy

[34] Cartinella JL, Cath TY, Flynn MT, Miller GC, Hunter KW, Childress AE. Removal of natural steroid hormones from wastewater using membrane contactor processes. Environmental Science and Technology. 2006;**40**:7381-7386

[35] Phuntsho S, Shon SK, Hong S, Vigneswaran S. A novel low energy fertilizer driven forward osmosis desalination for direct fertigation: Evaluating the performance of fertilizer draw solutions. Journal of Membrane Science. 2011;**375**:172-181

[36] Beaudry EG, Herron JR, Peterson SW. Direct osmosis concentration of waste water: Final report. Osmotek Inc., Corvallis; 1999

[37] Lee KL, Baker RW, Lonsdale HK. Membranes for power generation by pressure-retarded osmosis. Journal of Membrane Science. 1981;**8**:141-171

[38] Coday BD, Almaraz N, Cath TY. Forward osmosis desalination of oil and gas wastewater: Impacts of membrane selection and operating conditions on process performance. Journal of Membrane Science. 2015;**488**:40-55

[39] Duong PHH, Chung TS, Wei S, Irish L. Highly permeable double-skinned forward osmosis membranes for anti-fouling in the elmusified oil-water separation process. Environmental Science and Technology. 2014;**48**:4537-4545

[40] Zhao S, Zou L. Effects of working temperature on separation performance, membrane scaling and cleaning in forward osmosis desalination. Desalination. 2011;**278**:157-164

[41] Cath TY. Osmotically and thermally driven membrane processes for enhancement of water recovery in desalination processes. Desalination and Water Treatment. 2010;**15**: 279-286

[42] Bamaga OA, Yokochi A, Zabara B, Babaqi AS. Hybrid FO/RO desalination system: Preliminary assessment of osmotic energy recovery and designs of new FO membrane module configuration. Desalination. 2011;**268**:163-169

[43] Cath TY, Childress AE, Elimelech M. Forward osmosis: Principles, applications and recent developments. Journal of Membrane Science. 2006;**281**:70-87

[44] Hancock NT, Xu P, Roby MJ, Gomez JG, Cath TY. Towards direct potable reuse with forward osmosis: Technical assessment of long-term process performance at the pilot scale. Journal of Membrane Science. 2013;**445**:34-46

[45] van der Bruggen B, Luis P. Forward osmosis: Understanding the hype. Reviews in Chemical Engineering. 2014;**31**:1-12

[46] Thorsen T, Holt T. The potential for power production from salinity gradients by pressure retarded osmosis. Journal of Membrane Science. 2009;**335**:103-110

[47] Elimelech M, Phillip WA. The future of seawater desalination: Energy, technology, and the environment. Review Science. 2011;**333**:712-717

[48] McGinnis RL, Elimelech M. Global challenges in energy and water supply: The promise of engineered osmosis. Environmental Science and Technology. 2008;**42**:8625-8629

[49] Mohammadi T, Moghadam MK, Madaeni SS. Hydrodynamic factors affecting flux and fouling during reverse osmosis of seawater desalination. Desalination. 2002;**151**: 239-245

[50] Motsa MM, Mamba BB, Thwala JM, Verliefde ARD. Osmotic backwash of fouled FO membranes: Cleaning mechanisms and membrane surface properties after cleaning. Desalination. 2017;**402**:62-71

[51] Tu KL, Chivas AR, Nghiem LD. Effects of membrane fouling and scaling on boron rejection by nanofiltration and reverse osmosis membranes. Desalination. 2013;**310**:115-121

[52] Ge Q, Ling M, Chung TS. Draw solutions for forward osmosis processes: Developments, challenges, and prospects for the future. Journal of Membrane Science. 2013;**442**:225-237

[53] Singh R, Hankins NP. Introduction to Membrane Processes for Water Treatment, Emerging Membrane Technology for Sustainable Water Treatment. 1st ed. Amsterdam: Elsevier; 2016

[54] Chung TS, Li X, Ong RC, Ge Q, Wang HL, Han G. Emerging forward osmosis (FO) technologies and challenges ahead for clean water and energy applications. Current Opinion in Chemical Engineering. 2012;**1**:246-257

[55] Achilli A, Cath TY, Marchand EA, Childress AE. The forward osmosis membrane bioreactor: A low fouling alternative to MBR processes. Desalination. 2009;**239**:10-21

[56] McGinnis RL, McCutcheon JR, Elimelech M. A novel ammonia-carbon dioxide osmotic heat engine for power generation. Journal of Membrane Science. 2007;**305**:13-19

[57] Yong JS, Phillip WA, Elimelech M. Couple reverse draw solute permeation and water flux, in forward osmosis with neutral draw solutes. Journal of Membrane Science. 2012; **392-393**:9-17

[58] Xu Y, Peng X, Tang CY, QS F, Nie S. Effect of draw solution concentration and operating conditions on forward osmosis and pressure retarded osmosis performance in a spiral wound module. Journal of Membrane Science. 2010;**348**:298-309

[59] McCutcheon JR, Elimelech M. Desalination by ammonia-carbon dioxide forward osmosis: Influence of draw and feed solution concentration on process performance. Journal of Membrane Science. 2006;**278**:114-123

[60] Zou S, Gu Y, Xiao D, Tang CY. The role of physical and chemical parameters on forward osmosis membrane fouling during algae separation. Journal of Membrane Science. 2011;**366**:356-362

[61] Klaysom C, Cath TY, Depuydt T, Vankelecom IFJ. Forward and pressure retarded osmosis: Potential solutions for global challenges in energy and water supply. Chemical Society Reviews. 2013;**42**:6959-6989

[62] McCutcheon JR, Elimelech M, McGinnis RL. A novel ammonia-carbon dioxide forward (direct) osmosis desalination process. Desalination. 2005;**174**:1-11

[63] Ling MM, Chung TS, Lu X. Facile synthesis of thermosensitive magnetic nanoparticles as 'smart' draw solutes in forward osmosis. Chemical Communications. 2011;**47**:10788-10790

[64] Liu Z, Bai H, Lee J, Sun DD. A low-energy forward osmosis process to produce drinking water. Energy and Environmental Science. 2011;**4**:2582-2585

[65] Bacchin P, Aimar P, Field RW. Critical and sustainable fluxes: Theory, experiments and applications. Journal of Membrane Science. 2006;**281**:42-69

[66] Gkotsis PK, Banti DC, Pelleka EN, Zouboulis AI, Samaras PE. Review: Fouling issues in membrane bioreactors (MBRs) for wastewater treatment: Major mechanisms, prevention and control strategies. PRO. 2014;**2**:795-866

[67] Cath TY, Gormly S, Beaudry EG, Adams VD, Childress AE. Membrane contactor processes for wastewater reclamation in space. I. Direct osmotic concentration as pretreatment for reverse osmosis. Journal of Membrane Science. 2005;**257**:85-98

[68] She Q. Effect of hydrodynamic conditions and feed water composition on fouling of ultrafiltration and forward osmosis membranes by organic macro-molecules, M. Eng. Thesis. School of Civil and Environmental Engineering, Nanyang Technological University, Singapore; 2008

[69] Thelin WR, Silvertsen E, Holtv T, Brekke G. Natural organic matter fouling in pressure retarded osmosis. Journal of Membrane Science. 2013;**438**:46-56

[70] de Koning J, Bixio D, Karabelas A, Salgot M, Schäfer A. Characterization and assessment of water treatment technologies for reuse. Desalination 2008;**218**:92-104

[71] She Q, Wang R, Fane AG, Tang CY. Membrane fouling in osmotically driven membrane processes: A review. Journal of Membrane Science. 2016;**499**:201-233

[72] Xie M, Price WE, Nghiem LD, Elimelech M. Effects of feed and draw solution temperature and transmembrane temperature difference on the rejection of trace organic contaminants by forward osmosis. Journal of Membrane Science. 2013;**438**:57-64

[73] Tian E, Hu C, Qin Y, Ren Y, Wang X, Wang X, Xiao P, Yang X. A study of poly (sodium 4- styrenesulfonate) as draw solute in forward osmosis. Desalination. 2015;**360**:130-137

[74] Gou W, Ngo HH, Li J. A mini-review on membrane fouling. Bioresource Technology. 2012;**122**:27-34

[75] Katsoufidiou K, Yiantsios SG, Karabelas AJ. UF membrane fouling by mixtures of humic acids and sodium alginate: Fouling mechanisms and reversibility. Journal of Membrane Science. 2010;**264**:220-227

[76] Kim S, Hoek EMK. Interactions controlling biopolymer fouling of reverse osmosis membranes. Desalination. 2007;**202**:333-342

[77] Shirazi S, Li CJ, Chen D. Inorganic fouling of pressure-driven membrane processes – Critical review. Desalination. 2010;**250**:236-284

[78] Zhang J, Loong WLC, Chou S, Tang C, Wang R, Fane AG. Membrane biofouling and scaling in forward osmosis membrane bioreactor. Journal of Membrane Science. 2012; **403-404**:8-14

[79] Chen L, Tian Y, Cao CQ, Zhang J, Li ZN. Interactions energy evaluation of soluble microbial products (SMP) on different membrane surfaces: Role of the reconstructed membrane topology. Water Research. 2012;**46**:2693-2704

[80] Bhattacharjee S, Sharma A, Bhattacharya PK. Surface interactions in osmotic-pressure controlled flux decline during ultrafiltration. Langmuir. 1994;**10**:4710-4720

[81] Hong SK, Elimelech M. Chemical and physical aspects of natural organic matter (NOM) fouling of nanofiltration membranes. Journal of Membrane Science. 1997;**135**:159-167

[82] McCutcheon JR, Elimelech M. Influence of membrane support layer hydrophobicity on water flux in osmotically driven membrane processes. Journal of Membrane Science. 2008;**318**:458-466

[83] Phuntsho S, Kyong Shon H, Vigneswaran S, Kandasamy J, Hong S, Lee S. Influence of temperature and temperature difference in the performance of forward osmosis desalination process. Journal of Membrane Science. 2012;**415**:734-744

[84] Setiawan L, Wang R, Li K, Fane AG. Fabrication and characterization of forward osmosis hollow fibre membranes with antifouling NF-like selective layer. Journal of Membrane Science. 2012;**394**:80-88

[85] Yen SK, MehnasHaja F, NM S, Wang KY, Chung TS. Study of draw solutes using 2-methylimidazole-based compounds in forward osmosis. Journal of Membrane Science. 2010;**364**:242-252

[86] Gruber MF, Johnson CJ, Tang CY, Jensen MH, Yde L, Helix-Nielsen C. Computational fluid dynamics simulations of flow and concentration polarization in forward osmosis membrane systems. Journal of Membrane Science. 2011;**379**:488-495

[87] Su J, Chung TS, Helmer BJ, de Wit JS. Enhanced double-skinned FO membranes with inner dense layer for wastewater treatment and macromolecule recycle using sucrose as draw solute. Journal of Membrane Science 2012;**396**:92-100

[88] Li X, An G, Lin J, Li J. Monitoring of membrane scaling and concentration polarization in spiral wound reverse osmosis module using ultrasonic time-domain reflectometry with sound intensity calculation. Water Sustainability. 2014;**4**:167-180

[89] Lee S, Boo C, Elimelech M, Hong S. Comparison of fouling behaviour in forward osmosis (FO) and reverse osmosis (RO). Journal of Membrane Science. 2010;**365**:34-39

[90] Amy G. Fundamental understanding of organic matter fouling on membranes. Desalination. 2008;**118**:324-341

[91] Lu X, Castrillón SR-V, Shaffer DL, Ma J, Elimelech M. In situ surface chemical modification of thin-film composite forward osmosis membranes for enhanced organic fouling resistance. Environmental Science and Technology. 2013;**47**:12219-12228

[92] Motsa MM, Mamba BB, Verliefde ARD. Combined colloidal and organic fouling of FO membranes: The influence of foulant–foulant interactions and ionic strength. Journal of Membrane Science. 2015;**493**:539-548

[93] Mahlangu TO, Hoek EMV, Mamba BB, Verliefde ARD. Influence of organic, colloidal and combined fouling on NF rejection of NaCl and carbamazepine: Role of solute–foulant–membrane interactions and cake-enhanced concentration polarization. Journal of Membrane Science. 2014;**471**:35-46

[94] Tang S, Wang Z, Wu Z, Zhou Q. Role of dissolved organic matters (DOM) in membrane fouling of membrane bioreactors for municipal wastewater treatment. Journal of Hazardous Materials. 2010;**178**:377-384

[95] Subramani S, Badruzzaman M, Oppenheimer J, Jacangelo JG. Energy minimization strategies and renewable energy utilization for desalination: A review. Water Research. 2011;**45**:1907-1920

[96] Jones KL, O'Melia CR. Protein and humic acid adsorption onto hydrophilic membrane surfaces: Effects of pH and ionic strength. Journal of Membrane Science. 2000;**165**:31-46

[97] Xie M, Nghiem LD, Price WE, Elimelech M. Toward resource recovery from wastewater: Extraction of phosphorus from digested sludge using a hybrid forward osmosis–membrane distillation process. Environmental Science & Technology Letters. 2014;**1**:191-195

[98] Kim HC, Dempsey BA. Membrane fouling due to alginate, SMP, EfOM, humic acid and NOM. Journal of Membrane Science. 2013;**428**:190-197

[99] Jermann D, Pronk W, Meylan S, Boller M. Interplay of different NOM fouling mechanisms during ultrafiltration for drinking water production. Water Research. 2007;**41**:1713-1722

[100] Tang CY, Kwon YN, Leckie JO. Fouling of reverse osmosis and nanofiltration membranes by humic acid – Effects of solution composition and hydrodynamic conditions. Journal of Membrane Science. 2007;**290**:86-94

[101] Li Q, Xu Z, Pinnau I. Fouling of reverse osmosis (RO) membranes by biopolymers in wastewater secondary effluent: Role of membrane surface properties and initial permeate flux. Journal of Membrane Science. 2007;**290**:173-181

[102] Loh S, Beuscher U, Poddar TK, Porter AG, Wingard JM, Husson SM, Wickramasinghe SR. Interplay among membrane properties, proteins and operations on protein fouling during normal flow microfiltration. Journal of Membrane Science. 2009;**332**:93-103

[103] Yu S, Yao G, Dong B, Zhu H, Peng X, Liu J, Liu M, Gao C. Improving fouling resistance of thin-film composite polyamide reverse osmosis membrane by coating natural hydrophilic polymer sericin. Separation and Purification Technology. 2013;**118**:787-796

Applications of Reverse and Forward Osmosis Processes in Wastewater Treatment: Evaluation of Membrane Fouling

Achisa C. Mecha

Additional information is available at the end of the chapter

http://dx.doi.org/10.5772/intechopen.72971

Abstract

Although reverse osmosis (RO) process is widely used for wastewater reclamation, it requires high amount of energy that has a major effect on the economic effectiveness of the process. Furthermore, RO membranes are susceptible to fouling, which further limits their effectiveness and increases the costs due to the need for frequent cleaning. Consequently, the use of osmotically driven membrane separation processes such as forward osmosis (FO) has gained increasing consideration, although its uptake in wastewater remediation is still low. This is because the FO process, unlike the RO process, is operated by the osmotic gradient between the feed and draw solutions; therefore, it requires minimal or no hydraulic pressure. Hence, it has unique advantages, such as possibility of low fouling, and high water recovery. Nonetheless, the long-standing problem of membrane fouling still remains a major challenge even in the performance of FO processes especially when treating raw wastewaters, which have various contaminants. Furthermore, the mechanism of fouling in FO process has been found to be different from an RO process, and there is need for further studies to elucidate the differences of FO and RO fouling. These aspects are evaluated in this review.

Keywords: forward osmosis, membrane fouling, osmotic pressure, reverse osmosis, wastewater

1. Introduction

For many centuries, water has been considered a renewable, unlimited resource. However, in recent decades, the awareness that fresh water is not unlimited has arisen. The two major issues around water management are, thus, water scarcity and escalating pollution. Indeed, water pollution has put a potential strain on the existing water sources resulting in scarcity of fresh water. This has been occasioned by the rapid growth in global human population, thus increasing the demand; enhanced industrial and agricultural activity leading to rampant pollution of water sources; as well as climate change resulting in water scarcity through droughts. All these issues suggest the need for a more rational use of water resources [1]. The use of alternative sources of water such as seawater desalination and the reuse of wastewater after appropriate treatment is therefore necessary. Furthermore, the protection of natural water resources and development of new technologies for water and wastewater treatment for reuse are key priorities of the twenty-first century.

Wastewater reuse offers an opportunity to reduce demand on existing water resources [2]. This is because wastewater represents a suitable water source that can be used after appropriate treatment to reduce the fresh water demand and to lower the environmental impact of wastewater discharge [3]. Consequently, effluent from municipal wastewater treatment plants (MWWTPs) is a potential source of recycled water; however, to ensure its approval by the target population, microbial, physical, and chemical pollutants need to be removed using appropriate treatment technologies [4, 5].

Conventional municipal wastewater treatment processes rely on physicochemical and biological processes. However, with increasing contamination of wastewater by organic micropollutants and microbial contaminants, the current treatment technologies are often not successful in meeting the stringent standards. The reduction or complete removal of refractory organic contaminants from wastewater is important from the viewpoint of wastewater reclamation, recycling, and reuse [5]; however, conventional municipal wastewater treatment is inefficient especially in the removal of biorecalcitrant organic micropollutants and some resistant microorganisms.

There is therefore a pressing need to develop alternative wastewater remediation technologies that are capable of complete removal of organic micropollutants; have the provision of effective disinfection; are capable of utilization of minimum resources such as energy; are economically viable; and are environmentally friendly [6]. Suitable technologies should be able to enhance water recovery as well as extract biomass from the wastewater for reuse [7]. Membrane-based technologies have gained increasing prominence for wastewater remediation. Although low pressure processes such as microfiltration (MF) and ultrafiltration (UF) have been employed to treat secondary wastewater effluent, these technologies are not effective in removing emerging micropollutants and trace metals from wastewater, thus limiting the potential application of the reclaimed wastewaters. Consequently, the use of high pressure processes such as nanofiltration (NF) and reverse osmosis (RO) have been explored. However, they too suffer limitations such as high energy demand and severe membrane fouling, which ultimately increases the operating costs. This has prompted the exploration of osmotic pressure–driven membrane processes (ODMPs) such as forward osmosis (FO) as a suitable alternative to overcome these concerns [8]. This chapter presents the water scarcity and pollution challenge, applications of membrane-based processes (RO and FO) for wastewater remediation, and recent developments in addressing membrane fouling in RO and FO processes.

2. The RO and FO membrane processes

2.1. Principle of operation of RO and FO membranes

In the FO process, an osmotic pressure gradient across the semipermeable membrane drives water from a dilute feed solution (FS) to a concentrated draw solution (DS) [9]. In this way, the DS generates greater osmotic pressure and drives water from the feed through the membrane while rejecting solutes, thus separating the water from the diluted DS [10]. The RO process, on the other hand, employs hydraulic pressure to effect the permeation of water through a semipermeable membrane. The principle of operation of RO and FO processes is shown in **Figure 1**. The ideal semipermeable membrane for use in RO and FO processes should possess the following attributes: high water flux and salt rejection, less fouling propensity, and high chemical and thermal stability, among others [10]. The FO process has been shown to have a lower propensity to fouling and consequently, a higher reversibility of fouling than RO, and this is attributed to the lack of applied hydraulic pressure. Subsequently, FO can be used to treat low-quality feed waters such as municipal wastewater and landfill leachate, among

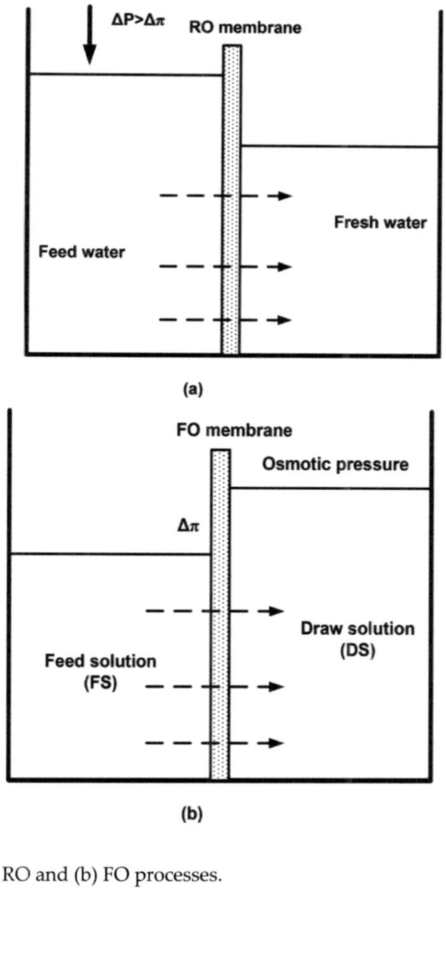

Figure 1. Working principle of (a) RO and (b) FO processes.

others [11]. Nevertheless, even in an FO-based separation process, energy is still required to extract clean water from the DS and to reuse the DS [12].

The general equation used to describe water flux across the RO and FO membrane (J_w) is calculated using Darcy's law [9]:

$$J_w = A_w \times (\sigma \Delta\pi - \Delta P) \quad (1)$$

where A_w is the membrane pure water permeability coefficient, ΔP is the applied hydrostatic pressure, $\Delta\pi$ is the differential osmotic pressure, and σ is the reflection coefficient indicating the rejection capability of a membrane (for an ideal membrane σ =1). Therefore, in FO, ΔP is zero thus making the water flux to be directly proportional to the difference in osmotic pressure, while for RO, $\Delta P > \Delta\pi$. This relationship is illustrated in **Figure 2**.

Despite not using hydraulic pressure, the FO process can produce permeate quality that is close to that produced by RO and superior permeate quality than that of microfiltration (MF) and ultrafiltration (UF) membranes [7]. Moreover, the FO process has benefits including high

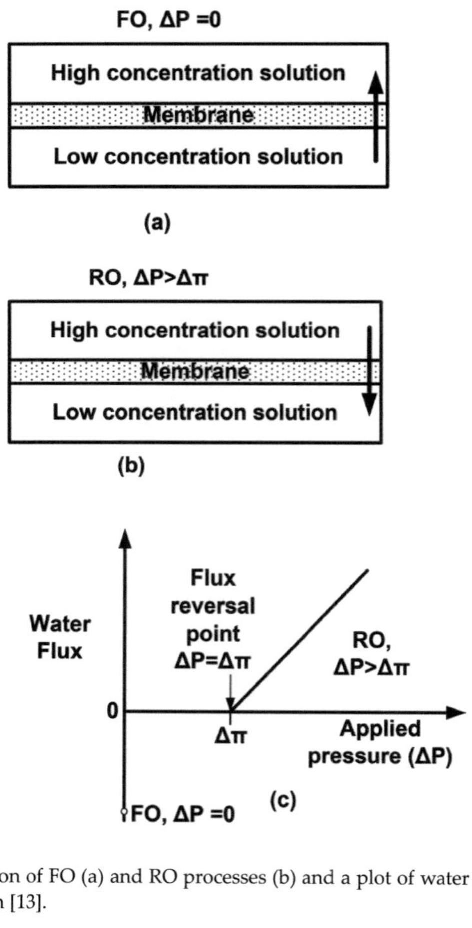

Figure 2. Schematic representation of FO (a) and RO processes (b) and a plot of water flux versus applied pressure for both processes (c). Adapted from [13].

rejection of a wide range of pollutants without using hydraulic pressure and hence the reduced energy expenditure and low membrane fouling tendency (more reversible fouling) [14]. For instance, a study by Altaee and colleagues [15] reported that the total power consumption by the FO process was 2–4% that of the RO-FO process, which shows that the use of FO can lead to significant reduction in energy expenditure. It is due to these unique advantages of FO membranes that they have been employed in many applications such as desalination of seawater, wastewater remediation, food and pharmaceutical processing, as well as renewable osmotic energy production [16].

However, notwithstanding these advantages of FO, it still suffers from the challenges faced by RO, mainly membrane fouling that results in reduced permeate quality and quantity as well as increased operational cost [17]. Developing an understanding of fouling behavior in FO is needed since it has been found that the fouling factors and mechanism of fouling in FO are different from those of an RO process [18]. Consequently, further research is required to understand the fouling behavior of FO and RO membranes to enable the development of tailored fouling controls [19].

2.2. Applications of RO and FO membranes in wastewater treatment

The FO and RO processes have been used to treat a variety of wastewaters such as municipal wastewater [14, 16, 18, 20], oily wastewater [21, 22], produced water [23], tannery wastewater [24], dairy wastewater [25], olive mill wastewater [26], as well as synthetic wastewater [8, 27]. In some of these studies, it has been reported that FO membranes could perform better than RO membranes. For instance, a comparative study by Cui and coworkers [28] on the removal of organic micropollutants (phenol, aniline, and nitrobenzene) reported that FO membranes achieved rejections of over 72%, which the authors observed that cannot be attained by commercial or lab-synthesized RO membranes. The FO and RO membranes can also be used in combination with other processes to increase the performance effectiveness. For instance, the use of combined MBR-RO and MBR-FO systems considerably improves the performance in wastewater treatment. Since the MBR alone is not effective in the removal of color and salts, the combination with RO and FO membranes allows for effective removal of these constituents [29]. Qui and colleagues [30] recently investigated the use of a biofilm-forward osmosis membrane bioreactor (BF-FOMBR) and reported that the process achieved very high removal efficiency of organic matter and nitrogen within a hydraulic retention time of 2 h. Furthermore, a significant reduction in FO membrane fouling was achieved (24.7–54.5%) due to decreased bacterial attachment and colonization of the membranes. A summary of the recent studies and the performance attained is shown in **Table 1**.

2.3. Limitations of RO and FO membranes

The use of membrane-based processes for wastewater treatment has been driven by the need to enhance water recovery, reduce energy consumption, and improve sustainability in application [31]. Consequently, membrane-based wastewater reclamation is considered a promising solution to supplement water supply and alleviate water shortage [18]. The RO process has received wide attention; however, it requires high hydraulic pressure, thus making it

Wastewater	Membrane type	Performance	Reference
Activated sludge	Cellulose triacetate Polyamide thin-film composite membranes (FO)	96% COD rejection.	[20]
Produced water	Cellulose triacetate Polyamide thin-film composite membranes (FO)	90% rejection of neutral hydrophobic compounds.	[23]
Oily wastewater	Hybrid forward osmosis membrane distillation (FO-MD) system	Water recovery of 90%. Almost complete rejection of oil and NaCl.	[21]
Soybean oil/water emulsion	Thin-film composite (TFC) FO membranes	Oil rejection of 99.9%.	[22]
Municipal wastewater	Superhydrophilic sulfonated poly-phenylenesulfone (sPPSU) polymer matrix TFC membranes (FO)	85% water recovery.	[16]
Municipal wastewater	FO membranes	A 5% flux decline in the absence of suspended solids and a 20% flux decline in the presence of suspended solids.	[18]
Synthetic urban runoff	Cellulose triacetate FO membrane	Rejection of trace metals (98–100%); phosphorus (97–100%); nitrate (52–94%). A 70% water recovery.	[27]
Synthetic wastewater	Cellulose triacetate (CTA) membranes (FO)	Rejection of pollutants in the wastewater (> 97%).	[8]
Municipal wastewater	Cellulose triacetate (CTA) membrane (FO)	89.2% removal efficiency of NH_4^+-N.	[14]
Tannery wastewater	ESPA-1 RO membranes	>98% rejection of COD and salts.	[24]
Dairy wastewater	TFC HR SW 2540 spiral RO membranes	99.9% TOC rejection and 99.5% conductivity reduction.	[25]
Olive mill wastewater	XLE and BW30 RO membranes	96.3% COD rejection.	[26]

Table 1. Studies on the application of RO and FO membranes in wastewater treatment.

energy intensive and costly due to the resulting membrane fouling and replacement. It is due to these concerns that in recent times the FO process has become an attractive alternative to RO due to the fact that it utilizes an osmotic pressure gradient as driving force for separation and also has additional merits such as lower energy consumption, less susceptibility to membrane fouling, and higher water recovery [32, 33]. Furthermore, FO membranes consistently reject a range of pollutants in municipal wastewater (chemical and biological contaminants), making FO an appropriate technology for wastewater remediation for reuse [20]; however, its application in wastewater treatment is still low [34]. Nevertheless, fouling still remains a formidable challenge even in FO processes limiting long-term operation, leading to flux decay and shortening of membrane lifespan [35].

3. Membrane fouling

3.1. Categories of membrane fouling

Membrane fouling arises from the accumulation of pollutants on the membrane surface leading to a reduction in flux. It has far-reaching implications since it affects the permeate quality and increases the operating costs such as process downtime leading to production losses, cleaning chemicals, energy and labor requirements, and eventually membrane replacements [36, 37]. The magnitude of membrane fouling depends on the physicochemical properties of the membrane and the wastewater composition. For instance, hydrophilic, low roughness, and neutral charge membranes present a high resistance to fouling [20]. In terms of location of foulants, fouling can be divided into surface fouling and internal fouling depending on the location of the foulants. Surface fouling is more frequent in high pressure membranes such as RO due to their compact and nonporous nature. On the other hand, based on foulant types, fouling can also be divided into biofouling, organic fouling, inorganic scaling, and colloidal fouling [20, 38].

a. Biofouling

This is the adhesion of microorganisms on the membrane surface leading to the formation of a biofilm. It occurs through the reversible attachment of planktonic bacteria, cell growth, and extracellular polymeric substance (EPS) production leading to the formation of biofilms [20]. Therefore, the two main components of biofilms are bacteria and EPS, which are excreted by bacteria. Biofouling is regarded as one of the most formidable forms of membrane fouling since bacteria reproduce on the membrane surface, thus enhancing the biofilm that leads to additional fouling [39]. This is because microorganisms are present in many water systems and they readily adhere to membrane surfaces and multiply.

b. Organic fouling

This arises from the adsorption or deposition of organic matter such as humic substances, polysaccharides, proteins, lipids, nucleic acids and amino acids, organic acids, and cell components on the membranes. It is the most common fouling experienced in wastewater treatment using membrane bioreactors (MBRs). The organics often become precursors of biofouling [40]. Effluent organic matter in wastewater arises from three sources: natural organic matters (NOMs), synthetic organic compounds (SOCs), and soluble microbial products (SMPs).

c. Inorganic scaling

This entails the chemical or biological deposition of inorganic substances on the membrane surface or within the pores, thus preventing permeation of water. It occurs when the concentration of some ions (such as metal sulfates and carbonates) in the water is high enough to exceed the equilibrium solubility product and hence become supersaturated leading to the deposition of the ions [13]. In fact, if the feed water is not well pretreated due to improper design of coagulation or oxidation processes, it may lead to the introduction of metal hydroxides into the fouling matrix, which causes significant challenges in chemical cleaning to enhance water flux.

d. Colloidal fouling

This refers to the deposition of fine suspended particles (colloids) on the membranes. Colloidal foulants can be divided into two types: inorganic foulants and organic macromolecules. Colloidal fouling leads to substantial flux decline resulting from the deposition of thick or less porous fouling layers composed of particulate matter. Consequently, this hinders back diffusion of salts that permeate water flux from the DS, thus increasing the salt concentration on the membrane surface.

A detailed analysis of the different modes of fouling in FO and RO membranes can be found in recent studies by Chun and colleagues [13] and by Jiang and coworkers [38], respectively. In addition, the following factors play a major role in fouling: the characteristics of the fouling matter, the chemistry of the DS and FS, the membrane properties (hydrophilicity and surface roughness), and hydrodynamic conditions, and they have been discussed in the literature [36].

3.2. Comparison of fouling in RO and FO membranes

Understanding the mechanisms of fouling is essential for improving membrane performance especially in FO membranes where very little has been done. For instance, the driving force for membrane separation plays a significant role in membrane fouling. It influences the fouling layer structure as well as the fouling reversibility. It has been reported that although the extent of compaction resulting from the permeate drag force is similar in FO and RO fouling layers, however, higher compressibility of foulants occurs under hydraulic pressure in RO processes. Therefore, in RO, there are two compaction mechanisms involved: compression of foulants and permeate drag force, whereas in FO, only the permeate drag force is predominant. These mechanisms reinforce one another, resulting in dense, compact, and irreversible fouling layers in RO [11].

Furthermore, in the RO processes, the hydraulic pressure–driving force remains constant during operation and hence the fouling effect can be readily determined. On the other hand, the fouling properties of FO process are different because of the changing osmotic pressure difference, accompanied by changes in concentration polarization. This makes it difficult to use the FO flux to accurately show the actual effect of membrane fouling [36]. Moreover, permeate flux and transmembrane pressure are commonly used to indicate membrane fouling in RO membranes, but these are not used in the FO process [36]. Additionally, in terms of transport, in the FO process, permeate water transports from the FS to DS; hence, the DS is diluted and FS concentrated steadily. Subsequently, the osmotic pressure decreases, leading to permeate flux decline along the membrane channel. However, in the RO processes, the concentration of the FS is only observed along the membrane channel [41]. Overall, studies have shown that the lack of hydraulic pressure in the FO system has a positive effect in that the membrane fouling generated is in most cases reversible and the water flux can be almost fully recovered using hydraulic washing, thus eliminating the use of chemical cleaning [37].

It has also been reported that membrane fouling in FO is less severe than in RO membranes. For instance, Yu and colleagues [42] compared the fouling propensity in RO and FO membranes treating activated sludge effluent and reported that the membrane fouling based on flux reduction was lower in FO membranes than in RO membranes. However, despite this, it is still necessary to pretreat the wastewater to prevent excessive fouling of FO membranes and decelerate membrane degradation [23]. A comparative study on the fouling of FO and RO membranes using polysaccharides (alginate, xanthan, and pullulan) depicted that alginate and xanthan resulted in more pronounced fouling in RO than in FO. Similarly, the study reported that polysaccharides naturally produced by marine bacteria improved the permeate flux instead of causing fouling in FO membranes [32]. Tow and coworkers, on the other hand, observed similarities in fouling in FO and RO membranes in terms of swelling and wrinkling of the fouling matter. They suggested that this could be leveraged to develop cleaning protocols for both FO and RO membranes [43]. In another study, Kwan and colleagues [44] evaluated biofouling in FO and RO membranes under similar hydrodynamic conditions and observed significant differences such as the following: (i) water flux decline was significantly lower in FO than in RO and (ii) biofilms in FO were loosely organized and in a thick layer, whereas in RO, they were tightly packed (due to hydraulic pressure). Consequently, the more packed biofilms in RO resulted in high resistance to water flow leading to higher flux decline. In another study, organic fouling has been reported to be dominant in RO membranes used for the treatment of municipal wastewater [45]. **Table 2** summarizes some of the recent studies on membrane fouling in RO and FO membranes.

Nevertheless, the fouling mechanism is complex and depends on numerous aspects such as water quality, process conditions, module design, and membrane properties, among others. It is therefore imperative to consider these factors in process design and development to mitigate fouling [9]. Moreover, the fouling behavior in the FO processes is unique because both sides of the FO membrane are involved [13], whereby there is membrane fouling and a drop in driving force [46]. A comprehensive evaluation of mass transport and fouling in FO and other ODMPs has been provided by She and colleagues [19].

3.3. Characterization of membrane foulants

Characterization of the fouling layer is important to enable the evaluation of membrane fouling especially the interaction of foulants with membranes and the composition of fouling matter.

Process	Water matrix	Type of fouling	Reference
FO, PFO, and RO	Sodium alginate	Organic fouling	[11]
FO and RO	Alginate, xanthan, and pullulan	Organic fouling	[32]
FO and RO	Activated sludge	Organic fouling	[42]
FO	Municipal wastewater	Cake layer formation	[46]
RO and FO	Alginate and methylene blue dye	Organic fouling	[43]
RO	Municipal wastewater	Organic fouling and inorganic scaling	[45]
FO and RO	Synthetic wastewater containing *Pseudomonas aeruginosa*	Biofouling	[44]

Table 2. Studies on membrane fouling in RO and FO membranes.

Process	Characterization technique	Water matrix	Reference
RO and FO	Fouling visualization apparatus	Alginate gel and methylene blue dye	[43]
RO and FO	CLSM	Sodium alginate	[11]
FO	SEM and LC-OCD	Synthetic wastewater	[33]
FO	SEM, FTIR, EDS	Oily wastewater	[37]
FO and OMBR	SEM, FTIR, EEM, EDX	Municipal wastewater	[14]
FO and RO	AFM and contact angle	Activated sludge	[42]
RO	FTIR, EEM	Municipal wastewater	[45]

Table 3. Studies on characterization techniques for RO and FO membranes.

This provides insight into the fouling mitigation strategies that can be adopted. Furthermore, a classification of fouling into chemical, physical, and microbiological enables also the identification of the appropriate techniques for characterization. Physical characterization can be performed by visual examination using environmental scanning electron microscopy (ESEM) and atomic force microscopy (AFM). Chemical characterization can be done using Fourier transform infrared (FTIR) and excitation emission matrix (EEM) analyses to determine the organic composition; energy dispersive X-ray spectroscopy (EDS) to determine the elemental composition of the fouling layer; evaluation of zeta potential to determine the surface charge and membrane hydrophilicity; and liquid chromatography with organic carbon detection (LC-OCD) to determine the different fractions of dissolved organic carbon. On the other hand, microbiological characterization can be accomplished using adenosine triphosphate (ATP) measurements, EPS quantification, and CLSM analysis for biofilm visualization and thickness estimation [8, 13, 14, 20]. More details can be found in a recent work by Li and coworkers [47] who reviewed the use of membrane fouling research methods to study fouling in RO and FO membranes. They also identified the main foulants involved in the various types of membrane fouling; however, they did not evaluate the mitigation strategies for membrane fouling. **Table 3** shows some of the studies that have been conducted and the characterization of membrane foulants.

4. Addressing membrane fouling

Municipal wastewater contains a variety of contaminants such as organic matter, inorganic matter, and microorganisms that can lead to membrane fouling [14]. Since membrane fouling is inevitable, it is imperative to develop strategies to address this challenge. Approaches for tackling fouling are twofold: (i) fouling mitigation through membrane and module development and optimization of hydrodynamic conditions and (ii) adapting cleaning approaches [48]. These strategies can further be broken down into the following: feed pretreatment, membrane monitoring and cleaning, membrane surface modification, or the use of novel membrane materials [38].

4.1. Feed pretreatment

It involves improving the feed water quality to minimize contaminant concentration prior to membrane filtration. It is aimed at ensuring reliable membrane operation and prolonging the membrane lifespan. Some of the most commonly used pretreatment technologies for RO include UF [49], coagulation/flocculation, and MF. In fact, FO can also be used as a pretreatment for RO because the former does not require hydraulic pressure and hence reduces the overall energy required and process costs by decreasing RO membrane fouling, minimizing the cleaning frequencies, and also increasing the water recovery [49, 50]. Nanofiltration has also been employed as a pretreatment for RO membranes. This is reported to have resulted in an increase in water recovery and water flux and also a reduction in RO membrane scaling and thus contributing to lowering the operating costs [51]. In another study on the treatment of geothermal water, NF was used as pretreatment for RO to reduce the concentration of divalent ions [52]. Combined pretreatment technologies have also been employed such as the use of ozonation, ceramic MF, and biological activated carbon (BAC) together as pretreatment for RO as reported by Zhang et al. [53]. In this combination, ozonation increased the oxidation of organic matter leading to its dissolution and facilitating removal by the ceramic MF and BAC prior to treatment by RO.

4.2. Membrane monitoring and cleaning

It entails the in situ monitoring of the membrane performance to evaluate the extent of fouling so as to conduct cleaning timeously. Some of the proven effective cleaning approaches of FO membranes include hydraulic cleaning and osmotic backwashing [23]. Osmotic backwash entails the reversed flow of water from the permeate side to the feed side based on the osmotic pressure difference. Lotfi and coworkers [33] observed that physical cleaning of FO membranes was effective leading to almost full restoration of the initial flux. In addition, treatment of oily wastewater using FO membranes indicated that osmotic backwashing resulted in over 95% water flux recovery and performed better than chemical cleaning using oxidants and acids [37]. Bell and colleagues employed chemically enhanced osmotic backwashing to clean FO membranes. The study showed that the cleaning removed cations and anions from the membrane surface but only slightly improving the water flux [23]. Similarly, Yu and colleagues demonstrated that during treatment of activated sludge using FO membranes, the flux was fully recovered using osmotic backwashing rather than cleaning by changing the cross-flow velocity or air scouring. They concluded that osmotic backwashing is a more efficient way to clean the FO membrane. A study by Wang and colleagues [54] investigated the chemical cleaning of FO membranes using different chemicals. They reported that disodium-ethylene-diamine-tetra-acetate (EDTA-2Na), sodium dodecyl sulfate (SDS), NaOH, HCl, and citric acid were not effective in removing the foulants after severe fouling; on the other hand, 0.5% hydrogen peroxide applied for 6 h at 25°C resulted in 95% recovery of permeability suggesting that almost all the foulants were removed. **Table 4** provides a summary of strategies employed in cleaning RO and FO membranes.

However, implementing costly cleaning protocols such as air scouring or chemical cleaning may be detrimental to the economic sustainability of the FO process. Therefore, it is necessary

Process	Cleaning strategy	Performance	Reference
TFC-FO	Water rinsing without using chemicals	97% water flux recovery.	[22]
FO	Hydraulic cleaning (cross-flow rate of 800 mL/min for 15 min)	90% water flux recovery.	[33]
FO	Hydraulic cleaning (cross-flow velocity 33 cm/s for 30 min) Osmotic backwash	75–80% flux recovery using hydraulic cleaning and 95% flux recovery using osmotic backwash.	[37]
FO OMBR	Hydraulic cleaning (cross-flow velocity of 10 cm/s for 60 min)	49.37% flux recovery in FO and 10.60% flux recovery in OMBR.	[14]
	Chemical cleaning (1% NaClO, 0.8% EDTA, and 0.1% sodium dodecyl sulfate (SDS) in sequence. Each lasted for 60 min.)	58–67% flux recovery in FO and 2–18.5% flux recovery in OMBR.	[14]
RO and FO	Hydraulic cleaning (cross-flow velocity of 25 cm/s for 60 min)	After hydraulic cleaning, the foulant peels off the membranes in both RO and FO.	[43]
FO, PFO, and RO	Hydraulic cleaning (cross-flow velocity 17 cm/s for 30 min)	Flux recovery: FO (99%); PFO (58%); and RO (10%).	[11]
FO and RO	Physical cleaning (cross-flow velocity of 8.5–25.5 cm/s for 1 min) Osmotic backwashing (1 min)	75% flux recovery by physical cleaning; 99.9% flux recovery by osmotic backwashing.	[42]
FO	Chemical cleaning (0.5% hydrogen peroxide for 6 h)	More than 95% recovery of permeability.	[54]

Table 4. Cleaning strategies employed for RO and FO membranes.

to explore proven strategies such as osmotic backwash, which has recently been demonstrated to successfully clean fouled FO membranes and has been extensively studied in the RO literature. This will allow for sustainable operation without use of chemicals [48]. In addition, real-time monitoring of the membrane process can provide useful information essential for efficient cleaning. To overcome the limitations of the individual cleaning methods, it is necessary to explore the use of multiple methods to take advantage of synergy in the use of multiple cleaning strategies such as a combination of osmotic backwashing and surface backwashing to further improve the performance of FO membrane [42]. For instance, a study by Sun and colleagues [14] showed that even in cases of severe membrane fouling, the use of hydraulic and chemical cleaning resulted in effective recovery of water permeability.

4.3. Membrane surface modification and the use of novel materials

It is based on the fact that membrane properties such as smoothness and hydrophilicity greatly influence performance. For instance, smooth surface and hydrophilic membranes are less prone to fouling compared to those with rough and hydrophobic surfaces. In addition to surface modification, the development of novel membrane materials with unique characteristics tailored to meet specific applications is another promising avenue. These novel materials include carbon nanotubes, zwitterionic materials, and metal oxide nanoparticles [38]. Li and coworkers [10] reviewed developments in materials and strategies for enhancing properties and performances of RO and FO membranes. They noted that surface modification of RO and FO membranes has received wide attention due to it being less costly and easy to perform compared to developing novel polymeric materials. However, surface modification may also have adverse effects such as pore blockage on the membrane active layer when some modifiers such as polyelectrolytes may promote concentration polarization and consequently reduce water flux. Asadollahi and colleagues [55] have recently also reviewed the enhancement of the performance of RO membranes through surface modification. They reported that the fact that membrane fouling has a strong dependence on membrane surface morphology and properties makes surface modification using physical and chemical methods a key tool to address membrane fouling. However, they also observed that surface modification has its demerits too such as the following: (i) it increases the membrane resistance, thus impeding permeation and reducing the water flux and (ii) the stability of surface modifiers during membrane cleaning and long-term operation has not been well studied. A study by Kochkodan and Hilal [56] evaluated the surface modification of polymeric membranes targeting the control of biofouling. The authors reported that generally high membrane hydrophilicity, smooth membrane surface, and the use of bactericidal or charged particles on the membrane surface result in a reduction in membrane biofouling. However, the challenge of developing membranes that can overcome the complexities of biofouling without having adverse effects still remains.

Therefore, understanding the mechanisms of fouling in membranes is paramount to develop the appropriate mitigation strategies. As an example, recently, Tow and colleagues [43] developed a fouling visualization apparatus to elucidate the mechanisms of organic fouling and cleaning in RO and FO processes. They identified one internal fouling mechanism that is unique to FO membranes based on vapor phase formation within the membrane. They further reported that although the use of feed spacers is advantageous in reducing the rate of fouling, it may also obstruct cleaning by preventing pieces of detached gel from flowing downstream.

5. Future perspectives

The performance of the FO process can be improved through its integration with other technologies to take advantage of the unique strengths of the individual processes. As an example, the FO-MD hybrid process has been employed for oily wastewater treatment [21]. The findings indicated that water recovery of greater than 90% was attained even at high salinities and also almost complete rejection of oil and sodium chloride. In another study [57], the FO-MD process was also applied for raw sewage; water recovery of 80% was achieved, and removal efficiency for trace organics was 91–98%. In addition, the use of FO-ED hybrid system for the treatment of secondary municipal wastewater resulted in treated water that met potable water standards (low concentration of TOC, carbonate, and low conductivity) [58]. Another promising hybrid process is the combination of FO and RO (FO-RO). Based on the unique advantages of RO and FO processes, it is important to exploit these to solve the challenges of wastewater remediation and even desalination. For instance, the potential of FO to reduce the

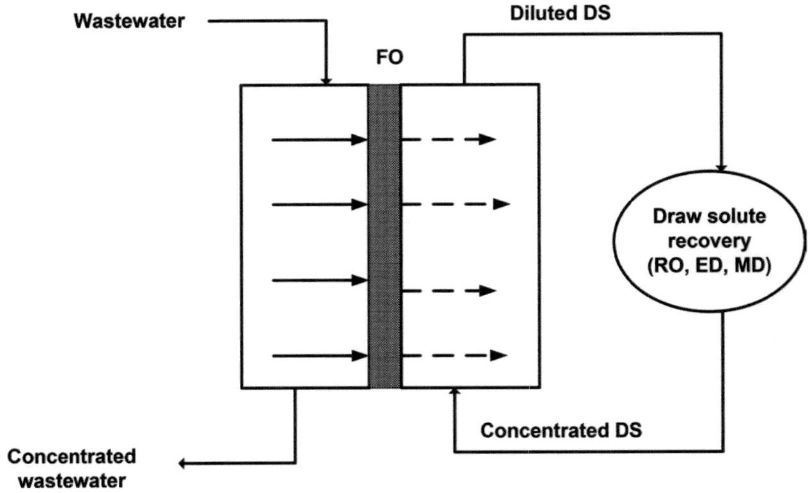

Figure 3. FO-based hybrid technologies (FO-MD, FO-RO, and FO-ED). Adapted from [63].

energy consumption of RO is very important. This can be done using an FO-RO hybrid process in which FO is implemented as a pretreatment step before RO. Furthermore, this FO-RO hybrid has the additional benefit of being a double-barrier protection leading to high-quality treated water [48]. Linares and coworkers [59] have recently shown that hybrid FO-RO systems are economically advantageous compared to other technologies for desalination or even wastewater treatment and recovery systems. Another integrated technology is the coupling of FO and microbial osmotic fuel cell (MOFC), which was performed by Werner and coworkers [60]. The key benefits reported were that the system could simultaneously treat wastewater treatment and desalinate seawater within the same reactor [60]. Furthermore, the integration of FO and conventional MBR can result in reduced energy consumption [61]. A coupled forward osmosis and microbial desalination cell (FO-MDC) was employed to simultaneously treat wastewater and desalinate seawater and the COD removals were satisfactory as well as high levels of desalination were achieved [62]. Therefore, these hybrid systems can greatly improve FO performance and increase its feasibility for commercial application. However, before the integrated processes can be implemented, there is a need for detailed studies on the energy consumption to determine their economic viability [34]. **Figure 3** shows a schematic representation of some of the FO-based hybrid technologies.

6. Conclusions

The review has provided insights into the use of forward osmosis either individually or in combination with other processes for wastewater treatment. Forward osmosis is gaining wide acceptability and application because of its unique advantages such as not requiring hydraulic pressure and less fouling propensity compared to conventional pressure-driven membrane processes. Inasmuch as the literature has indicated that the lack of hydraulic pressure in FO processes alters the extent of membrane fouling; further studies are required especially on how this influences the cleaning strategies to be adopted. Furthermore, it is necessary to develop new FO membranes taking into account the effect of the membranes on fouling and cleaning behavior. It is also imperative to explore the synergy in the use of multiple cleaning strategies such as a combination of osmotic backwashing and surface backwashing to further improve the performance of FO membrane.

Author details

Achisa C. Mecha

Address all correspondence to: achemeng08@gmail.com

Department of Chemical, Metallurgical and Materials Engineering, Tshwane University of Technology, Pretoria, South Africa

References

[1] De Sanctis M, Del Moro G, Levantesi C, Luprano ML, Di Iaconi C. Integration of an innovative biological treatment with physical or chemical disinfection for wastewater reuse. Science of The Total Environment. 2016;**543**(Part A):206-213

[2] Metcalf G, Pillay L, Murutu C, Chiburi S, Gumede N, Gaydon P. Wastewater reclamation for potable use. Vol. 1: Evaluation of membrane bioreactor technology for pretreatment. Water Research Commission 2014 WRC Report No. 1894/1/14. ISBN 978-1-4312-0564-6

[3] Qu X, Brame J, Li Q, Alvarez PJJ. Nanotechnology for a safe and sustainable water supply: Enabling integrated water treatment and reuse. Accounts of Chemical Research. 2012;**46**(3):834-843

[4] Puspita P, Roddick F, Porter N. Efficiency of sequential ozone and UV-based treatments for the treatment of secondary effluent. Chemical Engineering Journal. 2015;**268**:337-347

[5] Camel V, Bermond A. The use of ozone and associated oxidation processes in drinking water treatment. Water Research. 1998;**32**(11):3208-3222

[6] Singh S. Ozone Treatment of Municipal Wastewater Effluent for Oxidation of Emerging Contaminants and Disinfection. Windsor: University of Windsor; 2012

[7] Zhang H, Jiang W, Cui H. Performance of anaerobic forward osmosis membrane bioreactor coupled with microbial electrolysis cell (AnOMEBR) for energy recovery and membrane fouling alleviation. Chemical Engineering Journal. 2017;**321**:375-383

[8] Li S, Kim Y, Chekli L, Phuntsho S, Shon HK, Leiknes T, et al. Impact of reverse nutrient diffusion on membrane biofouling in fertilizer-drawn forward osmosis. Journal of Membrane Science. 2017;**539**:108-115

[9] Jasmina K, Subhankar B, Malini B, Claus H-N, Irena P. Forward osmosis in wastewater treatment processes. Acta Chimica Slovenica. 2017;**64**(1):83-94

[10] Li D, Yan Y, Wang H. Recent advances in polymer and polymer composite membranes for reverse and forward osmosis processes. Progress in Polymer Science. 2016;**61**:104-155

[11] Xie M, Lee J, Nghiem LD, Elimelech M. Role of pressure in organic fouling in forward osmosis and reverse osmosis. Journal of Membrane Science. 2015;**493**:748-754

[12] Cho M, Lee SH, Lee D, Chen DP, Kim I-C, Diallo MS. Osmotically driven membrane processes: Exploring the potential of branched polyethyleneimine as draw solute using porous FO membranes with NF separation layers. Journal of Membrane Science. 2016;**511**:278-288

[13] Chun Y, Mulcahy D, Zou L, Kim IS. A short review of membrane fouling in forward osmosis processes. Membranes. 2017;**7**(30):1-23

[14] Sun Y, Tian J, Zhao Z, Shi W, Liu D, Cui F. Membrane fouling of forward osmosis (FO) membrane for municipal wastewater treatment: A comparison between direct FO and OMBR. Water Research. 2016;**104**:330-339

[15] Altaee A, Zaragoza G, van Tonningen HR. Comparison between forward osmosis-reverse osmosis and reverse osmosis processes for seawater desalination. Desalination 2014;**336**:50-57

[16] Han G, Zhao B, Fu F, Chung T-S, Weber M, Staudt C, et al. High performance thin-film composite membranes with mesh-reinforced hydrophilic sulfonated polyphenylenesulfone (sPPSU) substrates for osmotically driven processes. Journal of Membrane Science. 2016;**502**:84-93

[17] Xie P. Simulation of Reverse Osmosis and Osmotically Driven Membrane Processes. South Carolina: Clemson University; 2016

[18] Kim S, Go G-W, Jang A. Study of flux decline and solute diffusion on an osmotically driven membrane process potentially applied to municipal wastewater reclamation. Journal of Industrial and Engineering Chemistry. 2016;**33**:255-261

[19] She Q, Wang R, Fane AG, Tang CY. Membrane fouling in osmotically driven membrane processes: A review. Journal of Membrane Science. 2016;**499**:201-233

[20] Bell EA, Holloway RW, Cath TY. Evaluation of forward osmosis membrane performance and fouling during long-term osmotic membrane bioreactor study. Journal of Membrane Science. 2016;**517**:1-13

[21] Zhang S, Wang P, Fu X, Chung T-S. Sustainable water recovery from oily wastewater via forward osmosis-membrane distillation (FO-MD). Water Research. 2014;**52**:112-121

[22] Han G, de Wit JS, Chung T-S. Water reclamation from emulsified oily wastewater via effective forward osmosis hollow fiber membranes under the PRO mode. Water Research. 2015;**81**:54-63

[23] Bell EA, Poynor TE, Newhart KB, Regnery J, Coday BD, Cath TY. Produced water treatment using forward osmosis membranes: Evaluation of extended-time performance and fouling. Journal of Membrane Science. 2017;**525**:77-88

[24] Fababuj-Roger M, Mendoza-Roca JA, Galiana-Aleixandre MV, Bes-Piá A, Cuartas-Uribe B, Iborra-Clar A. Reuse of tannery wastewaters by combination of ultrafiltration and reverse osmosis after a conventional physical-chemical treatment. Desalination. 2007;**204**(1):219-226

[25] Vourch M, Balannec B, Chaufer B, Dorange G. Treatment of dairy industry wastewater by reverse osmosis for water reuse. Desalination. 2008;**219**(1):190-202

[26] Coskun T, Debik E, Demir NM. Treatment of olive mill wastewaters by nanofiltration and reverse osmosis membranes. Desalination. 2010;**259**(1):65-70

[27] Li Z, Valladares Linares R, Abu-Ghdaib M, Zhan T, Yangali-Quintanilla V, Amy G. Osmotically driven membrane process for the management of urban runoff in coastal regions. Water Research. 2014;**48**:200-209

[28] Cui Y, Liu X-Y, Chung T-S, Weber M, Staudt C, Maletzko C. Removal of organic micro-pollutants (phenol, aniline and nitrobenzene) via forward osmosis (FO) process: Evaluation of FO as an alternative method to reverse osmosis (RO). Water Research. 2016;**91**:104-114

[29] Wenten IG, Khoiruddin. Reverse osmosis applications: Prospect and challenges. Desalination. 2016;**391**:112-125

[30] Qiu G, Zhang S, Srinivasa Raghavan DS, Das S, Ting Y-P. The potential of hybrid forward osmosis membrane bioreactor (FOMBR) processes in achieving high throughput treatment of municipal wastewater with enhanced phosphorus recovery. Water Research. 2016;**105**:370-382

[31] Cath TY. Osmotically and thermally driven membrane processes for enhancement of water recovery in desalination processes. Desalination and Water Treatment. 2010;**15**:279-286

[32] Xie Z, Nagaraja N, Skillman L, Li D, Ho G. Comparison of polysaccharide fouling in forward osmosis and reverse osmosis separations. Desalination. 2017;**402**:174-184

[33] Lotfi F, Chekli L, Phuntsho S, Hong S, Choi JY, Shon HK. Understanding the possible underlying mechanisms for low fouling tendency of the forward osmosis and pressure assisted osmosis processes. Desalination. 2017;**421**:89-98

[34] Lutchmiah K, Verliefde ARD, Roest K, Rietveld LC, Cornelissen ER. Forward osmosis for application in wastewater treatment: A review. Water Research. 2014;**58**:179-197

[35] Zhang S, Qiu G, Ting YP, Chung T-S. Silver–PEGylated dendrimer nanocomposite coating for anti-fouling thin film composite membranes for water treatment. Colloids and Surfaces A: Physicochemical and Engineering Aspects. 2013;**436**:207-214

[36] Majeed T, Phuntsho S, Jeong S, Zhao Y, Gao B, Shon HK. Understanding the risk of scaling and fouling in hollow fiber forward osmosis membrane application. Process Safety and Environmental Protection. 2016;**104**:452-464

[37] Lv L, Xu J, Shan B, Gao C. Concentration performance and cleaning strategy for controlling membrane fouling during forward osmosis concentration of actual oily wastewater. Journal of Membrane Science. 2017;**523**:15-23

[38] Jiang S, Li Y, Ladewig BP. A Review of reverse osmosis membrane fouling and control strategies. Science of The Total Environment. 2017;**595**:567-583

[39] Sun X-F, Qin J, Xia P-F, Guo B-B, Yang C-M, Song C, et al. Graphene oxide–silver nanoparticle membrane for biofouling control and water purification. Chemical Engineering Journal. 2015;**281**:53-59

[40] Le-Clech P, Chen V, Fane TAG. Fouling in membrane bioreactors used in wastewater treatment. Journal of Membrane Science. 2006;**284**(1-2):17-53

[41] Lee J, Kim B, Hong S. Fouling distribution in forward osmosis membrane process. Journal of Environmental Sciences. 2014;**26**(6):1348-1354

[42] Yu Y, Lee S, Maeng SK. Forward osmosis membrane fouling and cleaning for wastewater reuse. Journal of Water Reuse and Desalination. 2017;**7**(2):111-120

[43] Tow EW, Rencken MM, Lienhard JH. In situ visualization of organic fouling and cleaning mechanisms in reverse osmosis and forward osmosis. Desalination. 2016;**399**:138-147

[44] Kwan SE, Bar-Zeev E, Elimelech M. Biofouling in forward osmosis and reverse osmosis: Measurements and mechanisms. Journal of Membrane Science. 2015;**493**:703-708

[45] Tang F, Hu H-Y, Sun L-J, Wu Q-Y, Jiang Y-M, Guan Y-T, et al. Fouling of reverse osmosis membrane for municipal wastewater reclamation: Autopsy results from a full-scale plant. Desalination. 2014;**349**:73-79

[46] Zhang X, Ning Z, Wang DK, Diniz da Costa JC. Processing municipal wastewaters by forward osmosis using CTA membrane. Journal of Membrane Science. 2014;**468**:269-275

[47] Li L, Liu X-P, Li H-Q. A review of forward osmosis membrane fouling: Types, research methods and future prospects. Environmental Technology Reviews. 2017;**6**(1):26-46

[48] Blandin G, Vervoort H, Le-Clech P, Verliefde ARD. Fouling and cleaning of high permeability forward osmosis membranes. Journal of Water Process Engineering. 2016;**9**:161-169

[49] Khanzada NK, Khan SJ, Davies PA. Performance evaluation of reverse osmosis (RO) pretreatment technologies for in-land brackish water treatment. Desalination. 2017;**406**:44-50

[50] Zaviska F, Zou L. Using modelling approach to validate a bench scale forward osmosis pre-treatment process for desalination. Desalination. 2014;**350**:1-13

[51] Kaya C, Sert G, Kabay N, Arda M, Yüksel M, Egemen Ö. Pre-treatment with nanofiltration (NF) in seawater desalination—Preliminary integrated membrane tests in Urla, Turkey. Desalination. 2015;**369**:10-17

[52] Tomaszewska B, Rajca M, Kmiecik E, Bodzek M, Bujakowski W, Wątor K, et al. The influence of selected factors on the effectiveness of pre-treatment of geothermal water during the nanofiltration process. Desalination. 2017;**406**:74-82

[53] Zhang J, Northcott K, Duke M, Scales P, Gray SR. Influence of pre-treatment combinations on RO membrane fouling. Desalination. 2016;**393**:120-126

[54] Wang X, Hu T, Wang Z, Li X, Ren Y. Permeability recovery of fouled forward osmosis membranes by chemical cleaning during a long-term operation of anaerobic osmotic membrane bioreactors treating low-strength wastewater. Water Research. 2017;**123**:505-512

[55] Asadollahi M, Bastani D, Musavi SA. Enhancement of surface properties and performance of reverse osmosis membranes after surface modification: A review. Desalination. 2017;**420**:330-383

[56] Kochkodan V, Hilal NA. Comprehensive review on surface modified polymer membranes for biofouling mitigation. Desalination. 2015;**356**:187-207

[57] Xie M, Nghiem LD, Price WE, Elimelech M. A forward osmosis-membrane distillation hybrid process for direct sewer mining: System performance and limitations. Environmental Science and Technology. Australia, Australia/Oceania: Research Online. 2013;**47**(23):13486-13493

[58] Zhang Y, Pinoy L, Meesschaert B, Van der Bruggen B. A natural driven membrane process for brackish and wastewater treatment: Photovoltaic powered ED and FO hybrid system. Environmental Science & Technology. 2013;**47**(18):10548-10555

[59] Valladares Linares R, Li Z, Yangali-Quintanilla V, Ghaffour N, Amy G, Leiknes T, et al. Life cycle cost of a hybrid forward osmosis—Low pressure reverse osmosis system for seawater desalination and wastewater recovery. Water Research. 2016;**88**:225-234

[60] Werner CM, Logan BE, Saikaly PE, Amy GL. Wastewater treatment, energy recovery and desalination using a forward osmosis membrane in an air-cathode microbial osmotic fuel cell. Journal of Membrane Science. 2013;**428**:116-122

[61] Valladares Linares R, Li Z, Sarp S, Bucs SS, Amy G, Vrouwenvelder JS. Forward osmosis niches in seawater desalination and wastewater reuse. Water Research. 2014;**66**:122-139

[62] Yuan H, Abu-Reesh IM, He Z. Mathematical modeling assisted investigation of forward osmosis as pretreatment for microbial desalination cells to achieve continuous water desalination and wastewater treatment. Journal of Membrane Science. 2016;**502**:116-123

[63] Ansari AJ, Hai FI, Price WE, Drewes JE, Nghiem LD. Forward osmosis as a platform for resource recovery from municipal wastewater—A critical assessment of the literature. Journal of Membrane Science. 2017;**529**:195-206

Chapter 12

Membrane Gas Absorption Processes: Applications, Design and Perspectives

Julio Romero Figueroa and Humberto Estay Cuenca

Additional information is available at the end of the chapter

http://dx.doi.org/10.5772/intechopen.72306

Abstract

Membrane gas absorption (MGA) is one of the most attractive technologies among the osmotically driven membrane processes because of its configurational advantages with respect to the conventional absorption systems that use packed bed columns for different industrial applications. Nowadays, membrane gas absorption is used in industrial wastewater treatment, CO_2 absorption from greenhouse gases, treatment of flue-gas and off-gas streams, which contain SO_2, H_2S, NH_3 or HCl, upgrading and desulphurization of biogas from anaerobic digesters and landfills and acid gas removal of natural gas and olefin-paraffin separation in the petrochemical industry, among other applications. In this framework, the advantages of membrane gas absorption over packed bed processes are related to the decreasing of installation surface requirements through compact process design and easy operation modes. These aspects will increase the applications of these types of processes in the mid-term. Nevertheless, the main design criteria of this technology have been poorly addressed in the literature. This chapter summarizes the fundamental aspects of transport phenomena that drive these processes, as well as the main conceptual aspects, to propose a correct design through an overview of the current status of this technology and its potential applications, challenges and future trends.

Keywords: membrane gas absorption processes, gas-filled membrane absorption processes

1. Introduction

In a membrane contactor, the separation process integrates the mass transfer with the conventional phase contacting operation. Thus, membrane contactor operations can be designed with the same phenomenological approach of conventional extraction or absorption processes [1].

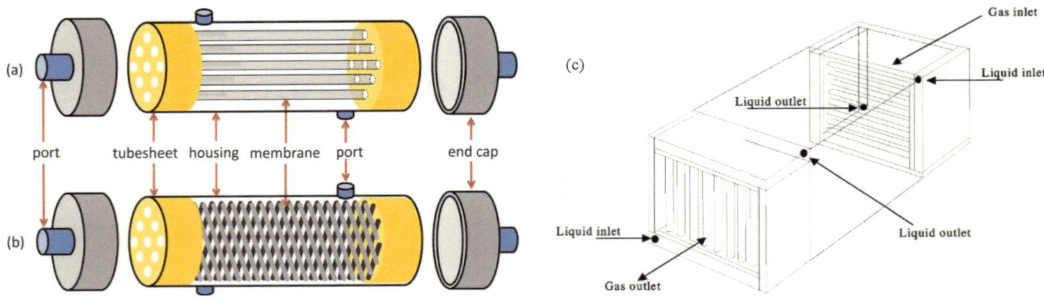

Figure 1. Hollow fiber membrane modules with (a) parallel, (b) crisscross and (c) transversal flow membrane arrangement [1, 2].

One of the most important aspects to be considered in the design of these membrane processes is the maximization of the contact surface area available for the mass transport through an interface, maintaining low pressure drop in the membrane modules. In this way, hollow fiber modules represent the most common geometrical configuration in membrane contactor processes because of their high value of contact surface area per volumetric unit, complemented with a relatively low pressure drop.

As phase contactors, these types of modules are conformed by a bundle of porous hollow fibers, which are arranged in a housing. Thus, one of the phases is circulated into the lumen side; meanwhile, the other phase flows through the shell side. However, the design of a membrane absorber cannot be based on the same hollow fiber modules used in filtration processes, which respond to other design criteria. **Figure 1** shows an outline of three different arrangements of hollow fiber modules [1, 2].

The geometrical arrangements described in **Figure 1** are not exclusive of membrane contactors, and it is used in other membrane processes such as filtration (MF, UF and NF), forward/reverse osmosis (FO, RO) and dialysis [2]. **Figure 1a** and **b** shows conventional arrangements designed from filtration applications. **Figure 1c** shows a transversal flow configuration specially designed for gas-liquid contactor duties [1]. This module involves a rectangular housing where the gas flow is perpendicular to the fibers, and the absorption liquid is circulated though the lumen.

The interface will be stabilized at the entrance of the pores on the lumen or on the shell side depending on the surface interaction between the membrane material and the contacted phases. Hollow fibers can be made in different types of materials such as hydrophobic and hydrophilic polymers [3, 4], ceramics [5] and metals [6]. Currently, hydrophobic membranes are widely used in gas-liquid contacting processes because of their larger contact area than the hydrophilic membranes [7].

2. Theory

Membrane gas absorption (and stripping) process is a gas-liquid contacting operation [8–10]. The core in the membrane gas absorption process is a microporous hollow fiber membrane.

The gas stream is fed along one side of the membrane; at the same time, absorption liquid is flowing at the other side of the membrane [1].

In the membrane absorption process, a hydrophobic or hydrophilic hollow fiber contactor is used to separate a feed solution containing a solute from the receiving gas phase. In the case of the stripping process, the solute to transfer is contained in the gas phase. The hydrophobic or hydrophilic character of the membrane determines the penetration of liquid solution or gas phase into the membrane pores, which are filled with liquid or gas. Thus, solute transfer through the membrane is achieved according to the following sequence of steps, which are presented in **Figure 2**:

1. Solute transfer through a boundary layer of gas phase at the membrane surface;
2. Solute gas transfer through the air gap that fills the pores;
3. Phase equilibrium between the feed solution at the membrane surface and the gas phase retained in the membrane pores for a hydrophobic membrane;
4. Mass transport of absorbed solute into the bulk receiving liquid phase.

For the stripping process, the solute will be transferred from the gas phase into the liquid phase. Moreover, two modes of operation are possible in gas/liquid contactors, according to the application: wetted mode and dry (or non-wetted) mode. Wetted mode occurs when the pores are filled with liquid, for example, if the liquid phase is aqueous and a hydrophilic membrane is used. Conversely, a hydrophobic membrane would operate in the dry mode in this case because the pores would be filled with gas. Dry mode is usually preferred in order to take advantage of the higher diffusivity in the gas; however, the wetted mode may be preferred if there is a fast or instantaneous liquid phase reaction; as a result, the gas phase resistance controls [10].

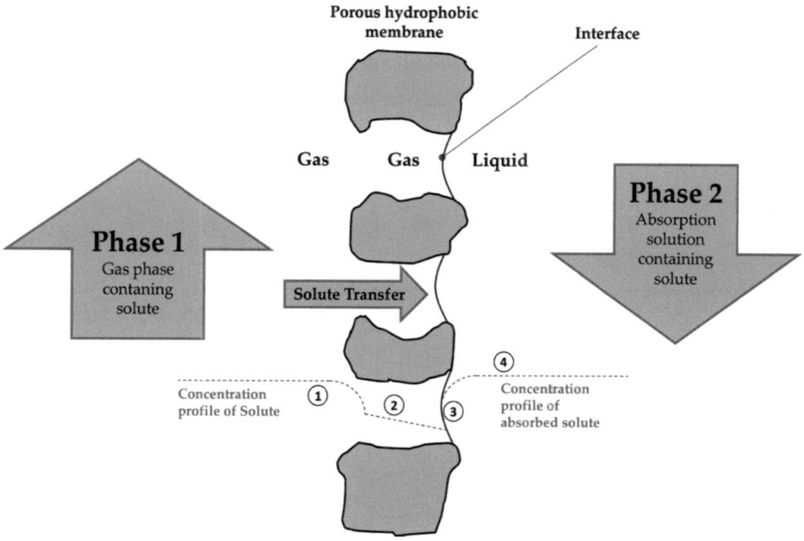

Figure 2. Outline of the membrane absorption process in a hydrophobic membrane.

This transfer of solute in the membrane absorption process can be described by means of a model based on a resistances-in-series approach applied on the proximities of the membrane [10, 11]. The overall solute transfer through the porous membrane can be described by the following equation.

$$N_i = KA\Delta C^i_{ml} \tag{1}$$

where N_i is the flux of solute transferred from the gas phase to the absorption phase, K is the overall mass transfer coefficient, A is the surface area available for mass transfer and ΔC_{ml}^{i} is the logarithmic mean driving force in the absorption phase expressed by:

$$\Delta C^i_{ml} = \frac{(C^i_L - C^{i*}_L)_1 - (C^i_L - C^{i*}_L)_2}{\ln\left[\frac{(C^i_L - C^{i*}_L)_1}{(C^i_L - C^{i*}_L)_2}\right]} \tag{2}$$

C_L^{i*} represents the pseudo-equilibrium concentration of solute in the absorption solution phase, which can be estimated by the following distribution equation:

$$C^{i*}_L = C_L \frac{y_i}{m_i} \tag{3}$$

In Eq. (3), C_L is the total concentration of the liquid phase and m_i is the partition constant (H_i/P) in mol of solute in the gas phase per mol of solute in the liquid phase, which represents the liquid feed-gas equilibrium that could be described by Henry's law for each solute transferred [11] as follows:

$$Py_i = H_i\ x_i \tag{4}$$

here, P is the total pressure, y_i is the mol fraction of solute in the gas phase, x_i is the mol fraction of solute in the liquid phase and H_i is the Henry's constant of solute i. The overall mass transfer coefficient can be represented as a global resistance, which involves the contribution of individual mass transfer steps [10, 12, 13]. Thus, the overall mass transfer coefficient K can be estimated by means of a resistances-in-series approach applied in the proximities of the membrane according to the following equation:

$$\frac{1}{K} = \frac{1}{k_L} + \frac{d_{in}}{m_i k_m d_{ml}} + \frac{d_{in}}{m_i k_G d_{out}} \tag{5}$$

For the driving force based on liquid phase and the gas phase flowing by the shell side and liquid phase by the lumen side. In the case of hydrophilic membranes, the overall mass transfer can be expressed by:

$$\frac{1}{K} = \frac{d_{out}}{k_L d_{in}} + \frac{d_{out}}{k_m d_{ml}} + \frac{1}{m_i k_G} \tag{6}$$

where k_L is the local mass transfer coefficient in the boundary layer of the liquid phase, k_m is the local mass transfer coefficient through the phase in the membrane pores, k_G is the local mass transfer coefficient in the boundary layer of the gas phase, d_{in} is the internal diameter of the fiber, d_{out} is the external diameter of the fiber and d_{ml} is the mean logarithmic diameter of the fiber. The resistances-in-series model is based on one-phase diffusion (i.e., liquid phase) with the assumption the overall mass transfer resistance only occurs in the liquid phase. This assumption is valid since the estimation of the mass transfer resistance in the absorption phase is lower than 4%, using the Hatta method [14]. The local mass transfer coefficient at both sides of the membrane (lumen and shell sides) can be estimated by means of a specific correlation [12, 15], which considers the geometry and the dimensionless Reynolds (Re), Schmidt (Sc) and Sherwood (Sh) numbers of the system. The correlation of mass transfer coefficient of each boundary layer depends on the circulation configuration of the phase in the membrane contactor. **Table 1** shows a summary for different correlations published in the literature.

On the other hand, inside the membrane pores, the local mass transfer coefficient for the retained phase can be described by molecular diffusion [18] according to the low estimated value of the dimensionless Knudsen number [21], close to 0.002. Thus, the local mass transfer coefficient in the gas pores can be estimated as follows:

$$k_m = \frac{D_{AB}\varepsilon}{\tau e} \tag{7}$$

Here, D_{AB} is the diffusion coefficient of the solute A in the phase B, which fills the membrane pores, ε is the porosity of the fibers, τ is the tortuosity of the fibers and e is the fibers thickness. The physical properties of this system, such as diffusion coefficients, viscosity and density of both phases must be established by using different theoretical or empirical relationships as function of system properties (absolute pressure, temperature and composition).

Correlation	Configuration	Observation	Reference	Eq. N°
$Sh = \alpha\left(\frac{d_{in}}{L}\mathrm{Re}Sc\right)^{0.33}$	Lumen side	The value of coefficient α can be 1.86, 1.64 (empirical) or 1.62 (theoretical). Characteristic length is d_{in}.	[16]	(7)
$Sh = 1.25\left(\frac{d_h}{L}\mathrm{Re}\right)^{0.93}Sc^{0.33}$	Shell side, parallel flow	$0 < \mathrm{Re} < 500$; $0.03 < \phi < 0.26$ Characteristic length is d_h.	[16]	(8)
$Sh = 0.022\mathrm{Re}^{0.6}Sc^{0.33}$	Shell side, parallel flow	Characteristic length is d_h.	[17]	(9)
$Sh = \beta(1-\varphi)\left(\frac{d_h}{L}\right)\mathrm{Re}^{0.6}Sc^{0.33}$	Shell side, parallel flow	β = 5.85 for hydrophobic membranes and 6.1 for hydrophilic membranes. $0 < \mathrm{Re} < 500$; $0.04 < \phi < 0.4$ Characteristic length is d_h	[18]	(10)
$Sh = 17.4(1-\varphi)\left(\frac{d_h}{L}\right)\mathrm{Re}^{0.6}Sc^{0.33}$	Shell side, parallel flow	$0 < \mathrm{Re} < 100$; $0.25 < \phi < 0.48$ Characteristic length is d_h	[19, 20]	(11)
$Sh = 0.9\mathrm{Re}^{0.4}Sc^{0.33}$	Shell side, cross flow	$1 < \mathrm{Re} < 25$; $\phi = 0.03$ Characteristic length is d_{out}	[16]	(12)

Note: d_h is the hydraulic diameter (4*[flow surface area]/[wetted perimeter]). Φ is the fiber packing fraction.

Table 1. Mass transfer correlations for local coefficients in different membrane module configurations.

3. Comparison between MGA and conventional packed columns

Mass transfer equipment can be sized as a relation between the number of transfer units (NTU) and the height (or length) of transfer units (HTU). The NTU value is determined by operational parameters such as stream flow rates, solutes concentration and equilibrium constant value of solutes, while HTU is defined by the equipment characteristics such as mass transfer area, stream velocities and mass transfer coefficients values. Thus, the height or length of a mass transfer equipment can be estimated as follows [13].

$$Z = HTU \cdot NTU \quad (8)$$

In terms of comparing the conventional packed columns and hollow fiber membrane contactors, the main difference of sizing will be HTU value, since this parameter depends on equipment dimensions and hydrodynamic characteristics. The HTU parameter can be estimated as shown in the following Eq. [22].

$$HTU = \frac{v}{Ka} \quad (9)$$

where v is the velocity of the stream flow rate and a is the specific transfer area (mass transfer area per equipment volume, m^2/m^3). Different studies have been conducted to compare Ka values for conventional mass transfer equipment and membrane modules applied to different absorption applications.

Table 2 shows that the Ka values for membrane modules can be 10 times higher than Ka values observed in conventional packed towers. Furthermore, the gas and liquid streams are independent in the membrane module; therefore, the gas flow can be increased without changing the liquid flow and vice versa. These altered flows will not cause flooding, as they might in a packed tower [8].

Application	*Ka* value for membrane module	*Ka* value for conventional absorption packed tower	Reference
Absorption of SO_2 in water from air	0.10–0.13 s^{-1}	0.01–0.04 s^{-1}	[22]
Absorption of CO_2 in water from air	0.12–0.25 s^{-1}	0.01–0.18 s^{-1}	[22]
Absorption of CO_2 in monoethanolamine aqueous solution from air	1.3–4.0 kmol/(m^3hkPa)	1.1–1.2 kmol/(m^3hkPa)	[23]
Absorption of CO_2 from flue gas	4.3 s^{-1}	0.47 s^{-1}	[1]
Absorption of CO_2 in monoethanolamine aqueous solution from flue gas	8.93×10^{-4}–7.53×10^{-3} mol/(m^3sPa)	2.25×10^{-4} mol/(m^3sPa)	[24]
Absorption of CO_2 in diethanolamine aqueous solution from air	0.126–0.43 s^{-1}	0.05 s^{-1}	[25]

Table 2. Comparison of *Ka* values between MGA modules and conventional absorption equipment.

4. MGA applications

There is a large body of literature on membrane absorption because this process can be applied to the same cases of most gas absorption processes with conventional dispersive contactors such as packed columns or spray towers. Thus, the use of membrane absorption can be justified when the use of membrane contactor modules involves clear operational and economic advantages over conventional dispersive contactors [1]. In some cases, this suitability is related to the treatment of smaller volume of gases.

Among the most studied cases are the absorption of CO_2 and its recovery from flue-, bio-, and off gases, the removal of SO_2, CO, H_2S, NH_3, HCN, HCl and VOCs from different streams, the upgrading and desulfurization of biogas produced from anaerobic digesters and landfills, the removal of acid gas from fuel gas mixtures and natural gas, the removal of mercury from natural gas, flue gas and glycol overheads, the separation of olefin-paraffin in petrochemical industry and the removal of specific compounds in indoor air [1, 4].

In the following sections, a summarized description of the main applications is presented in order to show the broad range of cases using different absorbents.

4.1. Absorption of CO_2 from flue gas

Nowadays, the reduction of greenhouse gases is probably the main challenge for scientists and engineers facing the unprecedented increase in the concentrations of these compounds, mainly represented by CO_2. In this framework, the absorption of CO_2 from flue gas becomes the most studied application of membrane gas absorption (MGA) processes because this process seems to be a promising alternative to the conventional dispersive absorption systems.

In this application, the selection of the membrane material represents a key parameter for the successful implementation of the process. Currently, typical membranes for gas–liquid contacting processes are prepared from polyethylene (PE), polypropylene (PP), polyvinylidene fluoride (PVDF), polytetrafluorethylene (PTFE) and polysulfone (PS). Among these materials, PTFE shows high hydrophobicity, good mechanical properties and high chemical stability [7, 26]. Different geometrical configurations of membrane contactors have been tested and reported in the literature [7, 27]. The performance of the CO_2 absorption will be more or less affected by the flow mode depending on the contactor geometry and the operation conditions. However, there are some issues that have to be taken into account in this application, when the gas mixture flows inside the lumen, because membrane pores can be plugged by the impurities present in the flue gas [28]. Thus, in the majority of studies, the absorbent flows inside the fibers and the flue gas stream flows in the shell side [7].

On the other hand, the major advancement in the CO_2 absorption has been carried out in the search of more efficient absorber solutions. Thereby, the main aspects that are to be taken into account in the selection of the absorber involve the nature of the process (physical or chemical) and its properties such as the regeneration capacity, viscosity, surface tension and its compatibility with the membrane material. The most commonly used absorber in membrane gas absorption of CO_2 is monoethanolamine (MEA), but there is a wide variety of absorbers such as amine solutions, alcohol-amine solutions as well as their blends [7], and more recently, other compounds such as ionic liquids [29]; the finding of a suitable CO_2 absorber has to match all these aspects. For the most common single absorbents, the CO_2 absorption performance order is NaOH > tetraethylenepentamine (TEPA) > triethylenetetramine (TETA) > diethylenetriamine (DETA) > amino acid potassium (GLY) > monoethanolamine (MEA) > diethanolamine (DEA) > diisopropanolamine (DIPA) > 2-amino-2-methylpropanol (AMP) > triethanolamine (TEA) > methyldiethanolamine (MDEA) > K_2CO_3. Meanwhile, the regeneration performance order is TEA > MDEA > DEA > AMP > DIPA > MEA > NaOH [7].

Recent studies [29, 30] involve the use of ionic liquids as absorbers in membrane absorption systems. Ionic liquids are salts that remain in liquid phase at temperatures lower than 100°C. These compounds are constituted by a relatively large organic cation and a smaller inorganic or organic anion, and they are considered a novel class of 'designer solvents', which show unique properties. Among these properties, their ionic nature and negligible vapor pressure are probably the most particular characteristics. These compounds, mainly based on imidazolium, ammonium, phosphonium, pyridinium, and pyrrolidinium cations, are being used as solvents, electrolytes and reaction media in different chemical processes. Ionic liquids have been studied for use as good gas absorbers [31], particularly of CO_2 [32]. 1-Butyl-3-methyl-imidazolium tetrafluoroborate ([bmim][BF_4]) and 1-(3-aminopropyl)-3-methyl-imidazolium tetrafluoroborate ([apmim][BF_4]) have been tested as absorbers of CO_2 in a membrane absorption system [29]. A much higher absorption was obtained with [apmim][BF4], but this ionic liquid was difficult to be regenerated under vacuum. Meanwhile, the less effective [bmim][BF_4] could be completely regenerated. More recent studies [30] involve the tests of membrane absorption using an amino acid-functionalized protic ionic liquid (monoethanolamine glycinate or [MEA][GLY]), which could be a potential substitute for the conventional chemical absorbent. Nevertheless, further research is necessary to find task-specific ionic liquids with lower viscosities and good absorption and regeneration capacities.

4.2. Removal of SO_2

The removal of SO_2 from gas streams was another pioneering application of hollow fiber absorption systems [22]. One of the first membrane absorption experiments using hollow fiber contactors for the simultaneous absorption of SO_2 and CO_2 considered the use of solutions of Na_2SO_3 [33]. The removal of SO_2 from flue gas has been intensively studied using different types of absorbers such as aqueous solutions of Na_2SO_3, Na_2CO_3, $NaHCO_3$ and NaOH [4]. Thus, different well-known chemical reactions can be considered depending on the absorber selected:

$$SO_2 + 2NaOH \rightarrow Na_2SO_3 + H_2O \quad (10)$$

$$SO_2 + Na_2CO_3 \rightarrow Na_2SO_3 + CO_2 \quad (11)$$

$$SO_2 + 2NaHCO_3 \rightarrow Na_2SO_3 + H_2O + 2CO_2 \quad (12)$$

$$SO_2 + Na_2SO_3 + H_2O \rightarrow 2NaHSO_3 \quad (13)$$

From these four chemical reactions, Park et al. [4] report that an aqueous solution of Na_2CO_3 proved to be the most efficient absorber when the feed SO_2 concentration was 400 ppm.

One of these alternatives is the dual alkali process [1], which involves the production of sodium bisulfite that can be reused on site. This process can be described by means of the reactions 19 and 20 that explain the absorption and regeneration step, respectively:

$$2NaHSO_3 + Na_2CO_3 \rightarrow 2Na_2SO_3 + H_2O + CO_2 \tag{14}$$

Klaassen et al. [1] described pilot-scale experiments in a potato starch production plant of AVEBE (the Netherlands) where the combustion of H_2S containing biogas in a steam boiler results in flue gas containing SO_2. Thus, sulfur dioxide was recovered as bisulfite from the flue gas and it can be reused in the starch production process according to the description given in **Figure 3**.

This installation was successfully tested with as capacity of 120 m^3/h obtaining a SO_2 recovery of over 95% for two production sessions of 6 months each. Problems related to the variation in the gas flow rate, changes in the SO_2 concentration or membrane fouling were not observed.

4.3. Absorption of CO

The absorption of CO from N_2-CO mixtures has been reported in the literature [34] using a hollow fiber module containing porous polypropylene fibers (Celgard X-20) and an ammoniacal cuprous chloride solution as receiving phase. Thus, the preferential absorption of carbon monoxide can be driven by the following reaction:

$$CO + CuCl \overset{NH_4Cl}{\Longleftrightarrow} [Cu(NH_4Cl)_3CO]^+ Cl^- \tag{15}$$

This process shows a very high selectivity and the permeation rate seems to be controlled by the mass transfer in the liquid phase at moderate liquid flow rates and by the chemical reaction

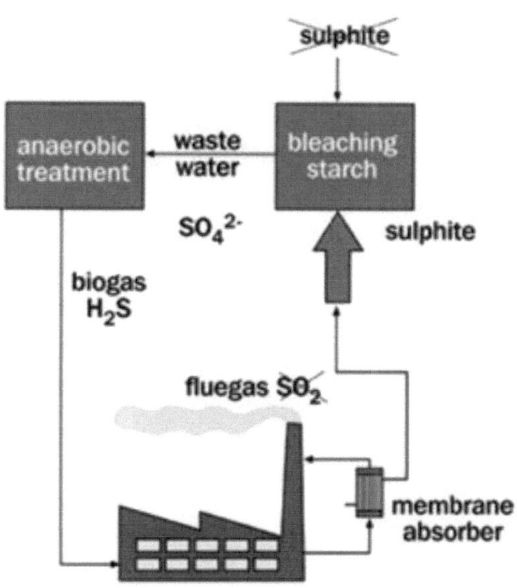

Figure 3. Scheme of the membrane gas absorption process for SO_2 removal and reuse [1].

at higher liquid flow rates. The selected polypropylene membranes seem to show a good chemical resistance to the solutes present in the absorption solution.

4.4. Elimination of H_2S

One of the major impurities of natural gas, refinery gas and coal gas is the hydrogen sulfide (H_2S) contained. Furthermore, it is an important indoor and outdoor air contaminant. This compound is toxic and corrosive and one of the main sources of acid rain [3]. Furthermore, this gas can be produced by sulfate-reducing bacteria under anaerobic conditions contributing to foul odors.

The most common processes to remove H_2S from gas streams are the gas absorption systems using water or different types of aqueous solutions such as sodium hydroxide, sodium carbonate [3], monoethanolamine (MEA) or diethaloamine [35] and ferric solutions of ethylenediaminetriacetic acid (EDTA) and hydroxyethylethylenediaminetriacetic acid (HEDTA) [36] and, more recently, ionic liquids [37].

The use of asymmetric hollow fiber membrane modules for absorption of H_2S has been studied by Li et al. [3], who tested two different hollow fiber membrane modules prepared from polysulfone or polyethersulfone hollow fibers with an outer edge thin layer and a 10% NaOH solution as absorbent. In this system, the presence of the membrane involves a significant increase of mass transfer resistance and the H_2S transfer could be increased if the structural membrane properties, such as porosity, are improved.

On the other hand, a further application considers the removal of H_2S from air using demineralized water (pH 7) for odor control [38]. Porous polypropylene hollow fiber modules with different geometrical parameters were used in this application obtaining 89% of removal for inlet concentrations of 100 ppmv when the gas stream was circulated through the lumen and the water through the shell of the membrane contactor. Fluid dynamic and geometrical aspects have to be considered to operate under the optimal conditions.

4.5. Removal of Hg from industrial gas streams

The removal of Hg from gas streams has also been analyzed using different types of hollow fiber membranes in transversal and shell-tube configurations and several oxidizing liquid solutions [39]. Mercury can be present in the atmosphere due to several industrial activities such as incineration of industrial and domestic waste and natural gas production, and its removal from gaseous streams can be complex because of its low concentration, which is common in the sub-ppm range. Thus, this application requires high gas/liquid flow ratio, and the liquid stream can be suitably circulated through the lumen of the hollow fibers.

Some oxidizing liquids tested to capture Hg from gas streams are H_2O_2/H_2SO_4, $K_2Cr_2O_7$, $K_2S_2O_8$, $Na_2S_2O_8$ + $AgNO_3$ as a catalyst, $KMnO_4$, NaClOx (saturated) and Cl_2 gas [39]. This oxidative membrane absorption process needs chemically resistant hollow fibers, and polytetrafluoroethylene (PTFE) seems to be the suitable membrane material because it shows stable behavior in contact with the oxidizing solutions [39].

4.6. Other applications

Other applications consider the use of a variant of the membrane absorption system called gas-filled membrane absorption. This configuration process is explained in detail in Section 6 and couples the stripping and absorption steps in a single membrane contactor. Thus, a compact design can be proposed, and the gas phase is confined into the membrane pores as an effective supported gas membrane. This system has been studied for the removal of NH_3 from wastewaters and aqueous streams [40], the extraction of SO_2 during the sulfite quantification in wines [41] and the elimination of HCN from pharmaceutical wastewaters [42], plating waters or its recovery from cyanidation solutions in the mining industry [11, 43]. **Figure 4** summarizes the treated and receiving streams in each one of these applications as well as the circulation configuration used in the abovementioned studies.

These three different applications of the GFMA process can involve the recovery of the specie transferred through the membrane and captured in the absorber phase. Thus, the NH_3 removal involves the saturation of the solute $(NH_4)_2SO_4$ or $(NH_4)Cl$ formed in the receiving solution to recover it as by-products. The SO_2 removal from wine involves the indirect quantification of the sulfite content in the absorber, and the elimination of HCN from a cyanidation solution involves the recovery of cyanide from the basic absorber to be reused in the same process.

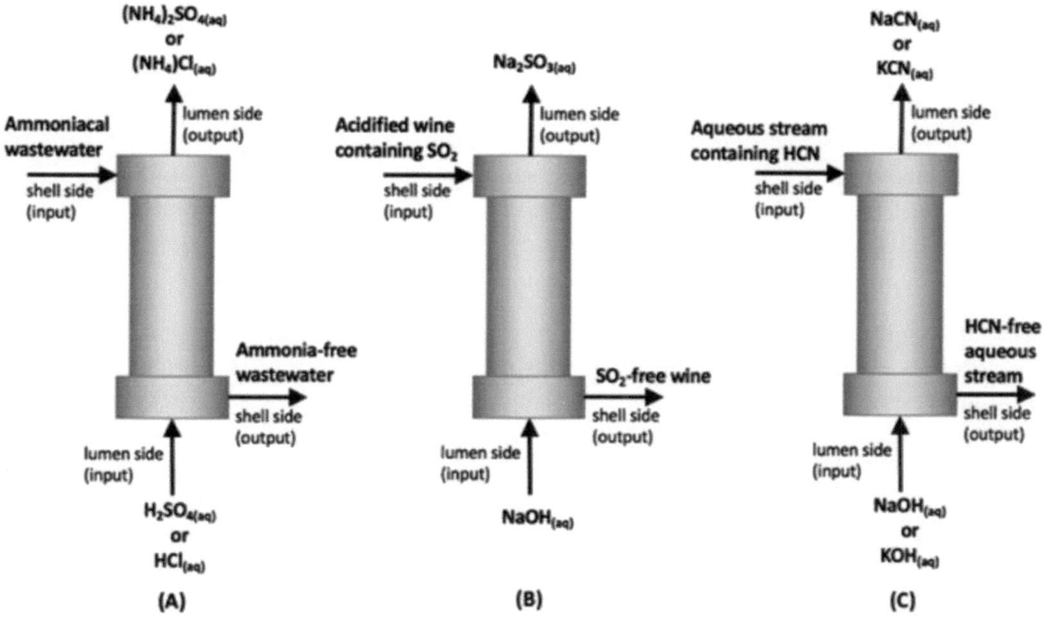

Figure 4. Outline of the input and output streams in GFMA processes for (a) NH_3 removal from wastewater [40], (b) absorption of SO_2 from wine samples [41] and (c) HCN recovery from cyanidation solutions [11].

5. Aspects of design

The design of a membrane gas absorption process is mainly focused on the mass transfer area required to ensure the absorption efficiency established. This area must be estimated by Eq. (1) according to the solute transferred from the gas phase into the liquid phase (N_i), which is defined by the mass balance (operational equation) in the process. In this context, the total area will be defined by the operational conditions (solute concentrations) and the mass transfer coefficient. When experimental results determine the absorption time, the total area required to transfer a solute flow can be estimated using Eq. (1). However, there are current limitations with respect to the modules size available, where the LiquiCel Extra Flow with center baffle module has the largest size, capable of treating a maximum liquid flow rate of 125 m^3/h, having a total transfer area of 373 m^2 [44]. In cases of large absorption times, the total area required could increase over the unitary area specified for one module forced to include an arrangement of hollow fiber membrane contactors modules in series and parallel configuration. In this scenario, one of the first analyses of the optimum hollow fiber membrane contactor arrangement was performed by Prasad and Sirkar [45], who estimated the number of membrane modules needed to treat 2 L/s of feed flow rate of an aqueous solution containing 4-cyanothiazole, which is treated with benzene to recover 98.3% of solute. For this purpose, the researchers proposed an arrangement using a LiquiCel hollow fiber membrane contactor module of 61 $\times$ 5.08 cm, 11,000 fibers and 4.6 m^2 of transfer area. Different arrangement of in series-parallel configuration was assessed in order to obtain a minimum number of total membrane modules needed to achieve the extraction efficiency of 4-cyanothiazole. Thus, the arrangement that determined the minimum number of membrane modules was 15 modules in series with 5 parallel configurations giving a total amount of membrane modules equal to 75. Nevertheless, the expected total pressure drop for each parallel configuration (containing 15 in series modules) was estimated on 3684 kPa; instead an arrangement of 5 in series modules with 34 parallel configuration minimizes the pressure drop (144 kPa), ensuring the 4-cyanothiazole extraction, although the total membranes modules required are 136. Hence, a conclusion of this study is that the optimum arrangement depends on a technical and economic analysis, considering the energy consumption determined by the pressure drop and the capital cost based on the total membrane modules defined.

Even though the arrangement analysis performed by Prasad and Sirkar [45] includes the pressure drop as a main parameter for design purposes, this study did not take into account the maximum permissible pressure by membrane module, having typical values around of 7.0 bar (700 kPa) at ambient temperature. In this regard, the first arrangement proposed by Prasad and Sirkar [45] involves to feed in the first membrane module of each parallel configuration at feed pressure of around 35 bar. This value is much higher than the maximum permissible pressure specified by commercial hollow fiber membrane contactors modules. This limitation of membrane contactors modules was included in a design analysis for a hydrogen cyanide recovery process using a gas-filled membrane absorption process (GFMA) [43]. In this study, the optimum configuration estimated was 39 hollow fiber membrane contactors [44] in-series to treat 60 m^3/h of cyanide solution to reach 90% of cyanide extraction. According to the maximum permissible feed pressure for membrane module (720 kPa) and the drop pressure for each membrane module

(27.58 kPa), the maximum stages of membrane modules-in-series were 16, forcing the inclusion of intermediate pumping stages. Therefore, the maximum permissible pressure for feed solution limits the total stages of membrane modules in-series, increasing the auxiliary equipment for an industrial plant design. In this scenario, the industrial modules available could limit the application of a membrane gas absorption process, especially for high flow rate requirements.

Summing up, the industrial design for a membrane gas absorption process must include an analysis of the optimum arrangement of in-series modules in parallel configuration, considering the pressure drop for each in-series circuit, the total membrane modules and the maximum permissible feed pressure for each module.

6. Gas-filled membrane absorption

The gas-filled membrane absorption process has been developed to perform the absorption and stripping stages in only one step of hollow fiber membrane contactor [28]. In this process, a hydrophobic hollow fiber contactor is used to separate a feed solution containing a volatile solute (stripping phase) from the receiving phase of absorption solution. The hydrophobic character of the membrane avoids the penetration of aqueous solutions into the membrane pores, which are filled with air. Thus, solute transfer through the membrane is achieved according to the following sequence of steps, which are presented in **Figure 5**:

1. Solute transfer through a boundary layer of feed solution to be treated at the membrane surface;
2. Phase equilibrium between the feed solution at the membrane surface and the gas phase (air) retained in the membrane pores;
3. Solute gas transfer through the air gap that fills the pores;
4. Phase equilibrium between the gas filling the pores and the receiving absorption solution at the membrane surface. In this step, the solute can be absorbed or can react into a new product;
5. Mass transport of absorbed solute into the bulk receiving solution.

The GFMA process has been applied to extract or recover solutes of interest, such as ammonia from wastewater [40, 46], SO_2 from wine [41, 47] and HCN from different wastewaters [11], [42, 48, 49]. These studies have shown high recoveries of volatile solutes (>90%), producing a concentrate product in the absorption solution. Moreover, a technical and economic study was carried out, comparing the GFMA process to recover HCN in gold mining and the conventional process, which uses stripping and absorption stage, separately, in packed towers [43]. This study estimated operational and capital cost reduction at about 10 and 20%, respectively, for the GFMA process, due to the saving on energy consumption (pumping vs. air blow in the towers) and footprint reduction.

Therefore, the GFMA process is an intensified membrane gas absorption process, which is capable of performing stripping and absorption stages in a single step. It is worth mentioning

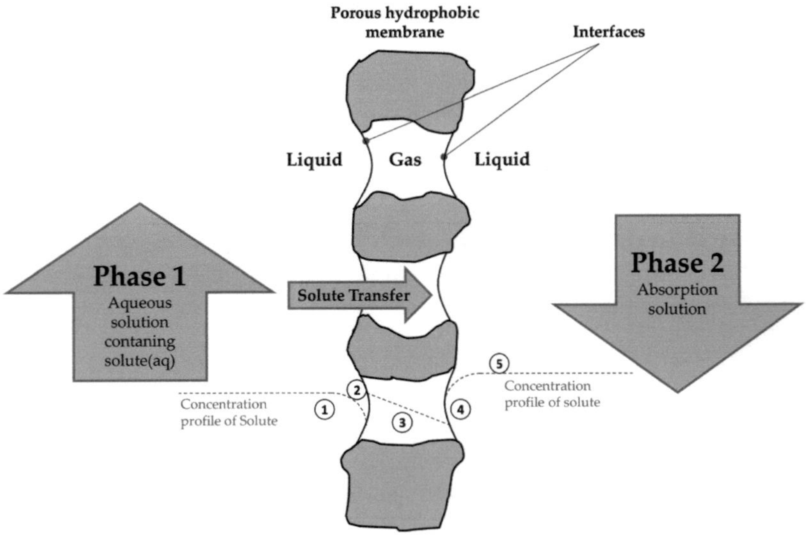

Figure 5. Scheme of the gas-filled membrane absorption process, which shows two gas-liquid interfaces at the pore entrances.

that the design aspects of a GFMA process are similar to the membrane gas absorption process commented upon earlier, taking into account the differences in physical properties on each phase.

7. Challenges and future trends

In this chapter, the most common applications of gas membrane absorption processes are described. Nowadays, these operations are applied in a wide range of fields and can be related to relevant environmental, technical and economic challenges. Nevertheless, the processes under study are currently using modules, which were originally designed for other purposes. Thus, the newest tools for industrial design such as 3D printing, the use of novel materials for membrane preparation and module fabrication, such as specific polymers or their blends, and the use of other absorbers, such as ionic liquids, could enhance the design of further operations according to the precepts of the process intensification; this would allow the design of safer, cleaner and cheaper operations, which are implemented in more efficient and compact units.

On the other hand, the well-known specific surface area into the membrane modules may enhance some procedures and processes at laboratory scale that need high reproducibility, such as analytical techniques [41] or the preparation of specific materials [50].

There is a broad spectrum of new applications, such as biorefineries or the production of bio-based materials that could require a major development of the membrane absorption processes as efficient separation techniques.

Acknowledgements

The authors gratefully acknowledge the financial support of the National Commission for Scientific and Technological Research (CONICYT Chile) through the research FONDECYT N° 1140208 and Project Fund No. FB0809 PIA CONICYT led by Prof. J. Romero and Dr. H. Estay, respectively.

Author details

Julio Romero Figueroa[1]* and Humberto Estay Cuenca[2]

*Address all correspondence to: julio.romero@usach.cl

1 Laboratory of Membrane Separation Processes - LabProSeM, Department of Chemical Engineering, University of Santiago de Chile, Chile

2 Advanced Mining Technology Center - AMTC, University of Chile, Chile

References

[1] Klaassen R, Feron P, Jansen A. Membrane contactors in industrial applications. Chemical Engineering Research and Design. 2005;**83**(A3):234-246

[2] Wan C, Yang T, Lipscomb G, Stookey D, Chung T. Design and fabrication of hollow fiber membrane modules. Journal of Membrane Science. 2017;**538**:96-107

[3] Li K, Wang D, Koe C, Teo W. Use of asymmetric hollow fibre modules for elimination of H_2S from gas streams via a membrane absorption method. Chemical Engineering Science. 1998;**53**(6):1111-1119

[4] Park H, Deshwal B, Kim I, Lee H. Absorption of SO_2 from gas using PVDF hollow fiber membranes in a gas-liquid contactor. Journal of Membrane Science. 2008;**319**:29-37

[5] Koonaphapdeelert S, Wu Z, Li K. Carbon dioxide stripping in ceramic hollow fibre membrane contactors. Chemical Engineering Science. 2009;**64**:1-8

[6] Luiten-Olieman M, Winubst A, Nijmeijer A, Wessling M, Benes N. Porous stainless steel hollow fiber membranes via dry-wet spinning. Journal of Membrane Science. 2011;**370** (1–2):124-130

[7] Zhang Z, Yan Y, Zhang L, Hollow JS. Fiber membrane contactor absorption of CO_2 from flue gas: Review and perspective. Global NEST Journal. 2014;**16**(2):354-373

[8] Qi Z, Cussler E. Microporous hollow fibers for gas absorption I. Mass transfer in the liquid. Journal of Membrane Science. 1985;**23**:321-332

[9] Qi Z, Cussler E. Microporous hollow fibers for gas absorption II. Mass transfer across the membrane. Journal of Membrane Science. 1985;**23**:333-345

[10] Gabelman A, Hwang S. Hollow fiber membrane contactors. Journal of Membrane Science. 1999;**159**:61-106

[11] Estay H, Ortiz M, Romero J. A novel process based on gas filled membrane absorption to recover cyanide in gold mining. Hydrometallurgy. 2013;**134–135**:166-176

[12] Cussler E. Diffusion: Mass Transfer in Fluid Systems. 2nd ed. Cambridge, UK: Cambridge University Press; 2009. 631 p

[13] Treybal R. Mass Transfer Operations. 3rd ed. Singapore: McGraw Hill; 1980. p. 784

[14] Perry R, Green D, Maloney J. Perry's Chemical Engineers' Handbook. 7th ed. New York: McGraw Hill; 1997

[15] Bird R, Stewart W, Lightfoot W. Transport Phenomena. 2nd ed. New York, USA: John Wiley and Sons; 2007

[16] Yang M, Cussler E. Designing hollow fiber contactors. AIChE Journal. 1986;**32**(11):1910-1916

[17] Knudsen J, Katz D. Fluid Dynamics and Heat Transfer. New York, USA: McGraw Hill; 1958

[18] Prasad R, Sirkar K. Dispersion-free solvent extraction with microporous hollow-fiber modules. AIChE Journal. 1988;**34**(2):177-188

[19] Basu R, Prasad R, Sirkar K. Nondispersive membrane solvent back extraction of phenol. AIChE Journal. 1990;**36**(3):450-460

[20] Yun C, Prasad R, Guha A, Sirkar K. Hollow fiber solvent extraction removal of toxic heavy metals from aqueous waste stream. Industrial and Engineering Chemistry Research. 1993;**32**:1186-1195

[21] Mulder M. Basic Principles of Membrane Technology. 1st ed. Netherlands: Kluwer Academic Publishers; 1996. 564 pp

[22] Karoor S, Sirkar K. Gas absorption studies in microporous hollow fiber membrane modules. Industrial and Engineering Chemistry Research. 1993;**32**:674-684

[23] deMontigny D, Tontiwachwuthikul P, Chakma A. Comparing the absorption performance of packed columns and membrane contactors. Industrial and Engineering Chemistry Research. 2005;**44**:5726-5732

[24] Nishikawa N, Ishibashi M, Ohta H, Akutsu N, Matsumoto H, Kamata T, Kitamura H. CO_2 removal by hollow Fiber gas-liquid contactor. Energy Coversion and Management. 1995;**36**(6–9):415-418

[25] Rangwala H. Absorption of carbon dioxide into aqueous solutions using hollow fibers membrane contactors. Journal of Membrane Science. 1996;**112**:229-240

[26] Cabezas R, Plaza A, Merlet G, Romero J. Effect of fluid dynamic conditions on the recovery of ABE fermentation products by membrane-based dense gas extraction. Chemical Engineering and Processing: Process Intensification. 2015;**95**:80-89

[27] Li J, Chen B. Review of CO_2 absorption using chemical solvents in hollow fiber contactors. Separation and Purification Technology. 2005;**41**:109-122

[28] Qi Z, Cussler E. Hollow fiber gas membranes. AIChE Journal. 1985;**31**(9):1548-1553

[29] Lu J, Lu C, Chen Y, Gao L, Zhao X, Zhang H, Xu Z. CO_2 capture by membrane absorption coupling process: Application of ionic liquids. Applied Energy. 2014;**115**:573-581

[30] Lu J, Ge H, Chen Y, Ren R, Xu Y, Zhao Y, Zhao X, Qian H. CO_2 capture using a funtional protic ionic liquid by membrane absorption. Journal of Energy Institute. **90**(6). DOI:10.1016/j.joei.2016.08.001

[31] Anderson J, Dixon J, Brennecke J. Solubility of CO_2, CH_4, C_2H_6, C_2H_4, O_2 and N_2 in 1-Hexyl-3-methylpyridinium Bis(trifluoromethylsulfonyl)imide: Comparison to other ionic liquids. Accounts of Chemical Research. 2007;**40**(11):1208-1216

[32] Cadena C, Anthony J, Shah J, Morrow T, Brennecke J, Maginn E. Why is CO_2 so soluble in Imidazolium-based ionic liquids? Journal of the American Chemical Society. 2004;**126**(16):5300-5308

[33] Nii S, Takeuchi H. Removal of CO_2 and/or SO_2 from gas streams by a membrane absorption method. Gas Separation and Purification. 1994;**8**(2):107-114

[34] Ghosh A, Borthakur S, Dutta N. Absorption of carbon monoxide in hollow fiber membranes. Journal of Membrane Science. 1994;**96**:183-192

[35] Vallée G, Mougin P, Jullian S, Furst W. Representation of CO_2 and H_2S absorption by aqueous solutions of Diethanolamine using an electrolyte equation of state. Industrial and Engineering Chemistry Research. 1999;**38**(9):3473-3480

[36] Wubs H, Beenackers A. Kinetics of H_2S absorption into aqueous ferric solutions of EDTA and HEDTA. AIChE Journal. 1994;**40**(3):433-444

[37] Huang K, Cai D, Chen Y, Wu T, Hu X, Zhang Z. Thermodynamic validation of 1-alkyl-3-methylimidazolium carboxylates as task-specific ionic liquids for H_2S absorption. AIChE Journal. 2013;**59**(6):2227-2235

[38] Boucif N, Favre E, Roizard D. Hollow Fiber membrane contactor for hydrogen Sulfide odor control. AIChE Journal. 2008;**54**(1):122-131

[39] van der Vaart R, Akkerhuis J, Feron P, Jansen B. Removal of mercury from gas streams by oxidative membrane gas absorption. Journal of Membrane Science. 2001;**187**:151-159

[40] Hasanoglu A, Romero J, Perez B, Plaza A. Ammonia removal from wastewater streams through membrane contactors: Experimental and theoretical analysis of operation parameters and configuration. Chemical Engineering Journal. 2010;**160**:530-537

[41] Plaza A, Romero J, Silva W, Morales E, Torres A, Aguirre M. Extraction and quantification of SO_2 content in wines using a hollow fiber contactor. Food Science and Technology International. 2014;**20**(7):501-510

[42] Shen Z, Han B, Wickramasinghe S. Cyanide removal from industrial prazinquantel wastewater using integrated coagulation-gas-filled membrane absorption. Desalination. 2006;**195**:40-50

[43] Estay H, Troncoso E, Romero J. Design and cost estimation of a gas-filled membrane absorption (GFMA) process as alternative for cyanide recovery in gold mining. Journal of Membrane Science. 2014;**466**:253-264

[44] LiquiCel. 14 x 40 Extra-flow product data sheet [Internet]. 2015. Available from: http://www.liquicel.com/uploads/documents/14x40-D102Rev4-10-15%20_ke2.pdf [Accessed: 21-07-2017]

[45] Prasad R, Sirkar K. Hollow Fiber solvent extraction: Performances and design. Journal of Membrane Science. 1990;**50**:153-175

[46] Zhu Z, Hao Z, Shen Z, Chen J. Modified modeling of the effect of pH and viscosity on the mass transfer in hydrophobic hollow fiber membrane contactor. Journal of Membrane Science. 2005;**250**:269-276

[47] Hasanoglu A, Romero J, Plaza A, Silva W. Gas-filled membrane absorption: A review of three different applications to describe the mass transfer by means of a unified approach. Desalination and Water Treatment. 2013;**51**(28–30):5649-5663

[48] Kenfield C, Qin R, Semmens M, Cussler L. Cyanide recovery across hollow Fiber gas membranes. Environmental science. Technology. 1988;**22**:1151-1155

[49] Short A, Haselmann S, Semmens M. The GM-IX process: A pilot study for recovering zinc cyanides. Journal of Environmental Science and Health. 1997;**32**:215-239

[50] Jia Z, Chang Q, Qin J, Mamat A. Preparation of calcium carbonate nanoparticles with a continuous gas-liquid membrane contactor: Particles morphology and membrane fouling. Materials and product. Engineering. 2013;**21**(2):121-126